# 20th Chemnitz Seminar on Materials Engineering



## 20. Werkstofftechnisches Kolloqim

IOP Conference Series: Materials Science and Engineering
Volume 373

Chemnitz, Germany
14 – 15 March 2018

ISBN: 978-1-5108-6486-3
ISSN: 1757-8981

**Printed from e-media with permission by:**

Curran Associates, Inc.
57 Morehouse Lane
Red Hook, NY 12571

**Some format issues inherent in the e-media version may also appear in this print version.**



Printed by Curran Associates, Inc. (2018)

For permission requests, please contact the Institute of Physics
at the address below.

Institute of Physics
Dirac House, Temple Back
Bristol BS1 6BE UK

Phone: 44 1 17 929 7481
Fax: 44 1 17 920 0979

techtracking@iop.org

**Additional copies of this publication are available from:**

Curran Associates, Inc.
57 Morehouse Lane
Red Hook, NY 12571 USA
Phone: 845-758-0400
Fax: 845-758-2633
Email: curran@proceedings.com
Web: www.proceedings.com

# Table of contents

## Volume 373

**20th Chemnitz Seminar on Materials Engineering - 20. Werkstofftechnisches Kolloquium**

**(previous editions: WTK-2016, WTK-2017)**
**14–15 March 2018, Chemnitz, Germany**

**Accepted papers received: 1 May 2018**
**Published online: 12 June 2018**

### Preface



### Articles

**Steels**

**Heat treatment**



**Brazing**

**Mechanical testing / Characterisation**

# Preface

This year, we celebrate an anniversary: the Institute of Materials Science and Engineering (Institut für Werkstoffwissenschaft und Werkstofftechnik, IWW) of Chemnitz University of Technology hosts the traditional scientific meeting, the Chemnitz Seminar on Materials Engineering – Werkstofftechnisches Kolloquium – for the 20th time.

We always strive to bring together scientists and engineers both from academia and industry, and this is again reflected by this year's different scientific sessions that cover various relevant topics, such as electroplating, composite materials, developments in steels, additive manufacturing and many more. Part of our scientific program is a special colloquium that marks the successful completion of the Collaborative Research Center SFB 692, which was launched in Chemnitz 12 years ago and has now reached the end of its regular funding period.

Like in the last years, we chose the IOP Conference Series: Materials Science and Engineering to provide all contributors with the opportunity to publish their papers in an international, peer-reviewed journal. This is understood and appreciated by nearly all participants of our meeting, and therefore this volume provides an excellent overview of the main topics of our Seminar.

Chemnitz, April 2018

Prof. Thomas Lampke

Prof. Guntram Wagner

Prof. Martin F.-X. Wagner

IOP Conf. Series: Materials Science and Engineering **373** (2018) 011002 doi:10.1088/1757-899X/373/1/011002

# PLD-Preparation of carbon based multilayered coatings

**R Bertram[1], T Lampke[2] and S Weißmantel[1]**

[1] Laserinstitut der Hochschule Mittweida

[2] Institut für Werkstoffwissenschaft und Werkstofftechnik der Technischen Universität Chemnitz

**Abstract.** Results obtained in the analysis of single layer and multilayered carbon stacks produced by PLD at low temperature will be presented. The mechanical properties of carbon single layers deposited at various laser pulse fluences, like hardness and elasticity, were determined by means of nano indentation measurements. It will be shown that a wide variety of mechanical properties can easily be adjusted by adapting the process parameters.

In addition, it will be demonstrated that the architecture of various pure carbon films strongly affects the coatings toughness. Scratch test measurements identified, therefore, that the cohesive break down of such layer systems can be hindered by refining the stack periodicity. This is along with no significant losses in hardness or elasticity if the stack architecture is suitable, as determined in nano indentation analyses. Furthermore, the resistance of different coating materials against abrasive wear were determined and compared. It was found that the carbon layer stacks again showed an outstanding performance.

In summary, the produced coating systems combine a very good adhesion on several metal and hard metal substrates, very high hardness and elasticity accompanied with low abrasive wear and high crack and spalling resistivity.

IOP Conf. Series: Materials Science and Engineering **373** (2018) 011002 doi:10.1088/1757-899X/373/1/011002

# Production of tungsten-fibre reinforced tungsten composite by a novel continuous chemical vapour deposition process

**H Gietl[1,*], J Riesch[1], J W Coenen[2], T Höschen[1] and R Neu[1]**

[1] Max-Planck-Institut für Plasmaphysik, Boltzmannstr. 2, 85748 Garching, Germany
[2] Forschungszentrum Jülich GmbH, Institut für Energie- und Klimaforschung – Plasmaphysik, 52425 Jülich, Germany

[*] e-mail: hanns.gietl@ipp.mpg.de

**Abstract.** For the use in a fusion reactor tungsten has unique properties such as low sputter yield, high melting point and low activation. The brittleness below the ductile-to-brittle transition temperature and the embrittlement during operation are the main drawbacks for the use of pure tungsten. Tungsten fibre-reinforced tungsten composites overcome this problem by utilizing extrinsic mechanisms to improve the toughness. The next step in the material development is the production of larger components and testing under cyclic high heat flux loading. A dense matrix produced with a reliable production routine is one of the major issues for the production of such components. In this contribution we present a novel continuous process of chemical vapour deposition. It was achieved by designing and testing a set up For Rotary Enhanced Deposition (FRED) in the already existing W InfiLtration MAchine (WILMA), fabricating first samples. These samples were examined by microstructural analysis and compared to material produced by standard techniques. It is shown that this advanced set up in combination with tungsten fabrics allows the productions of tungsten-fibre reinforced tungsten composite in a faster and more defined way.

IOP Conf. Series: Materials Science and Engineering **373** (2018) 011002 doi:10.1088/1757-899X/373/1/011002

# Additively manufactured isotropic and defect tolerant material by EBM processing – A new alloy design approach

**J Günther[1,*], F Brenne[1], M Droste[2], M Wendler[3], O Volkova[3], H Biermann[2] and T Niendorf[1]**

[1] Universität Kassel, Institut für Werkstofftechnik, Mönchebergstr. 3, 34125 Kassel, Deutschland
[2] TU Freiberg, Institut für Werkstofftechnik, Gustav-Zeuner Str. 5, 09599 Freiberg, Deutschland
[3] TU Freiberg, Institut für Eisen- und Stahltechnologie, Leipziger Str. 34, 09599 Freiberg, Deutschland

[*] e-mail: Guenther@uni-kassel.de

**Abstract.** Electron-beam melting (EBM) is an additive manufacturing (AM) technology allowing for the production of complex shaped components and efficient synthetization of advanced materials. Therefore, it recently gained high academic and industrial interest. However, the powder-bed based processes still face major challenges, e.g. evolution of strong textures and corresponding anisotropy imposed by epitaxial grain growth upon layer-wise consolidation of the initial powder. Furthermore, in most AM alloys process-induced defects detrimentally influence static and cyclic material properties. In the current study these issues are addressed by a new alloy designed for meeting the requirements set by current AM technologies. Firstly, the metastable austenitic steel in focus shows a distinct solidification mode and due to the presence of multiple solid state transformations induced by process-inherent intrinsic heat-treatment an equiaxed fine grained microstructure develops. Moreover, the alloy shows deformation induced phase transformation making it less susceptible to process-induced inhomogeneities. By controlling the process-parameters and the evaporation of volatile elements the deformation mechanisms and mechanical properties can be tailored for certain loading scenarios. For characterization X-ray diffraction, scanning electron microscopy and tensile testing have been employed. The presented alloy design could pave the way for a significant broadening of possible applications of powder-bed based AM components.

# Capability of 3D plasma metal deposition for additive manufacturing of multi-material components

**K Höfer[*], A Hälsig and P Mayr**

Chair of Welding Engineering, Chemnitz University of Technology, Reichenhainer Straße 70, 09126 Chemnitz

[*] e-mail: kevin.hoefer@mb.tu-chemnitz.de

**Abstract.** Within this work, the 3D plasma-metal deposition (3DPMD), based on a plasma powder deposition process is introduced as arc-based additive manufacturing technology. Various powders can be fed into a plasma arc and allow realising a layer-by-layer build-up of metallic or composite structures. 3DPMD has certain advantages compared to the established powder bed or wire based additive manufacturing processes. For example, up to four different powders can be mixed individually within one layer. This allows a targeted adaption of local properties in terms of microstructure, mechanical properties, wear resistance, corrosion resistance, etc. to the targeted load type for the component and the anticipated service conditions. The study aims to demonstrate the suitability of 3DPMD for the production of multi material components in layer-by-layer design for steels, titanium and metal matrix composites. Demonstrators of various metallic materials and metal-matrix-composites have been generated. Geometrical features, microstructure, homogeneity and mechanical properties of the different structures are discussed with a focus on process-structure-property relationships. In summary, 3DPMD offers the possibility to produce various multi-material structures. Using automated routines, it is possible to generate metallic or composite structures directly from CAD data using welding robots. Microstructures and properties are directly related to the process parameters and therefore knowledge about process-structure-property relationships allows manufacturing of tailored multi-material components.

# Highly sensitive thin film strain sensors for integration in thermoplastic-based hybrid laminates

**C Karapepas[1,*], D Nestler[2], D Wett[1] and G Wagner[1]**

[1] Technical University of Chemnitz, Chair of Composites and Material Compounds, Chemnitz, Germany
[2] Technical University of Chemnitz, Chair of Lightweight Structures and Polymer Technology, Chemnitz, Germany

* e-mail: christos.karapepas@mb.tu-chemnitz.de

**Abstract.** Lightweight components comprised of fibre-reinforced plastics have become more important for many different industrial applications in recent years. However, to take full advantage of this relatively new class of materials systems, Structural Health Monitoring of these materials is a necessity. The solution concept aims to integrate nickel-carbon nanocomposite layers by (DC) magnetron sputtering onto polyimide substrates to use as highly sensitive and temperature-compensated strain sensors in hybrid laminates. A suitable polyimide carrier substrate had already been determined which had satisfied the required layer deposition properties and could be processed by the hot-pressing process. A mosaic target was also developed for the physical vapor deposition system. The generated Ni-C sensor layers were characterized in terms of their coating rate in the PVD process, composition, structure and their temperature coefficient of electrical resistance. To determine the k-factor, meandering strain gauges have been produced by means of a mask and characterized using tensile testing module. To check the temperature stability of the Ni-C layers with respect to the hot pressing process, they have been thermally deposited and a temperature-dependent Raman and x-ray diffraction analysis have also been carried out for this purpose. In addition, the change in the TCR value has been determined.

IOP Conf. Series: Materials Science and Engineering **373** (2018) 011002 doi:10.1088/1757-899X/373/1/011002

# Laser hardening of high-alloy steel

**P Landgraf[1], T Grund[1], E Haack[2] and T Lampke[1]**

[1] Institute of Materials Science and Engineering, Chemnitz University of Technology
Chair of Materials and Surface Engineering Group
[2] LHW GmbH

**Abstract.** Laser hardening enable to heat the surface layer of the component and therefore to adjust the microstructure targeted and locally and thus the properties. The advantages of the process are short treatment times, a very good automation and easy integration in production lines as well as a high adaptability and flexibility. Target values of laser hardening are a requested hardness and the penetration depth of the laser track. The high heating rate and the short austenitization period, which is variable depending on the feed rate of the laser, makes the prediction of the laser treatment results difficult and requires an adaptation of the heat treatment parameters compared to conventionally heat-treated components. In high-alloy carbon-rich steels, such as the X153CrMoV12, different contents of carbon, chromium and other alloying elements are dissolved in the austenitic matrix during the laser treatment, depending on the austenitizing temperature. This complicates the process design and control in addition. During the investigations, the temperature and the feed rate of the laser and the surface condition (grounded and blasted) were varied. The laser tracks were analyzed by metallography and X-ray diffraction (XRD). Furthermore, the hardness profiles of the laser tracks were determined. The results show the potential influence of the different process parameters and surface conditions on the laser track in terms of hardness and penetration depths.

IOP Conf. Series: Materials Science and Engineering **373** (2018) 011002 doi:10.1088/1757-899X/373/1/011002

# Significance and characterization of complex formation for the galvanic deposition of cobalt

**M Müller*, I Scharf and T Lampke**

Institute of Materials Science and Engineering, Chemnitz University of Technology, 09125 Chemnitz, Germany

* e-mail: mueller.markus@mb.tu-chemnitz.de

**Abstract.** Almost every galvanic electrolyte contains complexing agents as additives. They have the task to lowering the free metal ion concentration and to prevent the precipitation of metalhydroxides. However, it is not possible to make an electrodeposition out of every metal complex. As a reason for this phenomenon, the dissociation rate of the metal complex was evaluated. The paper shows a methodical selection of the complexing agent for the electrodeposition of cobalt, based on thermodynamic and kinetic parameters. Thus, complexing agents from the substance group amino acids, carboxylic acids and inorganic compounds can be specifically characterized for their suitability. For each system of cobalt-complexing agent the necessary pH range and the excess of complexing agent to stabilize the electrolyte, can be calculated. This leads to an estimation of the applicability of a complexing agent for the galvanic process.

IOP Conf. Series: Materials Science and Engineering **373** (2018) 011002 doi:10.1088/1757-899X/373/1/011002

# Influence of Nano-Roughness on the Electrical Resistance of SiC micro Fibers

**S Palaniyappan, T Ega and G Wagner**

Institute of Materials Science and Engineering, Chemnitz University of Technology, Chemnitz, Germany

**Abstract.** In most of the fiber reinforced composites, the surface roughness of the reinforcing fiber has got quite a great impact on the amount of radial clamping stress being delivered between the fiber and the matrix. On the other hand, the change in average surface nano-roughness ($S_a$) of the fibers, which is caused by means of desizing, also influences their electrical resistance ($R_a$). In applications of composite materials, that demand the use of an embedded single fiber as a piezo-resistive sensor, the effect of this nano-roughness modification on their electrical resistance needs to be characterized. In this contribution, Silicon Carbide (SiC) ceramic fibers were desized using low temperature annealing in atmospheric air and the $S_a$ values were determined using Laser Scanning Microscopy (LSM). The $R_a$ measurements were carried out using 2-wire method connected to a Keithley source meter. The $S_a$ and $R_a$ of the desized fibers are found to be directly proportional to the time periods for the respective temperatures. However, while comparing between the two temperatures, the decrease in $S_a$ at the higher temperature actually lead to an increase in $R_a$ of the fibers. This phenomenon of investigating the effect of nano-roughness on the electrical resistance of SiC fibers paves the way to decide on the desired desizing temperature and time period for applications like fiber sensors, fiber coatings etc.

IOP Conf. Series: Materials Science and Engineering **373** (2018) 011002 doi:10.1088/1757-899X/373/1/011002

# Neural Network for Predicting Plasma Nitriding Results

**J Pribbenow[1], P Landgraf[1], M Mejauschek[2], T Grund[1] and T Lampke[1]**

[1] Institute of Materials Science and Engineering, Chemnitz University of Technology, 09125 Chemnitz, Germany
[2] Zentrum für Tribologische Schichten, Fraunhofer-Institut für Schicht- und Oberflächentechnik, Bienroder Weg 54 E, 38108 Braunschweig, DE

**Abstract.** An artificial neural network, particularly a backpropagation neural network (BPNN), has been developed to investigate the hardening behavior of different steel types according to various process parameters. Therefore, steels with different chemical compositions were plasma nitrated to study the hardness profile and surface hardness. The experimental data were used to train the neural network. For an optimal training, the weight distribution, hidden neurons and gradients of neuron activation functions were varied. This should facilitate a simulation of the hardness profile for types of steel with deviant chemical composition and nitriding parameters.

IOP Conf. Series: Materials Science and Engineering **373** (2018) 011002 doi:10.1088/1757-899X/373/1/011002

# Influence of Spray Angle and Spray Time on the Geometry of WC-Co HVOF Footprints

**W Tillmann, C Schaak* and O Khalil**

TU Dortmund University, Institute of Materials Engineering, Germany

* e-mail: christopher.schaak@udo.edu

**Abstract.** Currently, thermal spray guns mounted on and moved by robots are state of the art. Nevertheless, both the kinematics and dynamics of the robot are solely taken into account in the field of thermal spraying, although they both strongly influence the resulting coating properties. Manual teaching as well as automatic offline path planning are initial and simple steps to solve some of the movement problems with the robot to achieve better coating results. A second step is the optimization of a generated path with regard to the characteristics of the thermal spray process. For this purpose, a coating development simulation, which depends on the robot movement, is one approach to optimize the paths. Footprints / spray patterns can be used as input values for the coating development simulation. A footprint shows the mass flow of spray material at different lateral positions on the surface. However, the footprint geometry itself is influenced by different spray parameters. These deviations can lead to inaccuracies in the final simulation. Hence, the deviations of the footprint geometries for different spray angles and spray times were investigated and quantified. The obtained information will be used to improve the coating development simulation and to achieve a higher precision for the path optimization.

IOP Conf. Series: Materials Science and Engineering **373** (2018) 011002 doi:10.1088/1757-899X/373/1/011002

# Thermal fatigue testing of sand casted AlSi10Mg

**B Zillmann[*] and S Bucher**

CR/APM2, Robert Bosch GmbH, 71272 Renningen, Germany

[*] e-mail: benjamin.zillmann@de.bosch.com

**Abstract.** Sand casted aluminum is extensively used in heat exchangers for gas-fired applications. Consequently, a temperature gradient occurs over the wall thickness and causes transient thermal stresses. The thermal stresses are mainly influenced by the design, boundary conditions and the power of the heat exchanger. This thermal fatigue problem is well known and needs to be considered for the validation of new designs or if other changes in the system take place. So far, system tests are carried out to validate the required lifetime which is time consuming and therefore expensive due to the high cycles times for heating and cooling. The validation time can be drastically reduced by replacing the system tests to mechanical design element tests. Therefore, the transferability from thermal fatigue to the isothermal mechanical fatigue behavior needs to be considered. A new strategy for thermal fatigue testing will be introduced and finally compared to the isothermal mechanical fatigue behavior.

IOP Conf. Series: Materials Science and Engineering **373** (2018) 011003 doi:10.1088/1757-899X/373/1/011003

# Peer review statement

All papers published in this volume of *IOP Conference Series: Materials Science and Engineering* have been peer reviewed through processes administered by the proceedings Editors. Reviews were conducted by expert referees to the professional and scientific standards expected of a proceedings journal published by IOP Publishing.

IOP Conf. Series: Materials Science and Engineering **373** (2018) 012001 doi:10.1088/1757-899X/373/1/012001

# Influence of Si addition on the carbon partitioning process in martensitic-austenitic stainless steels

**Q Huang[1], O Volkova[1], BC De Cooman[2], H Biermann[3] and J Mola[1,*]**

[1] Institute of Iron and Steel Technology, Technische Universität Bergakademie Freiberg, Leipziger Str. 34, 09599, Freiberg
[2] Graduate Institute of Ferrous Technology, Pohang University of Science and Technology, Pohang, Gyeongbuk 790-784, South Korea
[3] Institute of Materials Engineering, Technische Universität Bergakademie Freiberg, Gustav-Zeuner-Straße 5, 09599 Freiberg, Germany

[*] e-mail: mola@iest.tu-freiberg.de

**Abstract.** The effect of Si on the efficiency of carbon partitioning during quenching and partitioning (Q&P) processing of stainless steels was studied. For this purpose, 2 mass-% Si was added to a Fe-13Cr-0.47C reference steel. The Si-free (reference) and Si-added steels were subjected to Q&P cycles in dilatometer. The carbon enrichment of austenite in both steels was evaluated by determining the temperature interval between the quench temperature and the martensite start temperature of secondary martensite formed during final cooling to room temperature. In Q&P cycles with comparable martensite fractions at the quench temperature, the carbon enrichment of austenite after partitioning was similar for both steels. To compare the mechanical stability of austenite, Q&P-processed specimens of both steels were tensile tested in the temperature range 20-200 °C. The quench and partitioning temperatures were room temperature and 450 °C, respectively. Si addition had no meaningful influence on mechanical stability of austenite. The results indicate that the suppression of cementite formation by Si addition to stainless steels, as confirmed by transmission electron microscopy examinations, has no noticeable influence on the carbon enrichment of austenite in the partitioning step.

## 1. Introduction

Introduction of retained austenite (RA) in the microstructure of martensitic steels has been one of the focuses in recent advanced high strength steel developments, with the aim to enhance the ductility of steels without severe sacrifice of the strength. This objective may be achieved by appropriate alloy and processing design. Due to its simplicity and efficiency in stabilizing austenite, a thermal processing method known as quenching and partitioning (Q&P) processing has received especial attention [1–5]. This heat treatment consists of quenching austenite to a temperature between martensite start ($M_s$) and finish ($M_f$) temperatures, followed by a partitioning treatment at higher temperatures to allow the diffusion of carbon from martensite to austenite, therefore stabilizing austenite at room temperature (RT) [6].

As the precipitation of carbides competes with the carbon enrichment of austenite, Q&P steels are usually alloyed with Si [7–9]. Si is known to suppress the formation of cementite, presumably due to its extremely low solubility in cementite [10]. Dilatometry experiments have indicated that the onset of cementite during continuous heating of low-alloy steels with 0.7 mass-% C is raised by approximately 200 °C by the addition of 2 mass-% Si [10]. Nevertheless, evidences of cementite formation in low-alloy martensitic steels containing up to 2 mass-% Si have been presented after tempering at relatively high temperatures [11,12]. In theory, if carbide precipitation is inhibited, more carbon is available for partitioning into RA. Accordingly, Si addition has been reported to increase the final RA fractions [13] or raise the carbon concentration of Q&P-processed carbon and low-alloy steels [14,15]. As an example, carbon partitioning between martensite and austenite was negligible in the Fe-1.1C-3Mn steel (concentrations in mass-%) partitioned at 300 °C, but it was noticeable in the steel containing 2 mass-% Si after applying the same Q&P routine [16].

The enhanced carbon enrichment of partitioned austenite upon Si addition occurs mostly during the formation of carbide-free bainite [17–19]. The latter is a result of isothermal decomposition of austenite in the partitioning step of carbon and low-alloy steels and is a common problem encountered during the Q&P processing of such steels [20–22]. In the case of stainless steels, on the other hand, the transformation of RA into bainite is inhibited by the presence of high Cr contents [23] which makes them ideal candidates for the Q&P processing. Nevertheless, the presence of high Cr contents is not effective against the formation of $M_3C$ cementite (M denotes Fe and substitutional elements) [24]. The precipitation of cementite in martensitic stainless steels can be retarded by the addition of Si [25]. Accordingly, Si addition to stainless steels is expected to enhance the carbon partitioning efficiency. Tobata *et al.* reported an increase in the fraction of RA in Q&P-processed martensitic stainless steels upon the addition of Si which was ascribed to the reduced $M_3C$ formation [26]. Nevertheless, although carbon concentration in the partitioned austenite was not given, the variation of austenite lattice parameter during partitioning was similar for the Si-free and Si-added steels.

The lack of convincing experimental proof for the theoretical expectation of enhanced carbon enrichment of austenite in the absence of $M_3C$ precipitates casts doubt on the efficiency of Si on the Q&P response of steels. The present study aims at clarifying the role of Si in the Q&P processing of stainless steels. For this purpose, Si-added and Si-free steels were subjected to Q&P cycles in dilatometry to evaluate the carbon enrichment of RA. Furthermore, the mechanical stability of austenite was compared by tensile testing of Q&P steels at temperatures ranging from RT to 200 °C.

## 2. Experimental procedure

Steels with the chemical compositions shown in Table 1 were produced in a vacuum induction and casting facility. The Si-free and Si-added steels are denoted as Cr and CrSi steels, respectively. The cast ingots were forged and caliber rolled to round bars with a diameter of 12 mm. A Bähr 805 pushrod dilatometer was used to simulate the Q&P cycles shown in Figure 1(a). Since the partitioning of carbon is more pronounced in the presence of martensite fractions above approximately 50 vol.% [27], the solution annealing temperature was 1050 °C where Cr-rich carbides are only partially in solution [28].The application of a high cooling rate of 20 °C/s between annealing temperature and 500 °C intended to prevent the growth of Cr-rich carbides upon cooling. Below 500 °C, cooling rate was reduced to 3 °C/s. The possibility of applying high quench temperatures enables to track changes in the secondary $M_s$ temperature without the need for subzero dilatometry. Quench temperature (QT) in dilatometry cycles was varied between 170 °C and 200 °C and the partitioning was done at 450 °C for 3 min.

In order to compare the carbon contents of austenite in both steels at 1050 °C, $M_s$ temperatures after annealing at 1180 °C and 1050 °C were determined by dilatometry. At 1180 °C, Cr-rich carbides are fully dissolved in the austenitic matrix [28]. Therefore, carbon concentrations of austenite at 1180 °C are equal to the nominal values in both steels. Considering that the addition of each 1 mass-% solute carbon in austenite reduces the $M_s$ temperature of stainless steels by approximately 510 °C [23], the carbon concentrations of austenite at 1050 °C can be estimated based on the rise in $M_s$ temperature caused by decreasing the annealing temperature from 1180 °C to 1050 °C.

The mechanical stability of RA in Q&P steels was evaluated by tensile tests. In order to enhance the stability of RA in comparison with dilatometry cycles, the carbon loss of austenite due to undissolved Cr-rich carbides was avoided by solution annealing at 1250 °C. According to [28], this temperature is sufficiently higher than the full dissolution temperature of Cr-rich carbides in the Si-free steel (1125 °C). As shown in Figure 1(b), the Q&P processing was simplified as the QT was equal to RT.

Quenching was done with He to achieve an average cooling rate of approximately 30 °C/s between 1250 °C and 400 °C. The cooling rate was lower at temperatures near RT. Subsequently, the specimens were partitioned at 450 °C for 5 min and air cooled to RT. Tensile tests were carried out in a Zwick/Roell-UPM 1476 type universal testing machine. The crosshead displacement speed was kept constant at 1.8 mm $min^{-1}$, corresponding to an initial strain rate of approximately $8 \times 10^{-4}$ $s^{-1}$. Tensile specimens with a diameter of 6 mm and a parallel length of 36 mm were machined according to the DIN 50125 standard. At least two tensile specimens from each steel were tested at each temperature. Furthermore, in order to evaluate the solid solution strengthening effect of Si, two Q&P tensile specimens of each steel were annealed at 700 °C for 5 h and then tensile tested at RT.
The martensite fraction of dilatometry specimens quenched directly from 1050 °C to RT and those of the tensile specimens before and after straining were measured by magnetic saturation measurements using a Metis MSAT device equipped with a Lakeshore 480 fluxmeter. As the only ferromagnetic phase in the microstructure, the fraction of martensite can be quantified by dividing the measured mass magnetization by the intrinsic mass magnetization of the martensitic constituent. The martensite fractions at different stages of dilatometry Q&P cycles were calculated based on the associated expansions. For this purpose, use was made of the lever rule in a manner similar to that described in [24]. The correlation between the magnitude of expansion and the fraction of martensite was established by quantifying the martensite fraction of dilatometry specimens by magnetic measurements. Transmission electron microscopy (TEM) examinations were done in a JEOL JEM-2100 type TEM with a $LaB_6$ gun operated at 200 kV. TEM samples were prepared by mechanical polishing to a thickness of approximately 100 μm followed by twin-jet polishing at RT with an electrolyte of 5 % perchloric acid in 95 % acetic acid.

**Table 1.** Chemical compositions of the studied steels in mass-%

| Steel ID | C | Cr | Si | Mn | N | Fe + others |
|---|---|---|---|---|---|---|
| **Cr** | 0.473 | 13.1 | 0.377 | 0.359 | 0.008 | Bal. |
| **CrSi** | 0.466 | 12.8 | 2.050 | 0.353 | 0.010 | Bal. |

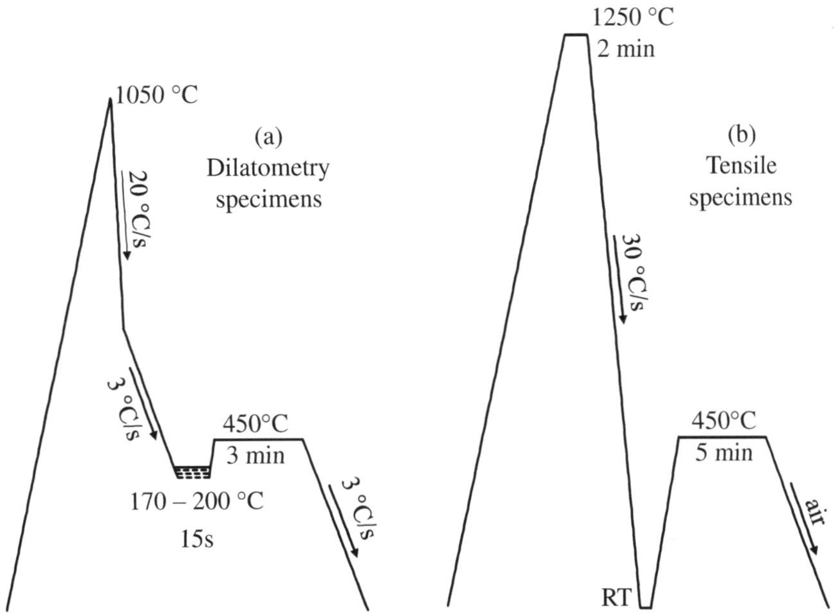


**Figure 1.** Schematic illustration of Q&P processing for (a) dilatometry specimens and (b) tensile specimens.

## 3. Results and discussion

Figure 2 shows the relative length changes during quenching of the steels from 1050 °C and 1180 °C. Relative length changes during quenching from 1050 °C are exemplified by two Q&P cycles with the routines shown in Figure 1(a). The QTs are 170 °C and 200 °C for the Cr and CrSi steels, respectively. After quenching from 1180 °C, due to the full dissolution of Cr-rich carbides at 1180 °C [28] , carbon concentrations of austenite and martensite are equal to the nominal values in both steels. The $M_s$ temperatures for the Cr and CrSi steels after annealing at 1180 °C were 124 °C and 98 °C, respectively. Due to the presence of undissolved Cr-rich carbides, the $M_s$ temperatures are higher for the annealing temperature of 1050 °C, namely they are 260 °C and 239 °C for the Cr and CrSi steels, respectively. As an increase of 1 mass-% solute carbon in austenite depresses the $M_s$ temperature of stainless steels by nearly 510 °C [23], the drops of 136 °C and 141 °C in the $M_s$ temperature can be correlated with reductions of 0.267 and 0.276 mass-% in the solute carbon content of Cr and CrSi steels, respectively. The comparable carbon losses of austenite upon the formation of Cr carbides implies that the solute carbon contents of austenite in the Cr and CrSi steels remain comparable after solution annealing at 1050 °C, which was also used as the annealing temperature for the dilatometry experiments represented in Figure 1(a).

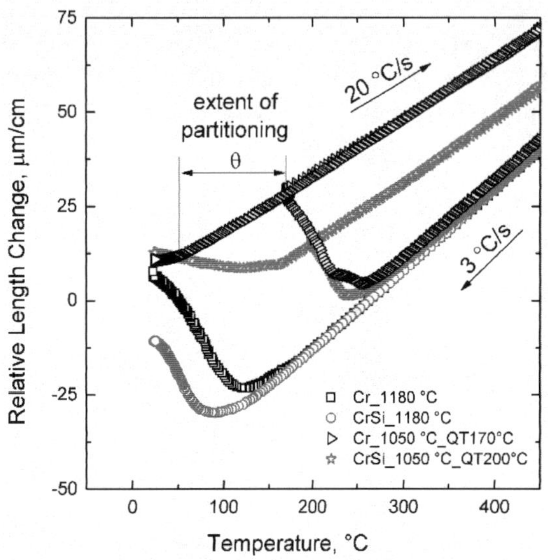


**Figure 2**. Relative length changes during cooling of Cr and CrSi steels from 1050 °C and 1180 °C. Cooling from 1050 °C was interrupted at 170 °C for the Cr steel and at 200 °C for the CrSi steel. Subsequent to interruption, specimens were reheated to 450 °C, held for 3 min, and finally cooled to RT at 3 °C/s.

The stabilization of RA during Q&P processing originates from its carbon enrichment during partitioning. As a result, the austenite cooled from the partitioning temperature transforms to secondary martensite at a temperature lower than the applied QT. In the present study, the temperature interval (θ) between QT and secondary $M_s$ temperature (marked in Figure 2) was used to quantify the extent of partitioning. To exclude the effect of Si addition on the $M_s$ temperature, θ is plotted against the primary martensite fraction ($f_{a'}$), namely the fraction of martensite present before partitioning. Presence of identical primary martensite fractions ensures that the amount of carbon available for partitioning into the austenite remains comparable. The results of dilatometry Q&P cycles according to Figure 1(a) are summarized in Figure 3. The secondary $M_s$ temperature for the CrSi steel quenched to 170 °C was below RT. Therefore, the associated θ could not be determined by dilatometry and is not given in Figure 3. As shown in Figure 3(a), the primary $f_{a'}$ at an identical QT is higher for the CrSi steel than it is for the Cr steel. At first glance, this might appear inconsistent since the martensitic expansions for both alloys are very similar in the vicinity of 200 °C (Figure 2). Nevertheless, the expansion caused by martensite formation is somewhat smaller for the CrSi steel than that for the Cr steel. According to correlative dilatometry and magnetic saturation measurements, the formation of each vol.% martensite in the Cr steel is associated with an expansion of nearly 0.0094 % at RT. For the CrSi steel, this value is reduced to 0.0076 % indicating that the CrSi steel undergoes a smaller expansion upon the formation of martensite. This result agrees with the dilatometry expansions for low-alloy steels with and without Si [10] and can be related to the lattice contraction of bcc steels upon alloying with Si [29].

For the primary $f_{a'}$ of nearly 44 vol.%, θ in the Cr steel is higher than that for the CrSi steel. The difference in θ values for the Cr and CrSi steels is much smaller when primary $f_{a'}$ exceeds 50 vol.%. The maximum difference in θ of 18 °C, observed in the presence of approximately 44 vol.% martensite, corresponds to a difference in the carbon enrichment of only 0.035 mass-%. In accordance with the θ values, the difference in the secondary $f_{a'}$ formed in the presence of a given primary $f_{a'}$ is negligible for both steels. The results indicate that, in contrast to the common belief, the carbon enrichment of austenite during Q&P processing of stainless steels is not much influenced by the addition of Si.

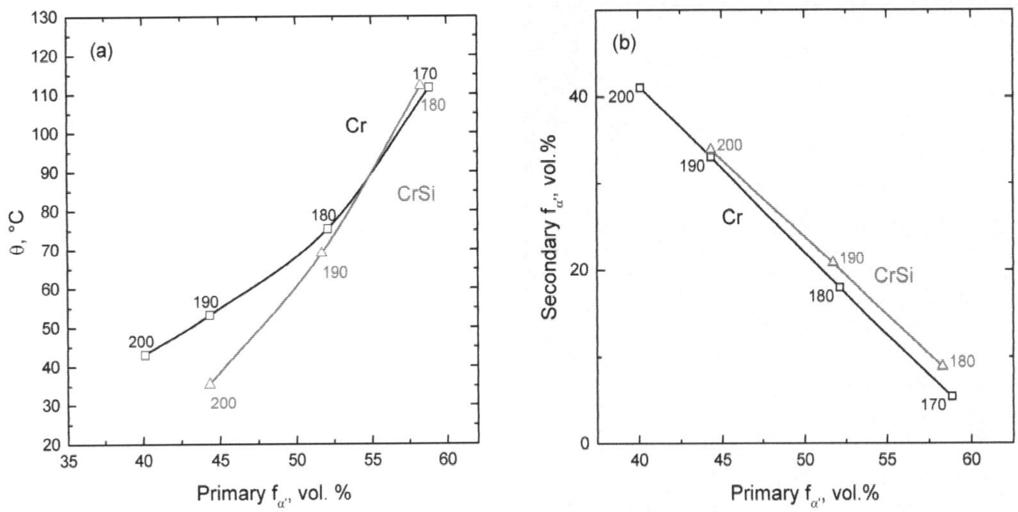


**Figure 3.** Variation of stabilization parameter θ (a) and secondary martensite fraction (b) with the primary martensite fraction. The numbers next to symbols denote the respective quench temperatures.

TEM investigations of the Cr steel quenched in liquid nitrogen from the solution annealing temperature of 1250 °C and tempered at 400 °C revealed the existence of $M_3C$-type carbides in the martensitic constituent (Figure 4(a)). As shown in Figure 4(b), such carbides do not form in the martensite phase of the CrSi steel heat treated in the same manner. The observation of similar degrees of stabilization for the austenite in the Cr and CrSi steels (dilatometry results) indicates that the suppression of $M_3C$ formation in the martensite does not promote the extent of carbon partitioning into the RA. Therefore, either more carbon remains as solute in the martensite phase of the CrSi steel or alternative carbon-consuming processes such as segregation to defects proceed to a larger extent for the CrSi steel.

The mechanical stability of RA in the steels treated by the simplified Q&P processing as shown in Figure 1(b) was determined by tensile testing at various temperatures. According to magnetic saturation measurements, the martensite fractions prior to tensile tests were 44 vol.% and 43 vol.% for the Cr and CrSi steels, respectively. The engineering stress-engineering strain curves are shown in Figures 5(a and b). In the studied temperature range, the CrSi steel shows both higher tensile strengths and higher tensile elongations. The temperature dependence of tensile elongation is summarized in Figure 5(c). Both steels exhibit poor ductility at RT and 80 °C. As the temperature increases, the tensile elongation increases in both steels which is due to the enhanced austenite stability and its delayed deformation-induced transformation to martensite [30,31]. The mechanical stability of austenite was quantified using the following equation as proposed by Sugimoto et. al. [32]:

$$k = -\ln((f_\gamma^0 - f_{\alpha'}^d) \cdot (f_\gamma^0)^{-1}) \cdot \varepsilon^{-1} \quad (1)$$

where $k$ is a parameter inversely proportional to the stability of austenite, $f_\gamma^\circ$ is the initial austenite fraction, $\varepsilon$ is the applied true strain in tension, and $f_{\alpha'}^d$ is the deformation-induced martensite fraction after applying the true strain $\varepsilon$. $f_\gamma^\circ$ was calculated from the martensite fraction in the grip section by assuming a carbide fraction equal to zero. The difference in the martensite fraction between the grip and gauge sections was taken as $f_{\alpha'}^d$, which is given in Figure 6(b). The local true strains shown in Figure 6(a) were obtained based on the reduction in area of nearly cylindrical specimens taken from the gauge sections of tensile-tested specimens. The calculated $k$ parameters are plotted in Figure 6(c) as a function of the tensile test temperature. At 20 °C and 80 °C, the CrSi steel exhibits somewhat smaller $k$ values, indicative of higher mechanical stabilities compared to the Cr steel. The difference in $k$ values becomes negligible at 160 °C and 200 °C. Due to the higher area reductions at these temperatures and the higher certainty of local strain readings based on the diameter change measurements, the $k$ values at 160 °C and 200 °C are expected to be better representatives of the mechanical stability of austenite. The negligible difference between the mechanical stability of austenite in the Cr and CrSi steels is in accordance with the dilatometry observations suggesting the inefficiency of Si addition in increasing the stability of austenite. Therefore, the importance of cementite suppression by Si addition on the efficiency of carbon partitioning might have been overestimated.

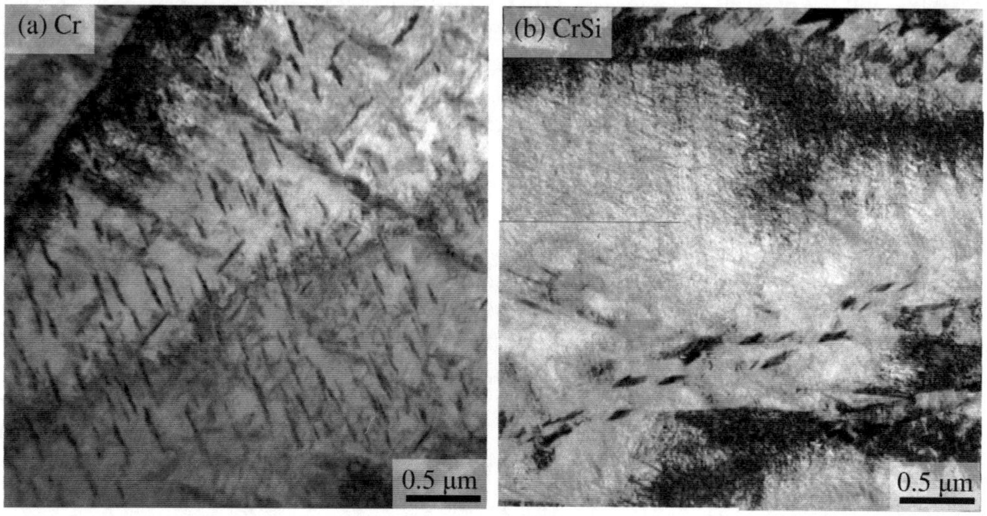


**Figure 4.** TEM bright-field micrographs of (a) Cr steel and (b) CrSi steel partitioned at 400 °C for 5 min subsequent to quenching from 1250 °C to -196 °C.

The higher strength of the CrSi steel in spite of having a martensite fraction comparable to that of the Cr steel was speculated to be related to the solid solution strengthening effect of Si. In order to verify this, tensile specimens Q&P processed according to Figure 1(b) were soft annealed at 700 °C for 5 h to obtain a low carbon ferritic matrix containing dispersed $M_{23}C_6$ carbides. The soft-annealed steels were then tensile tested at RT (Figure 7). The obviously higher tensile strength of the CrSi steel confirms that the higher strength of the Q&P tensile specimens of the CrSi steel may be attributed to the solid solution strengthening effect of Si.

IOP Conf. Series: Materials Science and Engineering **373** (2018) 012001 doi:10.1088/1757-899X/373/1/012001

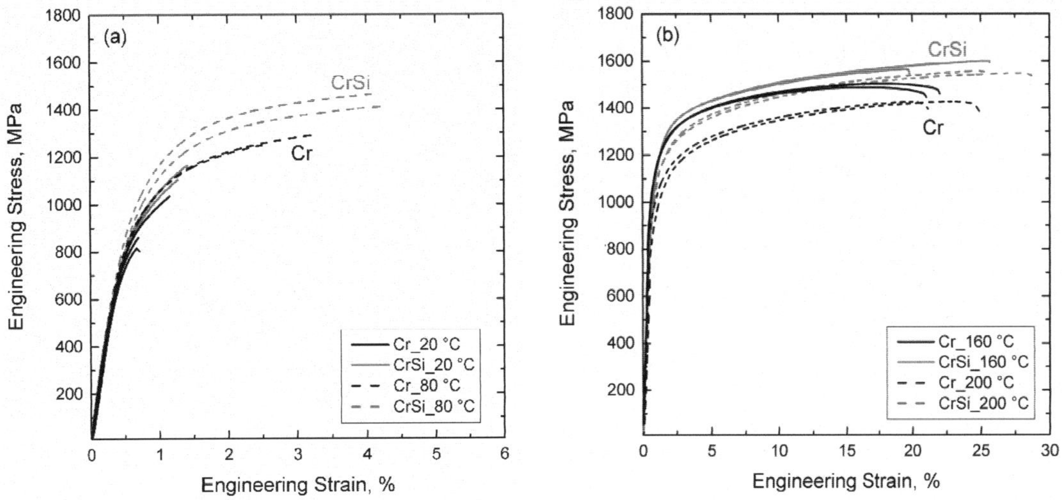


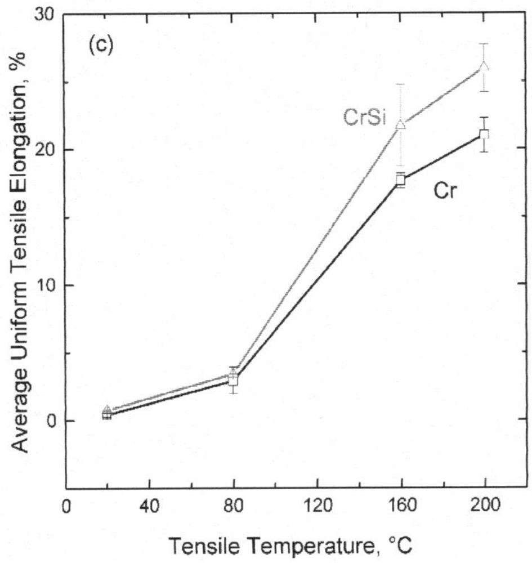


**Figure 5.** (a, b) Temperature dependence of engineering stress-strain curves for the steels treated by simplified Q&P processing according to Figure 1(b); (c) temperature dependence of uniform tensile elongation.

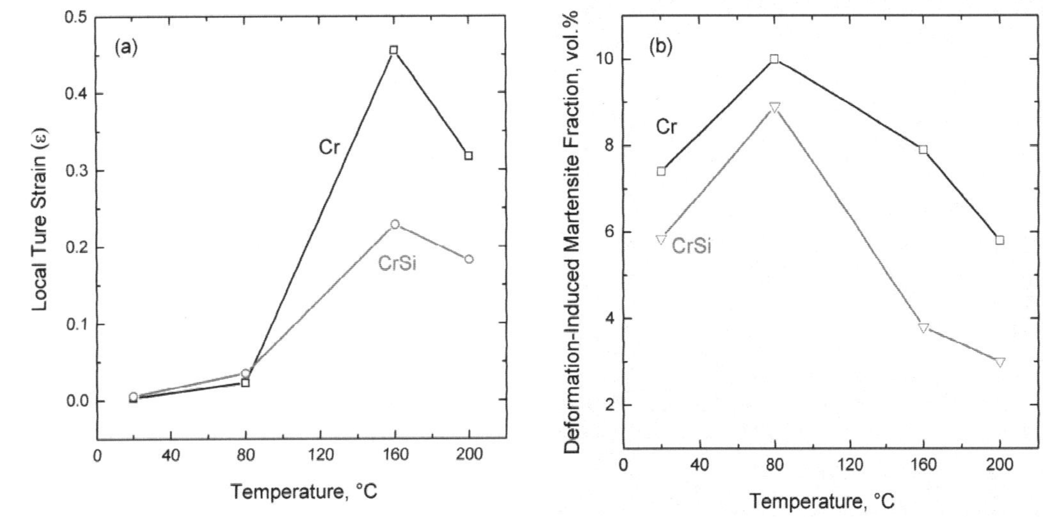

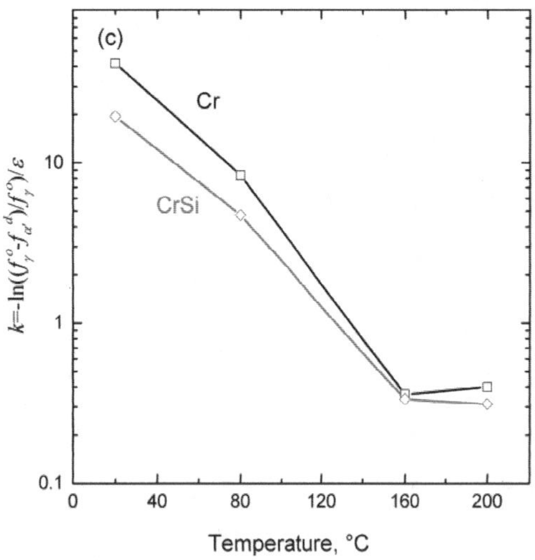


**Figure 6.** (a) Local true strains of the specimens taken from the gauge sections; (b) deformation-induced martensite fractions calculated from the martensite fractions of the specimens taken from the grip and the gauge sections of deformed tensile specimens; (c) Stability parameter $k$ at various tensile test temperatures calculated using equation (1) and the values in (a) and (b).

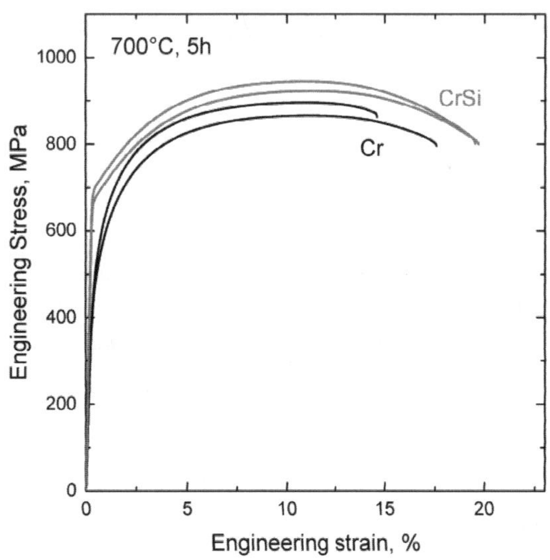


**Figure 7**. Flow curves of the steels soft-annealed at 700 °C for 5 h subsequent to Q&P processing according to Figure 1(b).

## 4. Conclusions

The influence of Si addition on the carbon partitioning during Q&P processing of stainless steels was investigated by the addition of 2 mass-% Si to the Fe-13Cr-0.47C steel. Q&P processing involving solution annealing at 1050 °C, quenching to various temperatures below $M_s$ temperature, and partitioning at 450 °C was performed in a dilatometer to evaluate the influence of Si on the thermal stability of austenite. A simplified Q&P processing involving solution annealing at 1250 °C, quenching to RT, and partitioning at 450 °C, on the other hand, was applied to tensile test specimens. The mechanical stability of RA was subsequently evaluated by tensile testing at temperatures between RT and 200 °C. The following conclusions were drawn:

1. According to dilatometry results, the carbon enrichment of austenite in the partitioning step was comparable for both steels.
2. Tensile specimens of Si-added and Si-free steels have nearly equal martensite fractions after the application of simplified Q&P. The difference in the mechanical stability of steels was negligible.

IOP Conf. Series: Materials Science and Engineering **373** (2018) 012001 doi:10.1088/1757-899X/373/1/012001

3. Since martensite in the Si-added steel was confirmed to be free of $M_3C$ cementite after partitioning, the importance of cementite suppression by Si addition on the efficiency of carbon partitioning might have been overestimated.
4. The higher tensile strengths of the Q&P processed specimens of the Si-added steel can be justified by the solid solution strengthening effect of Si.

## 5. References

[1] Clarke A J, Speer J G, Matlock D K, Rizzo F C, Edmonds D V and Santofimia M J 2009 Influence of carbon partitioning kinetics on final austenite fraction during quenching and partitioning *Scr. Mater.* **61** 149–52

[2] De Knijf D, Puype A, Föjer C and Petrov R 2015 The influence of ultra-fast annealing prior to quenching and partitioning on the microstructure and mechanical properties *Mater. Sci. Eng. A* **627** 182–90

[3] Gouné M, Danoix F, Allain S and Bouaziz O 2013 Unambiguous carbon partitioning from martensite to austenite in Fe–C–Ni alloys during quenching and partitioning *Scr. Mater.* **68** 1004–7

[4] Wendler M, Ullrich C, Hauser M, Krüger L, Volkova O, Weiß A and Mola J 2017 Quenching and partitioning (Q&P) processing of fully austenitic stainless steels *Acta Mater.* **133** 346–55

[5] Huang Q, Schröder C, Biermann H, Volkova O and Mola J 2016 Influence of Martensite Fraction on Tensile Properties of Quenched and Partitioned (Q&P) Martensitic Stainless Steels *Steel Res. Int.* **87** 1082–94

[6] Speer J G, Rizzo Assunção F C, Matlock D K and Edmonds D V 2005 The "quenching and partitioning" process: Background and recent progress *Mater. Res.* **8** 417–23

[7] Cooman B C D, Lee S J, Shin S, Seo E J and Speer J G 2016 Combined Intercritical Annealing and Q&P Processing of Medium Mn Steel *Metall. Mater. Trans. A* **48** 39–45

[8] Findley K O, Hidalgo J, Huizenga R M and Santofimia M J 2017 Controlling the work hardening of martensite to increase the strength/ductility balance in quenched and partitioned steels *Mater. Des.* **117** 248–56

[9] Ding R, Tang D, Zhao A, Dong R, Cheng J and Meng X 2014 Effect of intercritical temperature on quenching and partitioning steels originated from martensitic pre-microstructure *J. Mater. Res.* **29** 2525–2533

[10] Mola J, Luan G, Brochnow D, Volkova O and Wu J 2017 Tempering of Martensite and Subsequent Redistribution of Cr, Mn, Ni, Mo, and Si Between Cementite and Martensite Studied by Magnetic Measurements *Metall. Mater. Trans. A* 1–8

[11] Toji Y, Miyamoto G and Raabe D 2015 Carbon partitioning during quenching and partitioning heat treatment accompanied by carbide precipitation *Acta Mater.* **86** 137–47

[12] Caballero F G, Allain S, Cornide J, Puerta Velásquez J D, Garcia-Mateo C and Miller M K 2013 Design of cold rolled and continuous annealed carbide-free bainitic steels for automotive application *Mater. Des.* **49** 667–80

[13] Samanta S, Das S, Chakrabarti D, Samajdar I, Singh S B and Haldar A 2013 Development of Multiphase Microstructure with Bainite, Martensite, and Retained Austenite in a Co-Containing Steel Through Quenching and Partitioning (Q&P) Treatment *Metall. Mater. Trans. A* **44** 5653–64

[14] Kim B, Sietsma J and Santofimia M J 2017 The role of silicon in carbon partitioning processes in martensite/austenite microstructures *Mater. Des.* **127** 336–45

[15] Linke B M, Gerber T, Hatscher A, Salvatori I, Aranguren I and Arribas M 2017 Impact of Si on Microstructure and Mechanical Properties of 22MnB5 Hot Stamping Steel Treated by Quenching & Partitioning (Q&P)

[16] Toji Y, Matsuda H and Raabe D 2016 Effect of Si on the acceleration of bainite transformation by pre-existing martensite *Acta Mater.* **116** 250–62

[17] Luo Q, Kitchen M and Abubakri S 2017 Effect of Austempering Time on the Microstructure and Carbon Partitioning of Ultrahigh Strength Steel 56NiCrMoV7 *Metals* **7** 258

[18] Gui X, Gao G, Guo H, Zhao F, Tan Z and Bai B 2017 Effect of bainitic transformation during BQ&P process on the mechanical properties in an ultrahigh strength Mn-Si-Cr-C steel *Mater. Sci. Eng. A* **684** 598–605
[19] Huang Y, Li Q, Huang X and Huang W 2016 Effect of bainitic isothermal transformation plus Q&P process on the microstructure and mechanical properties of 0.2C bainitic steel *Mater. Sci. Eng. A* **678** 339–46
[20] Santofimia M J, Zhao L and Sietsma J 2009 Microstructural Evolution of a Low-Carbon Steel during Application of Quenching and Partitioning Heat Treatments after Partial Austenitization *Metall. Mater. Trans. A* **40** 46–57
[21] Babu S S, Specht E D, David S A, Karapetrova E, Zschack P, Peet M and Bhadeshia H K D H 2005 In-situ observations of lattice parameter fluctuations in austenite and transformation to bainite *Metall. Mater. Trans. A* **36** 3281–9
[22] Ghazvinloo H R, Honarbakhsh-Raouf A and Rashid A R K 2015 Mechanical Properties of a High Si and Mn Steel Heat Treated by One-Step Quenching and Partitioning *Metallurgist* **59** 90–6
[23] Tsuchiyama T, Tobata J, Tao T, Nakada N and Takaki S 2012 Quenching and partitioning treatment of a low-carbon martensitic stainless steel *Mater. Sci. Eng. A* **532** 585–92
[24] Mola J and Cooman B C D 2013 Quenching and Partitioning (Q&P) Processing of Martensitic Stainless Steels *Metall. Mater. Trans. A* **44** 946–67
[25] Mesquita R A and Kestenbach H J 2011 M3C and M7C3 Carbide Precipitation in Modified H11 Tool Steel *Solid State Phenom.* **172–174** 414–9
[26] Tobata J, Ngo-Huynh K-L, Nakada N, Tsuchiyama T and Takaki S 2012 Role of silicon in quenching and partitioning treatment of low-carbon martensitic stainless steel *ISIJ Int.* **52** 1377–1382
[27] Huang Q, Cooman B C D, Biermann H and Mola J 2016 Influence of Martensite Fraction on the Stabilization of Austenite in Austenitic–Martensitic Stainless Steels *Metall. Mater. Trans. A* **47** 1947–59
[28] Huang Q, Volkova O, Biermann H and Mola J 2017 Dilatometry Analysis of Dissolution of Cr-Rich Carbides in Martensitic Stainless Steels *Metall. Mater. Trans. A* **48** 5771–7
[29] Huyan F, Larker R, Rubin P and Hedström P 2014 Effect of Solute Silicon on the Lattice Parameter of Ferrite in Ductile Irons *ISIJ Int.* **54** 248–50
[30] Huang Q, Volkova O, Biermann H and Mola J 2017 Tensile elongation of lean-alloy austenitic stainless steels: transformation-induced plasticity versus planar glide *Mater. Sci. Technol.* **33** 1224–30
[31] Mola J, Ullrich C, Kuang B, Rahimi R, Huang Q, Rafaja D and Ritzenhoff R 2017 Austenitic Nickel- and Manganese-Free Fe-15Cr-1Mo-0.4N-0.3C Steel: Tensile Behavior and Deformation-Induced Processes between 298 K and 503 K (25 °C and 230 °C) *Metall. Mater. Trans. A* **48** 1033–52
[32] Sugimoto K-I, Kobayashi M and Hashimoto S I 1992 Ductility and strain-induced transformation in a high-strength transformation-induced plasticity-aided dual-phase steel *Metall. Trans. Phys. Metall. Mater. Sci.* **23 A** 3085–91

**Acknowledgments**
The financial support of German Research Foundation (DFG) under grant number MO 2580/1–1 is gratefully acknowledged. Authors extend their thanks to the technical staff at the Institute of Iron and Steel Technology of TU Bergakademie Freiberg for their assistance.

IOP Conf. Series: Materials Science and Engineering **373** (2018) 012002 doi:10.1088/1757-899X/373/1/012002

# Design of Fully Austenitic Medium Manganese Steels

**G Luan, O Volkova and J Mola***

Institute of Iron and Steel Technology, Technische Universität Bergakademie Freiberg, Leipziger St. 34, 09599 Freiberg, Germany

* e-mail: mola@iest.tu-freiberg.de

**Abstract.** Due to their higher ferrite potential compared to high Mn twinning-induced plasticity (TWIP) steels, medium Mn steels usually exhibit austenitic-ferritic microstructures, which makes them suitable for third-generation advanced high-strength steel applications. Nevertheless, the strain hardening characteristics of medium Mn steels are inferior to those of fully austenitic high Mn steels. The present work introduces alloy design strategies to obtain fully austenitic medium Mn steels capable of the TWIP effect. To achieve a fully austenitic microstructure, the martensite start temperature is reduced by raising the C concentration to above 1 mass-%, which in turn facilitates the formation of cementite. The formation of cementite during cooling from austenitization temperature is counteracted by alloying with Al. Microstructural examination of slowly-cooled Fe–Mn–Al–C and Fe–Mn–C steels indicated that Al changes the morphology of intergranular cementite from plate-shaped to equiaxed.

## 1. Introduction

Steels containing raised Mn contents receive an ever-increasing attention among the scientific community and steelmakers. The superior strength and ductility combination of such steels is of particular interest to the automotive industry, where enhanced passenger safety and reduced weight without substantial increase in manufacturing and materials costs drives the industry [1]. Steels with raised Mn levels can be classified as super high Mn steels if containing 22–30 mass-% Mn, high Mn twinning-induced plasticity (TWIP) steels with 12–22 mass-% Mn, and medium Mn transformation-induced plasticity (TRIP) steels with 4–8 mass-% Mn [2].

The occurrence of the TWIP effect in high Mn steels is responsible for their exceptional work hardening characteristics. High Mn steels can work harden to flow stresses in the range of 600–1100 MPa and show tensile elongations in the range of 60–95 %, indicative of their high formability [3]. The presence of high Mn concentrations in TWIP steels impairs their weldability and galvanizability and increases production costs, which are responsible for their limited industrialization [4]. Therefore, many researchers have attempted to decrease the Mn concentration of TWIP steels without significant sacrifice of the mechanical properties [4], 6,7]. Tsuchiyama et al. [7] reported a tensile strength of 1070 MPa and a total elongation of 33 % for an Fe–5Mn–1.2Si–0.1C steel (concentrations in mass-%). Zhang et al. [8] achieved similar tensile properties with an Fe–7Mn–0.23Si–0.14C steel. Tensile strengths and total elongations were lower for an Fe–5.3Mn–1.53Si–0.56Ni–0.01C–0.12(Nb+V+Ti) steel with a reduced austenite stability [9]. The preceding examples of medium Mn steels contain low C concentrations leading to the formation of austenitic-ferritic microstructures. As long as the carbide formation is suppressed and the fraction of austenite and ferrite is fixed, an increase in the C concentration can raise the stability of austenite and lead to enhanced strength-ductility combinations [10]. In order to retain a high fraction of austenite at room temperature (RT), these steels are generally intercritically annealed to enrich Mn (and C, if present) in the austenite [5,6,9]. When these steels are deformed, the occurrence of deformation-induced martensitic transformation results in high strain hardening rates at the expense of ductility. The stability of austenite and the type of deformation-induced processes can be controlled by adjusting the stacking fault energy (SFE) [11,12]. When the SFE is less than 15 $mJm^{-2}$, martensitic transformation and the TRIP effect are expected in the austenite. The TRIP and TWIP effects occur simultaneously in the SFE range 15–20 $mJm^{-2}$. The TWIP effect is dominant in the SFE range 20–35 $mJm^{-2}$ [2]. Since C and Mn both increase the SFE [13], it should be possible to substitute at least some of the Mn in high Mn steels with C and still retain the TWIP effect. Seo et al. [14] have demonstrated that twinning is indeed a dominant deformation mechanism during RT deformation of an Fe–7.0Mn–1.2C medium Mn high C steel. The present work discusses important considerations in the design of medium Mn high C steels with fully austenitic microstructures and adequate mechanical stabilities at RT.

## 2. Experimental procedure

Chemical compositions of the medium Mn high C steels studied in the present work are given in table 1. The ingots were produced using a vacuum induction melting and casting facility. The cast ingots were homogenized at 1250 °C for 2 hours under vacuum. The average cooling rate was 8 °C/s between 1250 °C and 650 °C and 4 °C/s between 650 °C and 400 °C. To eliminate carbides possibly formed during $N_2$ gas cooling to RT, small specimens taken from homogenized ingots were austenitized at 1150 °C for 30 min and then quenched in water. Ferromagnetic phase fractions after different mechanical and thermal treatments were quantified by magnetic flux density measurements in a magnetic field of nearly 3.5 kOe. These measurements were done using a Metis MSAT device equipped with a Lakeshore 480 fluxmeter. Specimens with an approximate length of 10 mm and cross sections of approximately 3.5 × 3.5 $mm^2$ were used to apply cooling rates of 0.5 °C/s and 0.1 °C/s in a Bähr-DIL805 dilatometer.

**Table 1**. Chemical composition of steels used in the present study (mass-%).

| Steel ID | C | Mn | Al | Fe |
|---|---|---|---|---|
| **8Mn2Al** | 1.44 | 8.0 | 1.9 | Balance |
| **8Mn3Al** | 1.47 | 7.8 | 3.3 | Balance |

## 3. Results and Discussion

After a discussion of alloy design strategies to obtain fully austenitic medium Mn steels exhibiting TWIP effect, the effect of Al addition to medium Mn high C steels on the microstructure after various thermal and mechanical treatments is discussed.

### *3.1. Design of fully austenitic medium Mn steels*

In order to ensure a fully austenitic microstructure in medium Mn steels, the formation of martensite, pearlite, and cementite must be inhibited by proper alloy design and thermal processing as outlined in the following.

*3.1.1. Inhibition of martensite.* To obtain a fully austenitic microstructure at RT, the martensite start ($M_s$) temperature must be reduced to below RT. Empirical relationships could be consulted to estimate $M_s$ temperature from the chemical composition, for instance [15]:

$$M_s = 539 - 423C - 30.4Mn - 7.5Si + 30Al \quad (1)$$

where $M_s$ temperature is in °C and elemental concentrations are in mass-%. According to equation (1), the addition of C is very effective in reducing the $M_s$ temperature and suppressing martensite formation. Carbon also has a substantial strengthening effect [16]. Experiences with stainless steels has demonstrated that the stability of austenite with an $M_s$ temperature equal to RT is too low for an appropriate formability at RT [17,18]. In such cases, maximum tensile elongation is obtained at temperatures higher than RT, where the formation of martensite during tensile deformation is almost entirely suppressed. In other words, to achieve steels with appropriate tensile elongation at RT, the $M_d$

temperature (the highest temperature associated with deformation-induced martensite formation) must be near RT [19] in which case the respective $M_s$ temperature would be subzero or non-existent. Figure 1 shows the variation of $M_s$ temperature for Fe–8Mn–2Al steels alloyed with up to 1.5 mass-% C. The $M_s$ temperature was calculated based on equation (1). The hatched area in figure 1 indicates the anticipated temperature range for $M_d$ assuming that it lies somewhere between 100 °C and 300 °C above $M_s$ temperature [17,18]. Although relationships such as the one given in equation (1) provide good approximations of $M_s$ for low-alloy steels containing common alloying elements, their applicability to high-alloy steels containing less common alloying elements such as Al is associated with a raised uncertainty. For instance, although Al is regarded in equation (1) as an $M_s$-raiser and it is therefore expected to raise the $M_d$ temperature, its addition to austenitic stainless steels has been found to decrease the $M_d$ temperature [20]. Based on figure 1, it is expected that the medium Mn high C steels in the present study (table 1) meet the condition of subzero $M_s$ temperature and $M_d$ temperatures in the vicinity of RT.

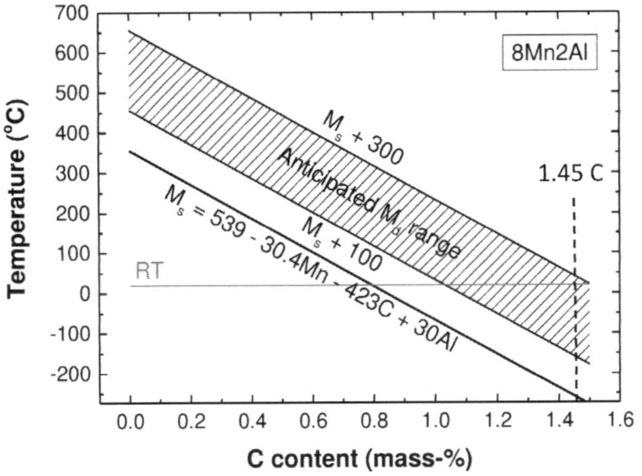


**Figure 1.** Calculated dependence of $M_s$ temperature on C content in Fe–8Mn–2Al–xC steels.

*3.1.2. Inhibition of pearlite.* Obtaining fully austenitic microstructures at RT requires that the formation of pearlite during cooling from austenite range is suppressed. This necessitates the addition of hardenability elements to decelerate the kinetics of pearlite formation. Mn has an obvious hardenability effect in steels [21] and its being an austenite stabilizer ensures that it does not impede the formation of austenite at the solution annealing temperature [22]. The microstructures in figure 2 demonstrate the effect of Mn addition to high C steels on the formation of pearlite during continuous cooling from 1150 °C at a rate of 0.5 °C/s. The microstructure of the binary Fe–1.4C steel consists of pro-eutectoid cementite and pearlite only. The reduced fraction of pearlite in the Fe–1.6C–3Mn steel indicates the decelerated pearlite formation upon alloying with 3 mass-%Mn. Increasing the Mn content to 6 mass-% (Fe–1.7C–6Mn steel) results in the full suppression of pearlite. The presence of Mn in medium Mn steels such as the ones dealt with in the present work (table 1) ensures their adequate resistance against the formation of pearlite.

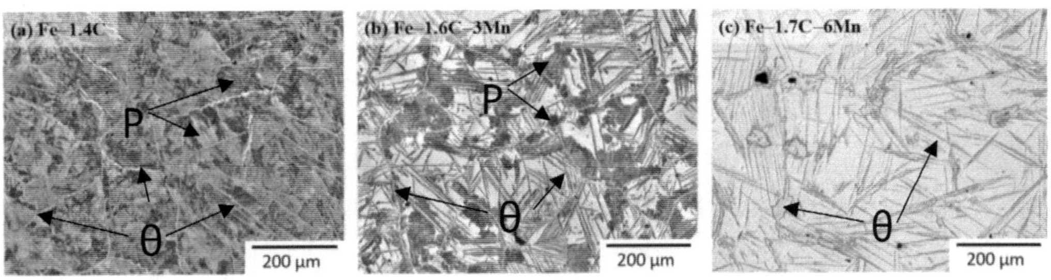


**Figure 2.** Microstructure of Fe–(1.4 - 1.7)C–(0 - 6)Mn steels after cooling from 1150 °C at a rate of 0.5 °C/s. Pikral was used as etchant. P and θ denote pearlite and cementite, respectively. Matrix with a bright contrast in (b) and (c) denotes austenite, possibly coexisting with bainite and martensite.

*3.1.3. Inhibition of cementite.* As discussed in 3.1.1, raised C concentrations are needed in order to achieve subzero $M_s$ temperatures in medium Mn steels. As figure 2 implies, the presence of high C concentrations in steels increases the risk of cementite formation during cooling from the austenite range. Figure 2 also shows that the formation of cementite is not inhibited by the presence of Mn although pearlite formation is clearly suppressed. Therefore, the addition of alloying elements such as Al and Si, which have a recognized decelerating effect on the precipitation of cementite [23,24], becomes necessary. The low solubility of Si in cementite might be responsible for the suppression of cementite during the austempering of Si-added steels [22]. Similar to Si, Al is also insoluble in cementite and therefore decelerates its formation [25]. Accordingly, as the Al content is increased, the C content of retained austenite in the Fe–C–Mn–Al TRIP-assisted steels increases as well [22].

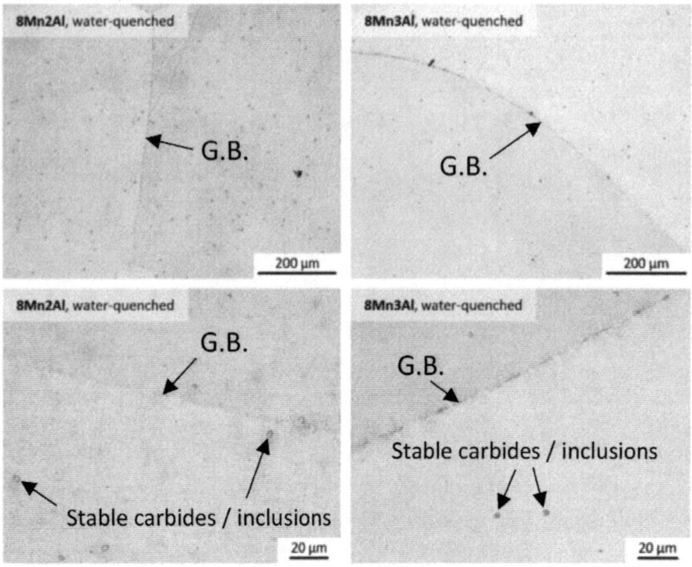


**Figure 3.** Microstructure of 8Mn2Al and 8Mn3Al steels after water quenching from 1150 °C. Nital was used as etchant. G.B. denotes grain boundaries. The matrix is austenite.

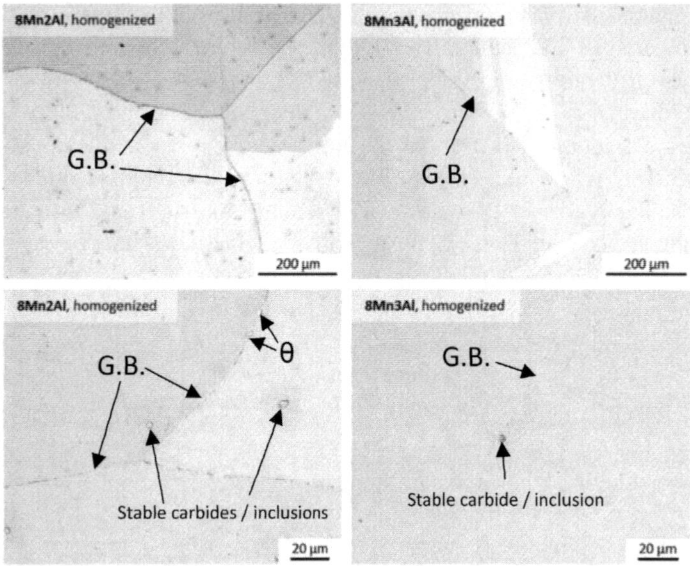


**Figure 4.** Microstructure of 8Mn2Al and 8Mn3Al steels after homogenization. Nital was used as etchant. G.B. and θ denote grain boundaries and cementite, respectively. The matrix is austenite.

In accordance with the preceding principles, the medium Mn high C steels used in the present study contain Al. The high affinity of Al for oxygen favors the formation of an oxide layer on the surface of steels. This oxide layer can in turn protect the bulk alloy from further oxidation and act as a barrier against the outward diffusion of C during high temperature solution annealing of high C steels. Furthermore, Al addition reduces the specific weight and can provide an additional tool to adjust the stability of austenite.

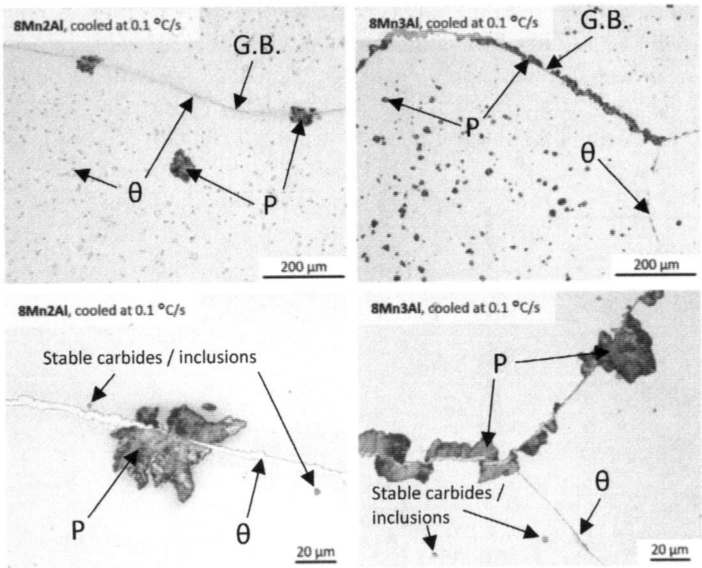


**Figure 5**. Microstructure of 8Mn2Al and 8Mn3Al steels after cooling from 1150 °C at a rate of 0.1 °C/s. Nital was used as etchant. P, G.B. and θ denote pearlite, grain boundaries and cementite, respectively. The matrix is austenite.

*3.2. Mechanical and thermal stability of austenite*

The microstructure of specimens after austenitization at 1150 °C and quenching in water are shown in figure 3. The microstructure of both steels is austenitic. With light optical microscopy, no obvious sign of cementite formation could be detected even at grain boundaries. The particles with an approximate diameter of 2 μm which are uniformly distributed in the microstructure are either inclusions or carbides of strong carbide-forming elements which might be present in the steels. Magnetic measurements using water-quenched specimens indicated near-zero magnetizations, thereby confirming the absence of pearlite and martensite. Magnetizations were also close to zero for the homogenized specimens which had been cooled at an average rate of 8 °C/s between 1250 °C and 650 °C and 4 °C/s between 650 °C and 400 °C. Small cementite particles were however visible at grain boundaries in this condition (figure 4). Cementite formation was more obvious in the 8Mn2Al than in the 8Mn3Al steel indicating the efficiency of Al in opposing cementite formation. When the cooling rate from 1150 °C was reduced to 0.1 °C/s, the fraction of cementite in both steels was increased dramatically (figure 5). In addition to cementite, some pearlite was also observed in the microstructure of both steels. The fraction of pearlite was lower in the 8Mn2Al steel than in the 8Mn3Al steel. In other words, an increase in the concentration of Al opposes the cementite formation but the simultaneous destabilization of austenite with respect to ferrite favors the formation of pearlite. A comparison of the microstructures in figure 5 with those of the Fe–1.7C–6Mn steel cooled at a higher cooling rate of 0.5 °C/s (figure 2) indicates the effectiveness of Al in inhibiting the cementite formation. Furthermore, Al addition changes the morphology of intergranular cementite from plate-shaped to equiaxed (figure 2 vs figure 5).

To compare the thermal stability of austenite in the water-quenched specimens, they were further cooled in liquid nitrogen. Magnetic measurements confirmed the absence of martensite, namely $M_s$ temperature cannot be reached at temperatures as low as -196 °C. These results qualitatively agree with the expectations based on figure 1. To assess the mechanical stability of austenite in the water-quenched specimens, they were given a cold rolling reduction of nearly 50 % at RT and the ferromagnetic phase

fraction was subsequently determined. The martensite fractions induced by cold rolling were 0.75 vol.% and 1.35 vol.% for the 8Mn3Al and 8Mn2Al steels, respectively. The formation of only small fractions of martensite by cold rolling at RT implies the high mechanical stability of austenite, in spite of the low substitutional content of alloys compared to high Mn steels. In other words, the high stability of austenite is mainly due to C, as if some of the Mn in high Mn steels is replaced by C.
The schematic time-temperature-transformation (TTT) diagram in figure 6 summarizes the effects of Mn and Mn+Al addition on the kinetics of phase transformations in high C steels. Although Mn addition significantly suppresses the kinetics of pearlite formation, mainly due to the destabilization of the ferritic constituent of pearlite, it is not effective against the precipitation of cementite. Therefore, obtaining fully austenitic microstructures in high C steels containing Mn as the only substitutional alloying element would not be possible unless very high cooling rates are applied. In order to reduce the critical cooling rate associated with a cementite-free microstructure, use should be made of alloying elements such as Al and Si. In the schematic of figure 6, the effect of Al on the kinetics of bainite formation was extracted from Ref. [26].

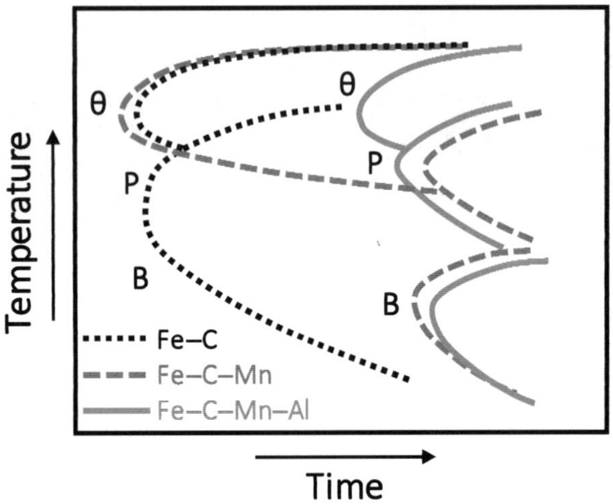


**Figure 6 .** Schematic TTT diagram representing how the kinetics of cementite (θ), pearlite (P), and bainite (B) formation in high C binary steels (dotted black) is influenced by the addition of Mn (dashed red) and Mn+Al (green solid).

## 4. Conclusions

1. The kinetics of pearlite formation in high C steels is decelerated by the addition of Mn.
2. Medium Mn steels can become fully austenitic at RT by raising the C content which reduces the $M_s$ temperature. To suppress the formation of cementite in high C steels, however, high cooling rates are necessary.
3. The addition of Al decelerates the kinetics of cementite formation in medium Mn steels and facilitates obtaining a fully austenitic microstructure at lower cooling rates compared to the ternary Fe–Mn–C steels. Furthermore, Al addition changes the morphology of intragranular cementite from plate-shaped to equiaxed.
4. Both of the Al-added medium Mn high C steels studied in the present work exhibited a high resistance against the formation of martensite by thermal and mechanical treatments. Both steels remained fully austenitic after cooling to -196 °C and formed only less than 2 vol.% martensite after applying a cold rolling reduction of nearly 50 % at RT.

## 5. References


[1] Ghassemieh E 2011 Materials in Automotive Application, State of the Art and Prospects *New Trends and Developments in Automotive Industry* (Croatia) pp 365–94

[2] Lee S and De Cooman B C 2014 Effect of the Intercritical Annealing Temperature on the Mechanical Properties of 10 Pct Mn Multi-phase Steel *Metall. Mater. Trans. A* **45** 5009–16

[3] De Cooman B C, Chen L, Kim H S, Estrin Y, Kim S K and Voswinckel H 2009 State-of-the-Science of High Manganese TWIP Steels for Automotive Applications *Microstructure and Texture in Steels* (Springer, London) pp 165–83

[4] Lee C-Y, Jeong J, Han J, Lee S-J, Lee S and Lee Y-K 2015 Coupled strengthening in a medium manganese lightweight steel with an inhomogeneously grained structure of austenite *Acta Mater.* **84** 1–8

[5] Lee S and De Cooman B C 2015 Tensile Behavior of Intercritically Annealed Ultra-Fine Grained 8% Mn Multi-Phase Steel *Steel Res. Int.* **86** 1170–8

[6] De Cooman B C, Lee S J, Shin S, Seo E J and Speer J G 2017 Combined Intercritical Annealing and Q&P Processing of Medium Mn Steel *Metall. Mater. Trans. A* **48** 39–45

[7] Tsuchiyama T, Inoue T, Tobata J, Akama D and Takaki S 2016 Microstructure and mechanical properties of a medium manganese steel treated with interrupted quenching and intercritical annealing *Scr. Mater.* **122** 36–9

[8] Zhang Y, Wang L, Findley K O and Speer J G 2017 Influence of Temperature and Grain Size on Austenite Stability in Medium Manganese Steels *Metall. Mater. Trans. A* **48** 2140–9

[9] Zou Y, Xu Y B, Hu Z P, Gu X L, Peng F, Tan X D, Chen S Q, Han D T, Misra R D K and Wang G D 2016 Austenite stability and its effect on the toughness of a high strength ultra-low carbon medium manganese steel plate *Mater. Sci. Eng. A* **675** 153–63

[10] Furukawa T, Huang H and Matsumura O 1994 Effects of carbon content on mechanical properties of 5%Mn steels exhibiting transformation induced plasticity *Mater. Sci. Technol.* **10** 964–70

[11] De Cooman B C, Estrin Y and Kim S K 2018 Twinning-induced plasticity (TWIP) steels *Acta Mater.* **142** 283–362

[12] Allain S, Chateau J-P, Bouaziz O, Migot S and Guelton N 2004 Correlations between the calculated stacking fault energy and the plasticity mechanisms in Fe–Mn–C alloys *Mater. Sci. Eng. A* **387–389** 158–62

[13] Saeed-Akbari A, Mosecker L, Schwedt A and Bleck W 2012 Characterization and Prediction of Flow Behavior in High-Manganese Twinning Induced Plasticity Steels: Part I. Mechanism Maps and Work-Hardening Behavior *Metall. Mater. Trans. A* **43** 1688–704

[14] Seo E J, Kim J K, Cho L, Mola J, Oh C Y and De Cooman B C 2017 Micro-plasticity of medium Mn austenitic steel: Perfect dislocation plasticity and deformation twinning *Acta Mater.* **135** 112–23

[15] Mahieu J, De Cooman B C and Maki J 2002 Phase transformation and mechanical properties of si-free CMnAl transformation-induced plasticity-aided steel *Metall. Mater. Trans. A* **33** 2573–80

[16] Bouaziz O, Zurob H, Chehab B, Embury J D, Allain S and Huang M 2011 Effect of chemical composition on work hardening of Fe-Mn-C TWIP steels *Mater. Sci. Technol.* **27** 707–9

[17] Huang Q, Volkova O, Biermann H and Mola J 2017 Tensile elongation of lean-alloy austenitic stainless steels: transformation-induced plasticity versus planar glide *Mater. Sci. Technol.* **33** 1224–30

[18] Mola J, Ullrich C, Kuang B, Rahimi R, Huang Q, Rafaja D and Ritzenhoff R 2017 Austenitic Nickel- and Manganese-Free Fe-15Cr-1Mo-0.4N-0.3C Steel: Tensile Behavior and Deformation-Induced Processes between 298 K and 503 K (25 °C and 230 °C) *Metall. Mater. Trans. A* **48** 1033–52

[19] Mola J 2017 Considerations in the Design of Formable Austenitic Stainless Steels Based on Deformation-Induced Processes *Austenitic Stainless Steels - New Aspects* (Croatia) pp 7–28

[20] Rahimi R, Ullrich C, Klemm V, Rafaja D, De Cooman B C, Biermann H and Mola J 2016 Influence of Al on the temperature dependence of strain hardening behavior and glide planarity in Fe–Cr–Ni–Mn–C austenitic stainless steels *Mater. Sci. Eng. A* **649** 301–12
[21] Calcagnotto M, Ponge D and Raabe D 2012 On the Effect of Manganese on Grain Size Stability and Hardenability in Ultrafine-Grained Ferrite/Martensite Dual-Phase Steels *Metall. Mater. Trans. A* **43** 37–46
[22] Kuziak R, Kawalla R and Waengler S 2008 Advanced high strength steels for automotive industry *Arch. Civ. Mech. Eng.* **8** 103–17
[23] Samek L, Moor E D, Penning J and De Cooman B C 2006 Influence of alloying elements on the kinetics of strain-induced martensitic nucleation in low-alloy, multiphase high-strength steels *Metall. Mater. Trans. A* **37** 109–24
[24] Kozeschnik E and Bhadeshia H K D H 2008 Influence of silicon on cementite precipitation in steels *Mater. Sci. Technol.* **24** 343–7
[25] Liu S, Zhang F, Yang Z, Wang M and Zheng C 2016 Effects of Al and Mn on the formation and properties of nanostructured pearlite in high-carbon steels *Mater. Des.* **93** 73–80
[26] Mertens A, Jacques P j., Sietsma J and Delannay F 2008 Relative Influences of Aluminium and Silicon on the Kinetics of Bainite Formation from Intercritical Austenite *Steel Res. Int.* **79** 954–9

**Acknowledgements**
G. Luan would like to thank the financial support of the China Scholarship Council (CSC). Thanks are also due to the technical staff at the Institute of Iron and Steel Technology of TU Bergakademie Freiberg.

IOP Conf. Series: Materials Science and Engineering **373** (2018) 012003 doi:10.1088/1757-899X/373/1/012003

# Corrosion characteristics of a quenching and partitioning steel determined by electrochemical impedance spectroscopy

**T Mehner[1,*], R Morgenstern[1], P Frint[2], I Scharf[1], M F-X Wagner[2] and T Lampke[1]**

[1] Materials and Surface Engineering Group, Institute of Materials Science and Engineering, Chemnitz University of Technology, 09107 Chemnitz, Germany
[2] Materials Science Group, Institute of Materials Science and Engineering, Chemnitz University of Technology, 09107 Chemnitz, Germany

[*] e-mail: thomas.mehner@mb.tu-chemnitz.de

**Abstract.** Among the advanced high-strength steels, quenching and partitioning (Q&P) steels provide excellent ultimate tensile strength in combination with good ductility. The mass fraction of martensite and austenite can be tailored by the Q&P treatment. In this study, the impact of the microstructure on the corrosion behaviour is investigated. For this, the microstructures of two Q&P processing routes and a quenched and tempered state are analysed using scanning electron microscopy and X-ray diffraction. Electrochemical impedance spectroscopy is conducted using two different media in order to determine the time evolution of corrosion rates during short-time corrosion. The results of this study demonstrate that correlations between Q&P microstructures and the corrosion behaviour can be analysed in detail with these experimental methods.

## 1. Introduction

Advanced high-strength steels (AHSS) provide excellent opportunities for light-weight design. Especially promising is the quenching and partitioning (Q&P) process, as Q&P steels combine high strength (e.g. yield stress > 1500 MPa) with a relatively large elongation to fracture > 20 % [1, 2]. The Q&P process consists of austenitising at a temperature $\vartheta_a$, quenching to a temperature $\vartheta_Q$ above the martensite finish temperature directly followed by a partitioning process at $\vartheta_P$. Afterwards, the Q&P steel is cooled down to room temperature, either in air or by quenching in oil or water [3, 4]. Such processes result in microstructures that consist mainly of martensite with a small volume fraction of stabilised, interlath retained austenite. This special microstructure gives rise to interesting mechanical properties – but it potentially also leads to corrosion susceptibility. Especially low-alloy steels cannot form passive layers and, thus, are prone to (uniform) corrosion. Despite the practical importance of corrosion resistance for future applications of Q&P steels, only few related studies can be found in the literature. The corrosion behaviour of a Q&P-treated medium carbon steel in NaCl solution was investigated by Yang et al. [5] using potentiodynamic polarisation curves and electrochemical impedance spectroscopy. They found similar corrosion properties and a corrosion rate that decreased slightly for the Q&P treated steel compared to the quenched and tempered (Q&T) state of the same steel using a maximum test duration of 350 h. Lu et al. [6] studied corrosion of a stainless steel in Q&T and Q&P states in NaCl solution. Depending on the partition temperature, they found that the pitting potential was increased by about 100 mV. Thus, the Q&P treatment reduced the susceptibility to pitting when partitioned at 350 °C or 450 °C.

Further research is required concerning the short-time corrosion behaviour of low-alloy Q&P steels. Whereas the diffusion of oxygen to the surface limits the corrosion rate for neutral pH values [7], especially the regime of the acid corrosion (pH < 4), in which microstructural effects gain importance, has not yet been investigated. To address this topic, this paper focusses on corrosion-rate determination by electrochemical impedance spectroscopy (EIS) of two Q&P states of a low-alloy steel in comparison with a Q&T state of the same alloy, and on their correlation with typical microstructural features in neutral and acidic media.

## 2. Material and Methods

Sheets with a composition of 0.38 C, 1.92 Si, 1.39 Cr, 0.66 Mn, 0.051 S, 0.038 Mo, 0.011 P (wt%) were conventionally hot rolled to a thickness of 10 mm. Three sheets were heat-treated using the parameters summarised in table 1. They were selected in order to achieve highest strength and moderate ductility based on previous work [8].

**Table 1.** Heat-treatment parameters for the Q&T and Q&P states. The partitioning time given is the time at $\vartheta_p$ and does not include the heating time (28 min or 35 min for Q&P 1 or 2, respectively).

| State | $\vartheta_a$ and austenitisation time (oven) | $\vartheta_Q$ (in water) | Molten salt temperature and holding time | $\vartheta_p$ and partitioning time (oven) | Post-annealing temperature and time |
|---|---|---|---|---|---|
| Q&T | 950 °C, 21 min | 60 °C | – | – | 250 °C, 10 min |
| Q&P 1 | 950 °C, 21 min | 400 °C | 200 °C, 5 min | 300 °C, 10 min | – |
| Q&P 2 | 950 °C, 21 min | 350 °C | 150 °C, 5 min | 250 °C, 10 min | – |

Specimens for EIS and scanning electron microscopy (SEM) analyses were cut from the sheets, ground and finally polished with 3 µm diamond polishing solution. The outer surface layer of about 300 µm was removed because it exhibited oxidation and decarburisation after the heat treatment. SEM was used to investigate the microstructure of the specimens. A NEON40EsB (Zeiss) microscope with a secondary-electron (SE) in-lens detector was applied with 5 kV acceleration voltage.
X-ray diffraction (XRD) measurements were performed utilising a D8 Discover diffractometer (Bruker AXS) with Co-K$_\alpha$ radiation (40 kV, 40 mA, point focus), a collimator with 1 mm diameter and an energy-filtered 1D Lynxeye XE detector. The diffraction diagrams were measured in the diffraction-angle ($2\theta$) range 10°–130° with a step size of 0.01° and a time of 3.4 s/step, which corresponds to 652.8 s/step due to the use of the 1D detector. The mass fractions, the lattice parameters ($a$ and $c$), microstrain (measure for the defect density, full width at half maximum (FWHM) of the Gaussian part of the line width over tan $\theta$) and the crystallite sizes of the phases austenite and martensite were determined using the Rietveld method, considering device-related line-broadening effects.

The carbon content $w_C$ (mass fraction in %) in the retained austenite was determined using the equation [9]

$$\boldsymbol{w}_{\mathrm{C}} = \frac{\boldsymbol{a}_{\gamma} - \mathbf{0.3573}\ \mathrm{nm}}{\mathbf{0.0033}\,\frac{\mathrm{nm}}{\mathrm{wt\%}}}, \tag{1}$$

as a function of the lattice parameter $a_\gamma$ of the retained austenite.

Corrosion testing by EIS was performed in diluted $H_2SO_4$ (0.25 mol/l) or NaCl solution (0.1 mol/l, pH 7) at 25 °C utilising an electrochemical work station IM6 (Zahner Elektrik GmbH) using a three-electrode setup with Ag/AgCl-saturated KCl reference electrode and a corroded area $A$ of 0.785 cm$^2$. For all specimens, 30 consecutive measurement cycles were performed, each one consisting of

- open circuit potential (OCP) stabilisation for 10 min,
- EIS measurement (amplitude 2 mV, 4 kHz–45 mHz, 5 frequencies per decade and 20 repetitions per frequency) for 23 min.

Thus, the total measurement time was 16.5 h per specimen. At least 4 specimens of each state were studied. The 30 EIS spectra for each specimen were interpolated linearly in time in order to extract spectra between 1.5 h and 14.5 h with a step size of 0.5 h. These spectra were averaged for each heat-treatment state. The equivalent circuit used for evaluation is shown in figure 1. It consists of a solution resistance $R_S$, charge-transfer resistance $R_{ct}$ and an imperfect capacitor, a so-called constant-phase element (CPE). $R_{ct}$ and the CPE represent the electrochemical behaviour of the specimen. The imperfection of the capacitor results mainly from the diffuse double layer and the surface roughness of the specimen.

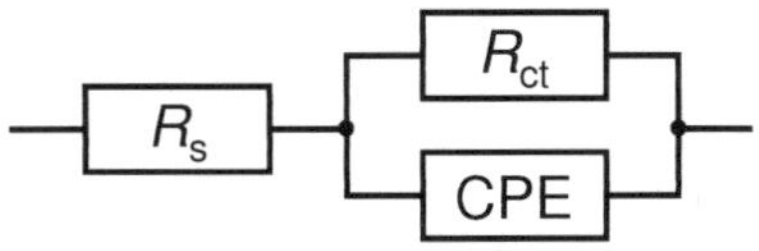


**Figure 1.** Equivalent circuit used for EIS.

The corrosion rate $k_{corr}$ is inversely proportional to $R_{ct}$. For Fe, it can be calculated by [10]:

$$\boldsymbol{k}_{\text{corr}} = \mathbf{0.9987}\ \frac{\text{mm} \cdot \text{cm}^2 \cdot \Omega}{\text{a} \cdot \text{K}} \cdot \frac{\boldsymbol{T}}{\boldsymbol{A} \cdot \boldsymbol{R}_{\text{ct}}} \text{ in } H_2SO_4 \text{ and} \quad (2)$$

$$\boldsymbol{k}_{\text{corr}} = \mathbf{0.9248}\ \frac{\text{mm} \cdot \text{cm}^2 \cdot \Omega}{\text{a} \cdot \text{K}} \cdot \frac{\boldsymbol{T}}{\boldsymbol{A} \cdot \boldsymbol{R}_{\text{ct}}} \text{ in NaCl,} \quad (3)$$

with the absolute temperature $T$.

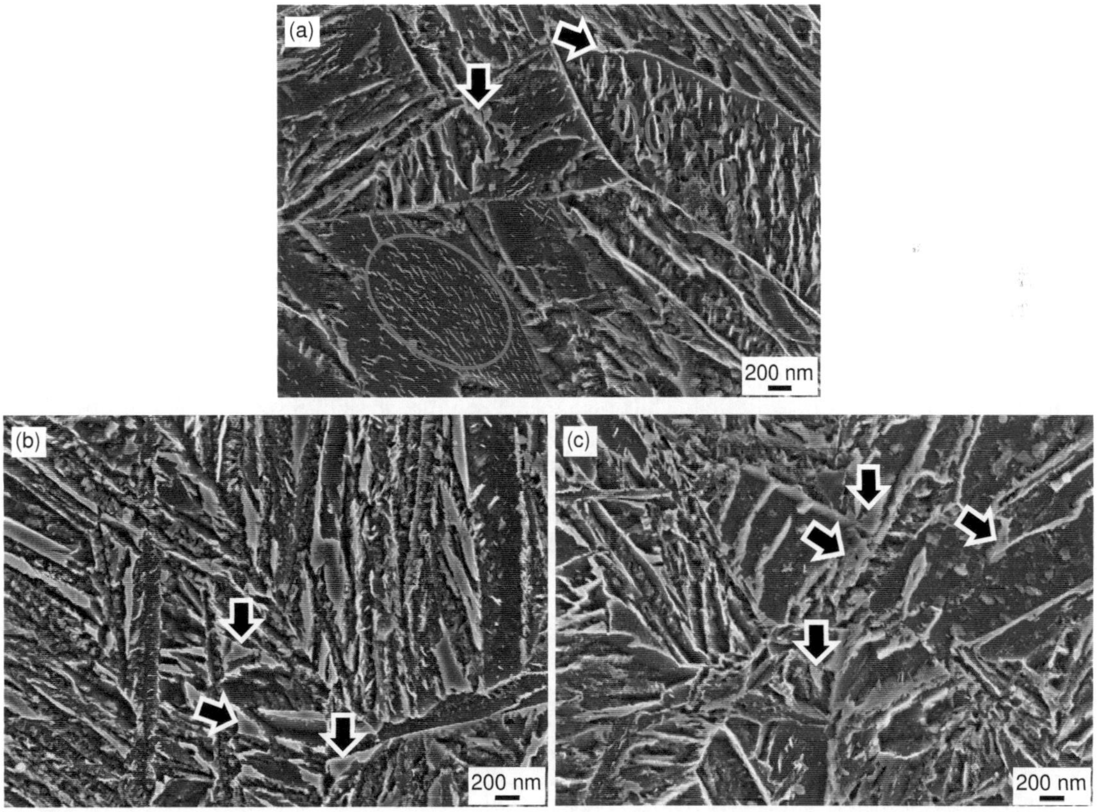


**Figure 2.** SEM micrographs of Q&T (a), Q&P 1 (b) and Q&P 2 (c). The red circles indicate examples of carbides and the arrows highlight retained austenite.

## 3. Results and discussion

### *3.1. Microstructural characterisation*

Typical SEM micrographs of the three different heat-treated states are shown in figure 2. Clear evidence of carbides in the Q&T martensitic areas can be found (see figure 2 (a)). Minor fractions of carbides rarely occur in the Q&P states as well (see figures 2 (b) and (c)). The microstructures of both Q&P processed conditions show similar features. Even though retained austenite and martensite cannot easily be distinguished microscopically, the Q&P states exhibit more areas in which austenite is likely present. Yang et al. found the retained austenite as interlath phase between the martensite laths in Q&P treated samples and almost no traces of inter-lath austenite in Q&T states [5]. A similar distribution can be assumed here. A quantification of the phase fractions was done by XRD. The diffraction diagrams revealed the presence of martensite and austenite only, whereas no carbide phases were detected. Thus, the mass fraction of these additional phases is below 1 %. Table 2 summarises the parameters determined by XRD and $w_C$ for the three states ($a$, $c$ – lattice parameters of martensite).

**Table 2.** Parameters of the phases martensite and austenite.

| | Q&T | Q&P 1 | Q&P 2 |
|---|---|---|---|
| | Martensite | | |
| Mass fraction (%) | 97.5 | 92.5 | 92.0 |
| $a$ (nm) | 0.2881 | 0.2882 | 0.2873 |
| $c$ (nm) | 0.2856 | 0.2857 | 0.2851 |
| Crystallite size (nm) | 57 ± 1 | 50 ± 1 | 66 ± 2 |
| Microstrain (°) | 0.793 | 0.766 | 0.560 |
| | Austenite | | |
| Mass fraction (%) | 2.5 | 7.5 | 8.0 |
| $a_\gamma$ (nm) | 0.3600 | 0.3606 | 0.3611 |
| Crystallite size (nm) | 17 ± 3 | 30 ± 3 | 37 ± 10 |
| $w_C$ (%) | 0.82 | 1.00 | 1.15 |

The Q&T state still contains a typical, small fraction of retained austenite, which is probably a consequence of stabilising effects of mechanical stresses and C diffusion during the cool-down processes. The two Q&P states possess almost the same amount of retained austenite (7.5–8.0 %), but show significant differences in terms of the other microstructural parameters: Q&P 2 has smaller lattice parameters, a larger crystallite size for martensite and larger $w_C$ in austenite. Due to the longer total time for partitioning and heating up, the recovery processes in the martensite phase can take place more intensely and thus reduce microstrains while also increasing the crystallite size. The lower quenching and holding temperature likely results in an even lower solubility of C in the martensite compared to Q&P 1, increasing the C diffusion rate into the austenite. As a consequence, $w_C$ is larger for Q&P 2. The similarity between the microstructural parameters of the martensite phases of Q&T and Q&P 1 is remarkable: As the lattice parameters of martensite are almost identical but the retained austenite contains more carbon for Q&P 1, the presence of carbide phases in Q&T is evident, which can be observed in the microstructural images in figure 2.

### *3.2. EIS characterisation*

In figure 3, typical Nyquist plots of EIS measurements of the three heat-treatment states in $H_2SO_4$ are shown. Clear differences between Q&T and Q&P can be identified. The diameter of the semicircle is significantly larger for Q&T compared with Q&P, indicating a lower corrosion rate for Q&T. The calculated $R_{ct}$ values for corrosion in both $H_2SO_4$ and NaCl were converted into corrosion rates (equations (2) and (3)), which are shown in figure 4 as a function of corrosion time. The corrosion appearance was uniform in both media.

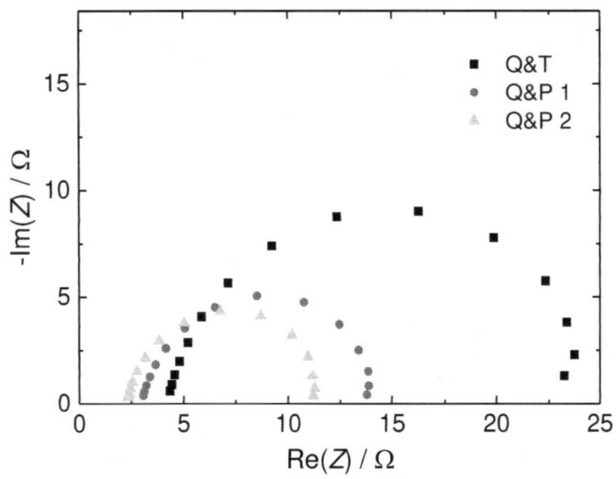


**Figure 3.** Exemplary Nyquist plots (impedance $Z$) measured during the corrosion of the three heat-treatment states in $H_2SO_4$ after 10 h.

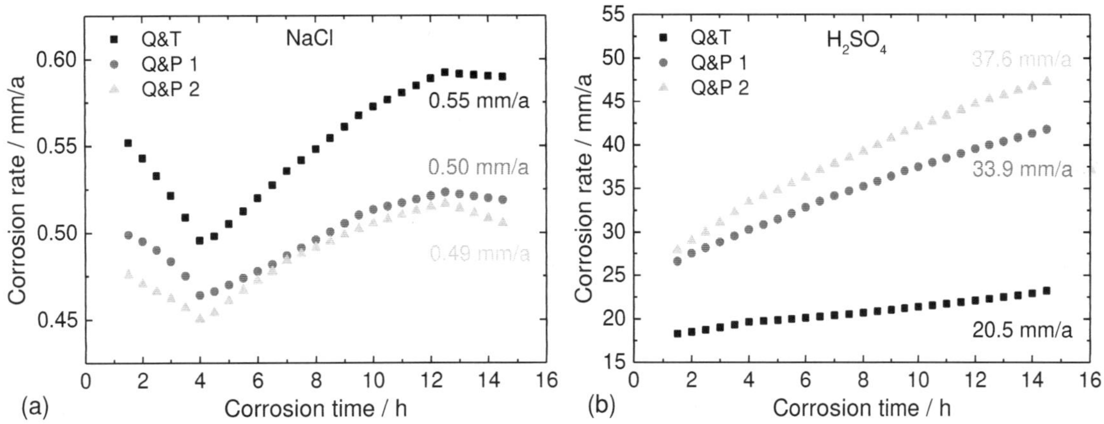


**Figure 4.** Time dependence of the corrosion rates of Q&T and Q&P heat-treatment states for the corrosion in NaCl (a) and $H_2SO_4$ (b). The numbers given are time averages of the corrosion rates.

For NaCl, all corrosion rates show a minimum after 4 h. This is a consequence of the formation of a passive layer due to the initial corrosion, which reduces the corrosion rate by about 10 %. The passive layer is dissolved in the further corrosion process leading to an increase of the corrosion rate, which reaches the initial value after about 8 h. After a corrosion time of about 12 h, the corrosion rate appears to stabilise. The differences between Q&P 1 and Q&P 2 are almost negligible. Another study with a low time resolution indicates an almost constant corrosion rate up to about 30 h, which is in agreement with the results in figure 4, followed by a significant increase [5]. For neutral pH values (7 for the NaCl solution), the differences of the microstructure (e.g. crystallite size or defect density) do not lead to differences of the corrosion rate, as the diffusion of oxygen to the surface limits the velocity of the cathode reaction and thus the corrosion rate [7]. However, significant differences of the phase composition can increase the corrosion velocity e.g. by increasing the cathodic area fraction. As the cathodic reaction is the reaction-rate limiting process for Fe [11], this will increase the corrosion rate: C atoms solved in martensite alter the potential of the areas to be more cathodic [12]. Figure 5 schematically shows the difference of the C distribution that may lead to an increase of the cathodic area for the Q&T state compared with Q&P: In Q&T, the C atoms are primarily solved in the martensite phase (only 2.5 % of retained austenite, black in figure 5 (a)). In proximity to the retained austenite, the C content in martensite drops because the C atoms can diffuse into the retained austenite during the tempering process (light grey areas). Due to the different concentration of C atoms, the martensitic areas close to the retained austenite will be more anodic compared to the other parts of the grain (dark grey areas). Thus, the grains mostly form cathodic areas with anodic parts near the retained austenite. The retained austenite is also cathodic to the martensite [13] in this medium. In Q&P (figure 5 (b)), in contrast, the C content and its difference within different parts of the martensitic grains is low and thus the differences between martensite and austenite predominantly determine the corrosion behaviour in neutral NaCl solution. The total cathodic area (retained austenite) in the Q&P state is smaller than in the Q&T state (C-rich martensite and retained austenite). Comparing Q&T with Q&P heat treated materials, this model rationalises the 10 % increase of the corrosion rate in Q&T.

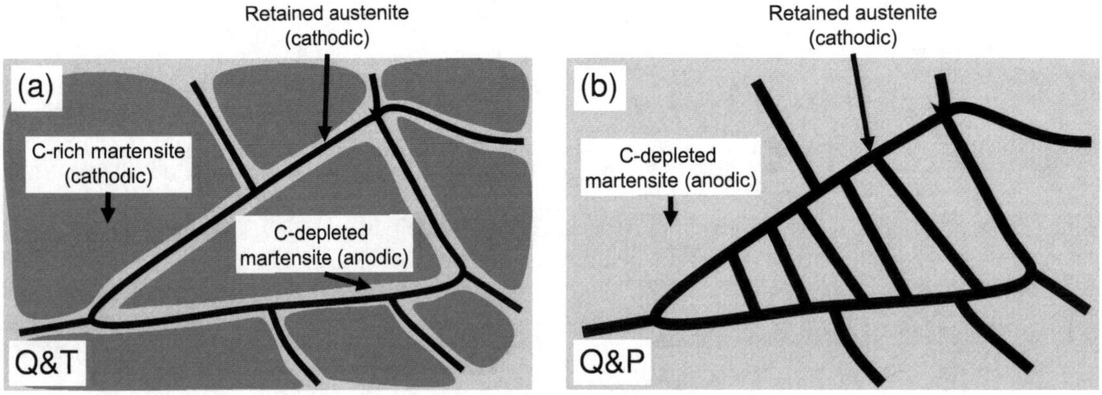


**Figure 5.** Schematic representation of the microstructure in Q&T (a) and Q&P (b) states. Black lines correspond to austenite and grey areas to martensite. In dark grey, cathodic C-rich martensite areas are shown.

The low pH value in the $H_2SO_4$ solution allows for hydrogen formation as the cathodic reaction and thus significantly increases the corrosion rate compared with the NaCl solution. In diluted $H_2SO_4$, the corrosion rates of the Q&P states are about a factor of 2 higher compared to that of the Q&T state. The dual-phase microstructure of the Q&P specimens can form a microscopic galvanic (bimetallic) corrosion element enhancing the velocity of the corrosion reaction. The highly-dispersed distribution of the austenite phase in the Q&P states results in the formation of a large number of such galvanic corrosion elements and thus in a significant increase of the corrosion rate compared with Q&T. The larger amount of carbon in the retained austenite in Q&P 2 reduces the hydrogen overpotential [12] leading to a further increase of the corrosion rate in comparison with Q&P 1. This also means that carbon atoms diffuse to defect sites (e.g. dislocations) in the retained austenite, forming additional corrosion elements and spots with reduced hydrogen overpotential, which both lead to a local favour of the cathodic reaction. With progressing time, the corrosion rate strongly increases (by almost a factor of 2), which is most likely related to the increase of the surface roughness and thus the size of the reactive area. This larger area provides even more cathodic areas and, in consequence, increases the corrosion rate.

## 4. Summary and conclusions

Depending on the corrosion system, Q&P treatments can improve or lower the corrosion resistance compared to a Q&T state of the same alloy. For neutral pH values with $Cl^-$ ions, the Q&P specimens show lower corrosion rates compared with Q&T specimens. During the corrosion in NaCl solution, a short-time passivation (< 4 h) of all specimens occurs. The stability of this layer is low and thus it quickly dissolves. The general time evolution of the three states is similar.

In diluted $H_2SO_4$, the average corrosion rate of the Q&P states is significantly higher (65 % for Q&P 1 and 80 % Q&P 2) than for the Q&T treatment at any time. In addition, within the first 16 h, it nearly doubles for the Q&P states. This increase is nearly linear in time with a slightly decreasing slope. Even though the state Q&P 1 shows a 10 % lower corrosion rate compared with Q&P 2, the overall very high corrosion damage of this steel in any heat-treatment condition in the acid medium proves the necessity for a corrosion protection in such aggressive environments. In the light of recent advances and potential applications of Q&P steels, there is a need for further, in-depths studies on the corrosion behaviour –

both on elementary corrosion mechanisms in multi-phase microstructures, and on engineering solutions to provide better corrosion protection to this promising class of structural materials.

## 5. References

[1] Yang J, Huang F, Guo Z, Rong Y and Chen N 2016 *Mat. Sci. Eng. A* **665** 76–85
[2] Jirková H, Masek B, Wagner M F-X, Langmajerová D, Kucerová L, Treml R and Kiener D 2014 *J. Alloys Compounds* **615** S163–68
[3] Sarikaya M, Thomas G, Steeds J W, Barnard S J and Smith G D W 1982 *Proc. Int. Conf. Solid to Solid Phase Transform. 1982* 1421–25
[4] Speer J, Matlock D K, De Cooman B C and Schroth J G 2003 *Acta Mater.* **51** 2611–22
[5] Yang J, Lu Y, Guo Z, Gu J and Gu C 2018 *Cor. Sci.* **130** 64–75
[6] Lu S-Y, Yao K-F, Chen Y-B, Wang M-H, Chen N and Ge X-Y 2016 *Cor. Sci.* **103** 95–104
[7] Whitman G W, Russell R P and Altieri V J 1924 *Ind. Eng. Chem.* **16** 665–70
[8] Kucerová L, Aisman D, Jirková H, Masek B and Hauserová D 2009 *An. DAAAM for 2009 & Proc. 20th Int. DAAAM Symposium* **20** 1035–36
[9] Ridley N, Stuart H and Zwell L 1969 *Trans. TMS-AIME* **245** 1834–36
[10] Mehner T 2017 *PhD thesis* Chemnitz University of Technology (in German)
[11] El Miligy A A, Geana D and Lorenz W J 1975 *Electrochimica Acta* **20** 273–81
[12] Uhlig H H 1971 *Corrosion and Corrosion Control: an Introduction to Corrosion Science and Engineering* vol 2 (New York: John Wiley & Sons)
[13] Francis R 2000 Bimetallic Corrosion: Guides to Good Practice in Corrosion Control (National Physical Laboratory)

## Acknowledgments

The authors gratefully acknowledge the experimental support of Dr. Harry Podlesak (SEM) and Marc Pügner (XRD).

# On the hardness of high carbon ferrous martensite

**J Mola[*] and M Ren**

Institute of Iron and Steel Technology, Technische Universität Bergakademie Freiberg, Leipziger St. 34, 09599 Freiberg, Germany

[*] e-mail: mola@iest.tu-freiberg.de

**Abstract.** Due to the presence of retained austenite in martensitic steels, especially steels with high carbon concentrations, it is difficult to estimate the hardness of martensite independent of the hardness of the coexisting austenite. In the present work, the hardness of ferrous martensite with carbon concentrations in the range 0.23-1.46 mass-% was estimated by the regression analysis of hardnesses for hardened martensitic-austenitic steels containing various martensite fractions. For a given carbon concentration, the hardness of martensitic-austenitic steels was found to increase exponentially with an increase in the fraction of the martensitic constituent. The hardness of the martensitic constituent was subsequently estimated by the exponential extrapolation of the hardness of phase mixtures to 100 vol.% martensite. For martensite containing 1.46 mass-% carbon, the hardness was estimated to be 1791 HV. This estimate of martensite hardness is significantly higher than the experimental hardness of 822 HV for a phase mixture of 68 vol.% martensite and 32 vol.% austenite. The hardness obtained by exponential extrapolation is also much higher than the hardness of 1104 HV based on the rule of mixtures. The underestimated hardness of high carbon martensite in the presence of austenite is due to the non-linear dependence of hardness on the martensite fraction. The latter is also a common observation in composite materials with a soft matrix and hard reinforcing particles.

## 1. Introduction

The high strength and toughness of martensitic steels make them suitable candidates for a variety of structural and engineering applications [1]. The higher strength of martensite compared to other constituents such as ferrite, austenite, pearlite, and bainite is the reason behind the increasing use of martensite-containing steels. For instance, low carbon dual-phase and hot stamping steels with partially and almost fully martensitic microstructures, respectively, are being increasingly used for structural automobile parts [2,3]. Medium and high carbon martensitic steels, on the other hand, are commonly used in engineering applications such as gears, shafts, and bearings [4,5].

The strength of ferrous martensite arises from factors such as dislocation hardening [6,7], grain boundary hardening as described by the Hall-Petch relation [8], and interstitial strengthening [9]. The strengthening effect of carbon has been proposed to be a consequence of the concurrent increase in the density of dislocations [6]. The dependence of hardness on the carbon concentration has been established in numerous investigations [4,10–12]. For low carbon steels, hardness values reported in various investigations are largely consistent and indicate significant strengthening of martensite by an increase in the solute carbon concentration [12]. The rate of increase in the hardness of martensite decreases as the carbon concentration increases [4,12]. The hardness may even decrease as the carbon content exceeds 0.8-1 mass-% [13]. The latter observation may be justified by the presence of large quantities of austenite in the microstructure of as-quenched high carbon steels. Therefore, the hardness of martensite in high carbon steels cannot be estimated unless the softening effect of retained austenite is properly excluded.

By adjusting martensitic-austenitic microstructures containing various retained austenite contents, the present work enables to determine the dependence of hardness on the fraction of martensite. Establishing an appropriate fitting and extrapolation scheme in turn enables to obtain a realistic estimation of the hardness of martensite, especially in the case of high carbon martensitic steels where the coexistence of high austenite fractions with the martensite is inevitable.

## 2. Experimental procedure

Ingots with an approximate weight of 300 g and chemical compositions as given in table 1 were prepared in a cold-crucible induction melting facility. Specimens with approximate thicknesses of 4.5-5 mm were austenitized at 1150 °C for 5 min and subsequently quenched to room temperature (RT) using brine (10 NaCl in water). Subsequently, various martensite fractions were adjusted by resumed cooling to temperatures as low as -196 °C, approximately 3 sec after brine quenching. To ensure the absence of decarburization and oxidation in hardened specimens, an outermost layer of specimens with an approximate thickness of 0.8 mm was removed by mechanical grinding. Bulk martensite fractions were then quantified by magnetic measurements at room temperature using a Metis MSAT device equipped with a Lakeshore 480 fluxmeter. For pure iron, the equipment returns a mass magnetization of 213.5 emu/g. Given that the mass magnetization of iron decreases by 1.18 emu/g per mass-% solute Mn [14], the expected magnetization of specimens in the fully martensitic condition ($\sigma_{\alpha'}$) was calculated using the following equation:

$$\sigma_{\alpha'}\ (\text{emu/g}) = 213.5 - (1.18 \times C_{Mn}) \quad (1)$$

Martensite fractions of ternary Fe−Mn−C steels were quantified by dividing measured magnetizations by the magnetization based on equation (1). To obtain hardness values representative of all phases, Vickers hardness measurements with a load of 98 N (HV10) were done using a Wolpert 430-SVD indenter. To estimate the hardness of austenite in the steel 1.4C, a small load of only 49 mN (HV0.005) was applied using a Shimadzu HMV-2000 indenter. To obtain reliable average hardness values, five indentations were performed per specimen. To minimize the occurrence of aging/tempering reactions prior to hardness measurements, the specimens were mounted using cold resin. Mechanical grinding was done using SiC papers with grit sizes between 220 and 2000. Subsequently, specimens were polished using diamond suspensions with particle sizes of 3 µm and 1 µm for a total duration of 20 min. Final polishing for 2 min was done using a suspension of colloidal silica with a particle size of 0.04 µm. The deformation-induced transformation of austenite during the polishing steps was minimized by grinding under the flow of water at 80 °C. Light optical microscopy examinations using unpolarised light were done using a Neophot 30 microscope.

**Table 1.** Chemical composition of ternary steels in mass-%.

| Steel ID | Mn | C | Fe |
|---|---|---|---|
| **0.2C** | 1.49 | 0.23 | bal. |
| **0.6C** | 1.49 | 0.64 | bal. |
| **1.0C** | 1.33 | 1.04 | bal. |
| **1.4C** | 1.48 | 1.46 | bal. |

## 3. Results and discussion

Table 2 summarizes the martensite contents associated with various quench temperatures. The choice of low quench temperatures aimed to eliminate austenitic regions with a low stability at RT by inducing their thermal transformation to martensite. In general, the highest quench temperature was raised as the carbon concentration increased. This reduced the likelihood of deformation-induced martensite formation during the surface grinding and polishing steps. For the 1.4C steel, RT was used as the highest quench temperature. Figure 1 shows light optical micrographs for the 1.4C steel quenched to three different temperatures, indicating the increase in the fraction of martensite at lower quench temperatures. To obtain at least two different martensite contents in each steel, quenching to RT was also done for the 0.2C steel. The fraction of retained austenite after cooling to RT was only 3.5 vol.%. The retained austenite was therefore present as film-like regions between martensitic laths. This retained austenite morphology is associated with a much higher mechanical stability compared to the blocky retained austenite present in steels with higher retained austenite contents [15]. Therefore, quenching to RT was also done for the 0.2C steel. Given the influence of C on the martensite start temperature [16]. As table 2 shows, the martensite fraction obtained by quenching to -196 °C decreased as the carbon concentration increased.

**Table 2**. Martensite fractions after quenching to indicated temperatures

| Steel ID | Martensite fractions after quenching, vol.% | | | | |
|---|---|---|---|---|---|
| | **20 °C** | **0 °C** | **-50 °C** | **-100 °C** | **-196 °C** |
| **0.2C** | 96.5 | - | - | - | 98.6 |
| **0.6C** | - | - | - | 92.4 | 94.4 |
| **1.0C** | - | - | 83.8 | 85.7 | 87.3 |
| **1.4C** | 8.1 | 30.2 | 54.3 | 64.9 | 68.4 |

Due to the presence of untransformed austenite in the microstructure of all steels even the 0.2C steel with the lowest carbon content, the true hardness of martensite cannot be determined independently of the hardness of austenite. A possible way of excluding the effect of austenite is to extrapolate hardness values for the specimens with various martensite contents. In figure 2, hardness values are plotted as a function of the martensite fraction. The hardness values were measured using an applied load of 98 N (10 kgf). Under this load, the mean diagonal length of indentations would be at least 136 µm for hardness values up to 1000 HV. Therefore, the indentations are large enough to ensure the contribution of both phases to the measured values. Linear extrapolation has been previously used to estimate the hardness of fresh and tempered martensites in the microstructure of a quenched and partitioned Fe−13Cr−0.3C stainless steel [17]. When martensite fractions are varied in a narrow range close to 100 vol.%, different extrapolation schemes lead to similar hardness approximations for martensite. Nevertheless, linear extrapolation of hardness to 0 vol.% martensite might lead to negative hardness values for austenite [17], indicating the inappropriateness of linear extrapolation. In the present case too, linear fitting to experimental hardness values for 0.2C, 0.6C, and 1.0C steels would lead to negative hardness values for austenite. To overcome this obvious drawback of linear fitting, a non-linear fit was made to the experimental data. Hardness values for the 1.4C steel with the broadest variation of martensite fraction indicate the appropriateness of an exponential fit to the data.

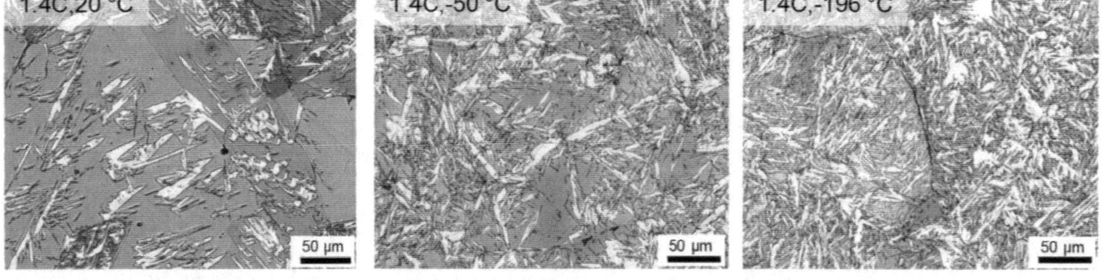


**Figure 1.** Microstructure of the 1.4C steel after quenching to the marked temperatures. Nital was used as etchant. Martensite and austenite appear bright and dark, respectively.

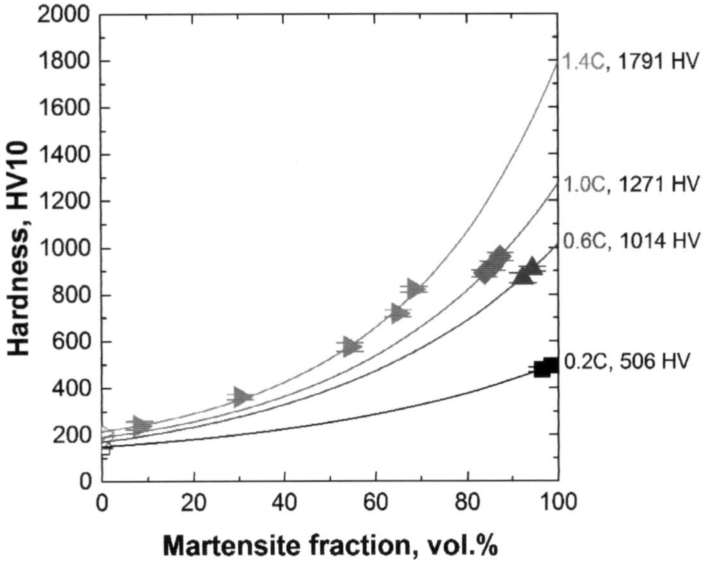


**Figure 2.** Variation of hardness for hardened steels with different martensite fractions. Error bars denote standard deviation.

To enhance the regression analysis, the hardness of austenite in the experimental steels was approximated using the following procedure (figure 2, open symbols at 0 % martensite). By indenting a large austenitic region with a small load of 49 mN (5 gf), the hardness of austenite in a specimen of 1.4C steel containing a low martensite fraction was determined to be 211 HV. The hardness of austenite in steels containing lower carbon concentrations was assumed to decrease linearly to that for an SUS316L austenitic stainless steel (140 HV) [18] which was assumed to also represent the hardness of interstitial-free austenite in the present steel. The effect of carbon concentration on the hardness of austenite in the experimental steels ($HV_\gamma$) was subsequently expressed using the following equation:

$$HV_\gamma = 140 + (48.6 \times \text{mass-\% C}) \quad (2)$$

The extrapolated hardness of martensite based on exponential fits to the hardness values is marked in figure 2 and summarized using the data denoted $HV_{\alpha'}^{\text{exponential}}$ in figure 3. The results reveal that the hardness of martensite is close to 1014 HV for the 0.6C steel and increases to 1271 HV and 1791 HV for the 1.0C and 1.4C steels, respectively. These hardness levels are significantly higher than the measured hardness values for the specimens quenched to -196 °C ($HV_{\alpha'+\gamma}^{\text{measured}}$ in figure 3). The hardness values denoted $HV_{\alpha'}^{\text{RoM}}$ in figure 3 represent the hardness of martensite if the softening effect of austenite is excluded by applying the rule of mixtures to the hardness values for the specimens quenched to -196 °C ($HV_{\alpha'+\gamma}^{\text{measured}}$). This was done by means of the following equation:

$$HV_{\alpha'}^{\text{RoM}} = \frac{HV_{\alpha'+\gamma}^{\text{measured}} - (f_\gamma \times HV_\gamma)}{1 - f_\gamma} \quad (3)$$

where $f_\gamma$ denotes austenite fractions after quenching to -196 °C. The dependence of $f_\gamma$ on the carbon concentration is also represented in figure 3. The $HV_\gamma$ values were estimated from equation (2). As shown in figure 3, $HV_{\alpha'}^{\text{exponential}}$ values are well above the $HV_{\alpha'}^{\text{RoM}}$ values as well as the literature results in which the effect of austenite has been excluded by the rule of mixtures [13]. Consideration of the hardness values for metal-matrix composites with various fractions of hard reinforcing particles confirms the non-linearity of the relationship between the hardness and the second phase volume fraction [19–21]. In other words, plastic deformation due to indentation is mainly accommodated by the soft matrix phase. In such cases, the measured hardness values remain below the straight line drawn between the hardness values for the soft and hard constituents.

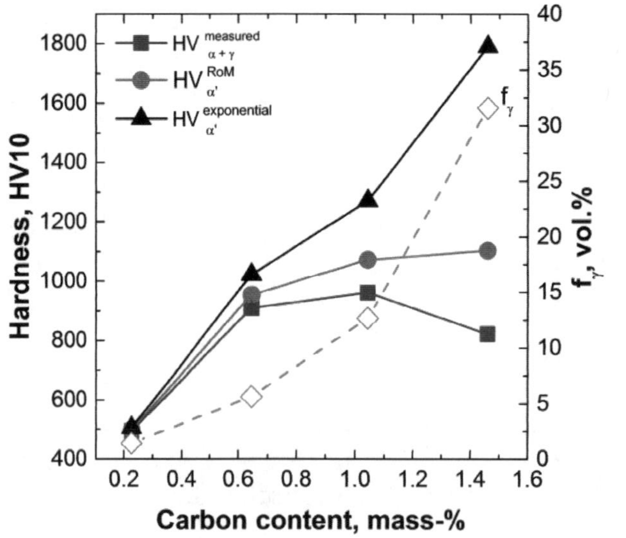


**Figure 3.** Austenite fraction ($f_\gamma$) and hardness of steels with various carbon contents after quenching to -196 °C ($HV^{measured}_{\alpha'+\gamma}$). The extrapolated hardness of martensite based on exponential fits to the hardness values in the presence of various martensite fractions is marked $HV^{exponential}_{\alpha'}$. The hardness of martensite estimated based on the rule of mixtures is marked $HV^{RoM}_{\alpha'}$.

Nano-indentation measurement have indicated that the hardness is almost equal for hardened steels containing 0.6 mass-% and 0.8 mass-% carbon [22]. Invalid assumptions on the material response, excessive surface roughness, and insufficient penetration depths are among common sources of error in nano-indentation experiments [23,24]. Although increasing the penetration depth during the nano-indentation of hardened steels with martensitic-austenitic microstructures leads to an enhanced reliability of derived hardness values, it increases the contribution of retained austenite to the estimated hardness values for martensite. The difficulty of excluding the effect of austenite without compromising the reliability of nano-indentation measurements emphasizes the usefulness of regression analysis for the estimation of the true hardness of martensite.

## 4. Conclusions

In summary, the dependence of hardness on the martensite fraction of hardened martensitic-austenitic steels with various phase balances was described using exponential fits to the hardness data. The hardness of martensite containing 1.46 mass-% carbon was estimated to be 1791 HV which is significantly higher than the measured hardness of 822 HV for the same steel containing nearly 68 vol.% martensite and the hardness of 1104 HV obtained after excluding the effect of austenite by the rule of mixtures.

## 5. References

[1] De Cooman B C and Speer J G 2011 *Fundamentals of Steel Product Physical Metallurgy* (Warrendale: Association for Iron and Steel Technology)

[2] de la Concepción V L, Lorusso H N and Svoboda H G 2015 Effect of Carbon Content on Microstructure and Mechanical Properties of Dual Phase Steels *Procedia Mater. Sci.* **8** 1047–56

[3] Fan D W, Kim H S and De Cooman B C 2009 A Review of the Physical Metallurgy related to the Hot Press Forming of Advanced High Strength Steel *Steel Res. Int.* **80** 241–248

[4] Krauss G 1999 Martensite in steel: Strength and structure *Mater. Sci. Eng. A* **273–275** 40–57
[5] Bhadeshia H K D H 2012 Steels for bearings *Prog. Mater. Sci.* **57** 268–435
[6] Kehoe M and Kelly P M 1970 The role of carbon in the strength of ferrous martensite *Scr. Metall.* **4** 473–6
[7] Chilton J M and Kelly P M 1968 The strength of ferrous martensite *Acta Metall.* **16** 637–56
[8] Hansen N 2004 Hall–Petch relation and boundary strengthening *Scr. Mater.* **51** 801–6
[9] Grange R A, Hribal C R and Porter L F 1977 Hardness of tempered martensite in carbon and low-alloy steels *Metall. Trans. A* **8** 1775–85
[10] Litwinchuk A, Kayser F X, Baker H H and Henkin A 1976 The Rockwell C hardness of quenched high-purity iron-carbon alloys containing 0.09 to 1.91% carbon *J. Mater. Sci.* **11** 1200–6
[11] Ohmura T, Tsuzaki K and Matsuoka S 2001 Nanohardness measurement of high-purity Fe–C martensite *Scr. Mater.* **45** 889–94
[12] Sherby O D, Wadsworth J, Lesuer D R and Syn C K 2008 Revisiting the Structure of Martensite in Iron-Carbon Steels *Mater. Trans.* **49** 2016–27
[13] Galindo-Nava E I and Rivera-Díaz-del-Castillo P E J 2016 Understanding the factors controlling the hardness in martensitic steels *Scr. Mater.* **110** 96–100
[14] Yamauchi H, Watanabe H, Suzuki Y and Saito H 1974 Magnetization of α-Phase Fe-Mn Alloys *J. Phys. Soc. Jpn.* **36** 971–4
[15] Xiong X C, Chen B, Huang M X, Wang J F and Wang L 2013 The effect of morphology on the stability of retained austenite in a quenched and partitioned steel *Scr. Mater.* **68** 321–4
[16] Grange R A and Stewart H M 1946 The Temperature Range Of Martensite Formation *AIME Tech. Publ.* **1996** 1–24
[17] Mola J and De Cooman B C 2013 Quenching and Partitioning (Q&P) Processing of Martensitic Stainless Steels *Metall. Mater. Trans. A* **44** 946–67
[18] Suzuki K and Shobu T 2010 Residual Microstress of Austenitic Stainless Steel Due to Tensile Deformation *Mater. Sci. Forum* **652** 7–12
[19] Levin B F, DuPont J N and Marder A R 2000 The effect of second phase volume fraction on the erosion resistance of metal-matrix composites *Wear* **238** 160–7
[20] Sambathkumar M, Navaneethakrishnan P, Ponappa K, Sasikumar K S K, Sambathkumar M, Navaneethakrishnan P, Ponappa K and Sasikumar K S K 2017 Mechanical and Corrosion Behavior of Al7075 (Hybrid) Metal Matrix Composites by Two Step Stir Casting Process *Lat. Am. J. Solids Struct.* **14** 243–55
[21] Sivasankaran S and Alaboodi A S 2016 Structural Characterization and Mechanical Behavior of Al 6061 Nanostructured Matrix Reinforced with TiO2 Nanoparticles for Automotive Applications
[22] Ohmura T and Tsuzaki K 2003 Evaluation of matrix strength of Fe-C as-quenched and quench-tempered martensite using nanoindentation techniques *J. Phys. IV Proc.* **112** 267–70
[23] Menčík J 2012 Uncertainties and Errors in Nanoindentation *Nanoindentation in Materials Science* (InTech)
[24] Šmeļova V, Schwedt A, Wang L, Holweger W and Mayer J 2017 Electron microscopy investigations of microstructural alterations due to classical Rolling Contact Fatigue (RCF) in martensitic AISI 52100 bearing steel *Int. J. Fatigue* **98** 142–54

**Acknowledgements**
Support of the technical staff at the Institute of Iron and Steel Technology of TU Bergakademie Freiberg is gratefully acknowledged.

IOP Conf. Series: Materials Science and Engineering **373** (2018) 012005 doi:10.1088/1757-899X/373/1/012005

# On the origin of subgrain boundaries during conventional solidification of austenitic stainless steels

**R Rahimi[1], H Biermann[2], O Volkova[1] and J Mola[1,*]**

[1] Institute of Iron and Steel Technology, Technische Universität Bergakademie Freiberg, Leipziger Str. 34, 09599, Freiberg

[2] Institute of Materials Engineering, Technische Universität Bergakademie Freiberg, Gustav-Zeuner-Straße 5, 09599 Freiberg, Germany

[*] e-mail: mola@iest.tu-freiberg.de

**Abstract.** The origin of subgrain formation during conventional casting and solidification of stainless steels was studied using two austenitic stainless steels with 0 and 4 mass-% Al. Whereas the Al-free alloy showed no subgrain formation, the Al-added alloy developed a high density of subgrains separated by low-angle grain boundaries. The occurrence of subgrains in the Al-added alloy was justified by its ferritic mode of solidification as predicted by thermodynamic calculations and confirmed by dynamic scanning calorimetry measurements. The subgrains might be a consequence of the plastic deformation of soft primary ferrite dendrites by the fluid flow and their subsequent inheritance by the austenite. Alternatively, they might have been induced during the austenite formation from delta ferrite, most likely via a peritectic reaction. The absence of subgrains in the Al-free alloy was justified by its austenitic mode of solidification.

## 1. Introduction

Bending and fragmentation of dendrite arms are well demonstrated to happen during solidification processes like stir casting [1] and spray forming [2] which are inherently associated with external forces. These can give rise to the formation of low-angle grain boundaries (LAGBs) inside the grains. In the absence of external forces during conventional casting, LAGBs are apparently not expected. However, internal forces caused by fluid-dendrite interactions, specifically in the mushy zone, may bring about mechanisms by which such a deformation could arise [3]. Two possible mechanisms are the rotation of the principal growth direction by thermo-solutal advection and the direct mechanical damage to the growing dendrite by the fluid flow [4]. In the thermo-solutal advection mechanism, a deviation in isotherms caused by the fluid flow results in the bending of dendrites. Direct mechanical deformation in the latter mechanism arises when the liquid flows next to the dendrite and causes friction-induced stresses.

Deformation of dendrite during solidification is theoretically and experimentally investigated is some studies [5,6]. For instance, Dragnevski et al. [5] modeled the dendrite-fluid flow interactions during solidification of Cu-O and Cu-3wt.%Sn melts in a variety of supercoolings in the approximate range of 47-160 K. After a series of calculations, they concluded that the thermo-solutal advection cannot account for the deformation of dendrites in the studied supercooling range. Within a certain range of supercoolings (50 K in Cu–O and 75–85 K in Cu–3wt%Sn), however, the required stress for the bending of dendrite arms was supplied by the fluid flow. At these supercoolings, the fine size of dendrites and the high melt velocities supply the required stress for the mechanical damage of dendrites. Due to the small supercoolings associated with conventional solidification, the mechanical deformation of dendrites is unlikely during conventional solidification processes. This belief, however, was challenged by Doherty later [7]. In his critical article, he listed a couple of studies in which the formation of in-grain misorientations is reported for conventional casting processes. For example, in a statistical work, a higher than usual frequency of LAGBs is seen during direct chill (DC) casting of commercial pure Al [8]. It should be noted that high cooling rates as well as relative movement of mold wall with respect to the casting are the factors which may influence the solidification process. When the casting is still solidifying, these external forces can extend into the semi-solid mushy zone and slightly bend the dendrite arms [6]. In at least one of the references cited by Doherty, namely [9], it seems that misorientation substructures have truly formed during directional solidification of 99.9% pure Al. The development of grain boundaries with misorientations less than 15° is clearly shown in that work using electron backscattered diffraction (EBSD) method. Additionally, it is certainly possible that large solidification contractions can lead to significant internal stresses, particularly when dendrites impingement occurs. The extreme case of internal stress development is the occurrence of hot tearing where the lack of mushy zone feeding can lead to the cracking of the sample [10].

In the present study, differences in the tendency of austenitic stainless steels to form in-grain misorientations are discussed in view of the differences in their solidification modes.

## 2. Experimental methods

Two austenitic stainless steels with the chemical compositions shown in Table 1 were cast in a vacuum induction melting (VIM) furnace. Electron backscattered diffraction (EBSD) and electron channeling contrast imaging (ECCI) investigations were conducted on as-cast samples using a Zeiss LEO-1530 GEMINI field emission scanning electron microscope (FESEM). Transmission electron microscopy (TEM) observations were carried out in a JEOL field emission TEM (FETEM) at an operating voltage of 200 kV. TEM foils were prepared using mechanical grinding to a thickness of below 100 μm followed by twin-jet electrolytic polishing using a solution of 5% perchloric acid in 95% acetic acid at room temperature. The equilibrium phase diagram of the alloy system was calculated using the Thermo-Calc software. Differential scanning calorimetry (DSC) experiments were done in a 404 C Pegasus calorimeter with heating and cooling rates of 50 K/min and 10 K/min, respectively.

**Table 1.** Chemical composition of the cast alloys in mass- %.

| Alloy | C | Al | Mn | Cr | Ni | Si |
|---|---|---|---|---|---|---|
| 0Al | 0.43 | 0.02 | 6.63 | 17.76 | 9.27 | 0.40 |
| 4Al | 0.42 | 3.95 | 6.24 | 16.91 | 8.69 | 0.39 |

## 3. Results and discussion

### *3.1. Subgrains characterization*

Figures 1(a,d) show ECC images of the alloys in the as-cast condition. The corresponding inverse pole figure (IPF) maps and grain boundary misorientation maps obtained by EBSD are shown in Figures 1(b-c,e-f). The results indicate significant differences in the microstructures of the alloys. Based on the ECC images alone, the 4Al alloy appears to have a significantly finer grain size than the 0Al alloy. Based on the EBSD maps, however, the grain size of both alloys is relatively comparable if the small island grains in the 4Al alloy and the in-grain small misorientations are neglected. According to the boundary map in Figure 1(f), the island grains in the 4Al alloy are twin-related to the surrounding matrix. The most obvious difference between the microstructures is the massive development of subgrains in the 4Al alloy and their absence in the Al-free variant. The subgrains observed in the 4Al alloy arises from the presence of a high density of LAGBs (Figure 1(f)). LAGBs are normally observed after annealing of cold-worked metals at temperatures below the recrystallization temperature, where recovery is the dominant restoration mechanism [11]. By contrast, the subgrains in the 4Al alloy have already formed in the as-cast condition prior to any deformation and annealing.

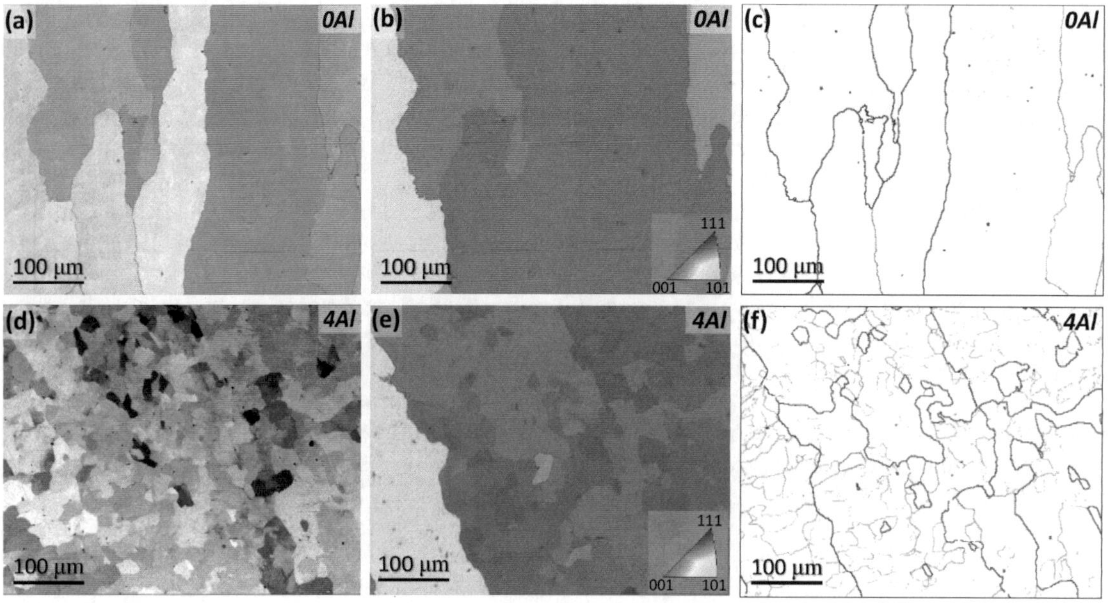


**Figure 1.** Microstructure of as-cast alloys; (a,d) ECC images; (b,e) corresponding EBSD IPF maps; (c,f) corresponding grain boundary distribution maps. Colors in IPF maps denote the crystallographic directions parallel to the plane of view. High-angle grain boundaries (HAGBs), LAGBs, and Σ3 twin boundaries are represented by black, gray, and red lines, respectively.

The LAGBs, which are otherwise difficult to observe by optical microscopy, become visible after soaking at temperatures where precipitates form [12]. This is due to the formation of precipitates at LAGBs. To enable the precipitation of Cr-rich carbides in both alloys, they were aged at 650 °C for 8 hours. The optical micrographs of the aged specimens (Figure 2) indicate the existence of subgrains only in the 4Al alloy. Optical micrographs also indicate that the substructures are not uniformly distributed throughout the γ grains and that there also exist regions free of subgrains. Some regions without subgrains are indicated by arrows in Figure 2(b).

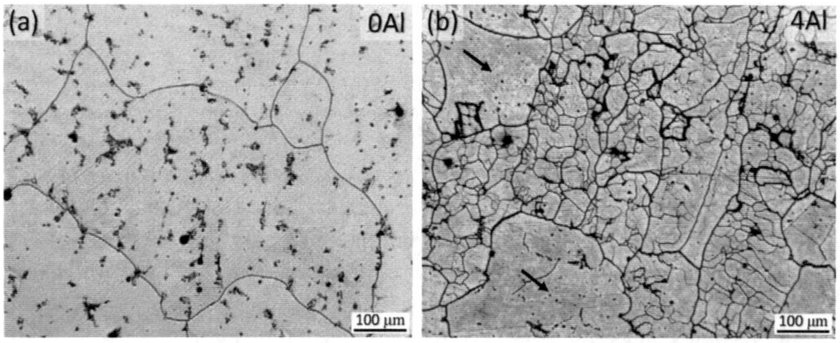


**Figure 2.** Optical micrograph of (a) 0Al and (b) 4Al alloys after annealing at 1250 °C for 5 minutes and heat treating at 650 °C for 6 hours. Samples were etched with V2A agent. Before annealing, samples were homogenized at 1200 °C for 1 h.

The TEM micrograph in Figure 3 shows a subgrain boundary in the Al-added steel. Depending on the nature of dislocations, the boundary can be regarded as a tilt or twist boundary for edge and screw

IOP Conf. Series: Materials Science and Engineering **373** (2018) 012005 doi:10.1088/1757-899X/373/1/012005

dislocations, respectively [13]. The similarity of the selected area electron diffraction (SAED) patterns taken from the regions marked A and B in Figure 3(b) and the similarity of dislocation arrangements on both sides of the boundary indicates that this boundary is a LAGB. Accordingly, the pattern of LAGBs in the 4Al alloy does not act as effective obstacles to dislocation motion as both alloys exhibit comparable flow stresses [14,15]. Similarly, the presence of solidification-induced LAGBs with misorientations below 6 ° has been reported to have a negligible influence on the tensile properties of Ni-based single crystal superalloys [16].

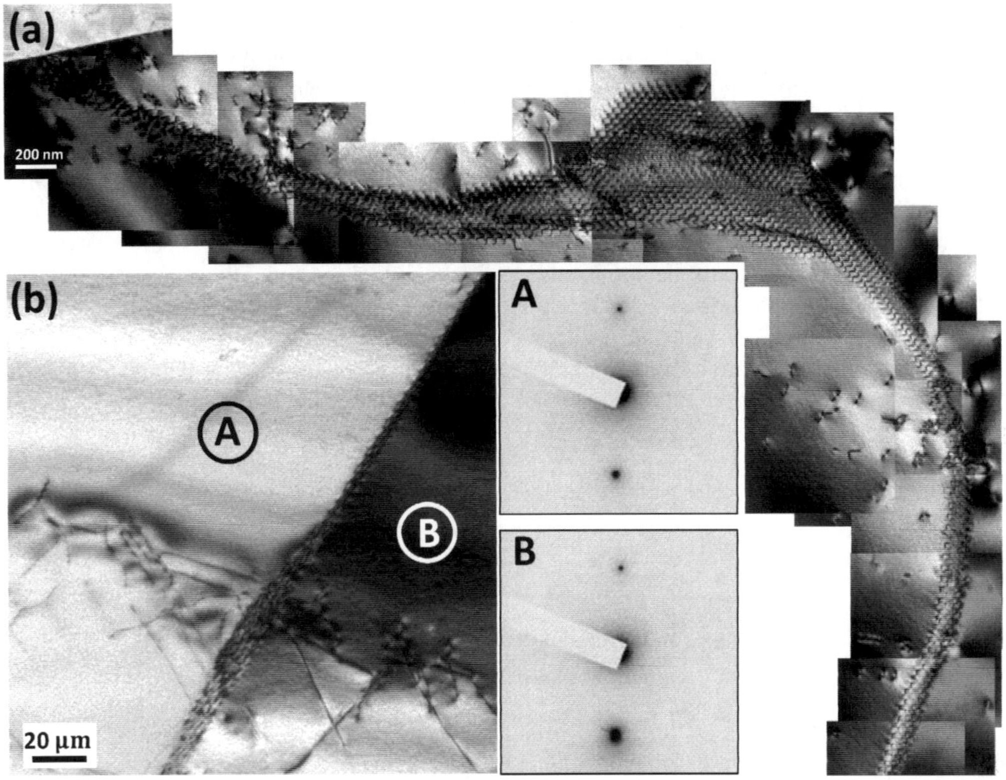


**Figure 3.** TEM bright field micrograph of 4Al (a) subgrain boundary (b) interaction of a boundary and moving dislocations. SAD patterns are taken from spots A and B in two adjacent austenite grains.

The most common justification for the occurrence of in-grain features is the bending of dendrite arms by the fluid flow in the mushy zone. Nevertheless, since both alloys were cast and fabricated using the same processing conditions (i.e. the same casting and heat treatment conditions), the occurrence of the subgrains only in 4Al alloy is rather an inherent characteristic of the 4Al alloy. In the following, we discuss two possibilities for the origin of in-grain boundaries in the 4Al alloy.

*3.2. Dendrite arm bending during solidification*

As discussed before, subgrain formation in as-cast alloys has often been linked to the dendrite arm bending during solidification. The required force for this deformation could be provided by external or internal forces, depending on casting and solidification conditions. Although there is no consensus on whether the internal forces in conventional solidification suffice to deform dendrites, it appears that no better alternative is available. According to Doherty [7], "*It would seem very difficult to account for these small ingrain misorientations in conventionally solidified dendrites, if they did not arise by dendrite arm bending*". In this regard, the misorientation substructures observed in the 4Al alloy may be attributed to the slight bending of dendrite arms. The absence of in-grain misorientations in the 0Al alloy might then be justified by differences in the solidification mode of the steels. The pseudo-binary equilibrium phase diagram in Figure 4 predicts that the solidification of 0Al and 4Al alloys begins with the formation of austenite and ferrite dendrites, respectively. To confirm the validity of Thermo-Calc predictions, specimens of both alloys were continuously heated in DSC to study their transformation/melting behavior. Both steels exhibit small endothermic peaks at slightly below 1300 °C, which mark the dissolution of Cr-rich carbides. At temperatures above approximately 1339 °C, the 0Al alloy exhibit a large endothermic peak which marks the onset of melting (solidus temperature). Melting of austenite is complete at approximately 1480 °C. Due to the absence of any phase transformation between the melting start and finish points, the austenite in the 0Al alloy is stable up to the liquidus temperature, namely it melts without preceding or simultaneous formation of ferrite. In the case of the 4Al alloy (Figure 5(b)), in contrast, a slope change and an extra endothermic peak occurred between the solidus and liquidus temperatures. The slope change at ~1367 °C and the peak at ~1397 °C can be attributed to the transformation of austenite into ferrite during melting of the 4Al alloy, namely the higher thermodynamic stability of ferrite compared to austenite just below the liquidus temperature. Based on the DSC results, the first phase to form during solidification of the 4Al alloy is ferrite, although it transforms to austenite before the solidus temperature is reached. In contrast, the solidification of the 0Al alloy starts with the formation of austenite which remains stable down to RT. Due to the high C concentration of the alloys and the likelihood of cooling curves being influenced by severe decarburization in the liquid state, interpretations regarding the sequence of transformations during solidification were made based on the sequence of transformations during the reverse process, rather than that during cooling from the liquid range.

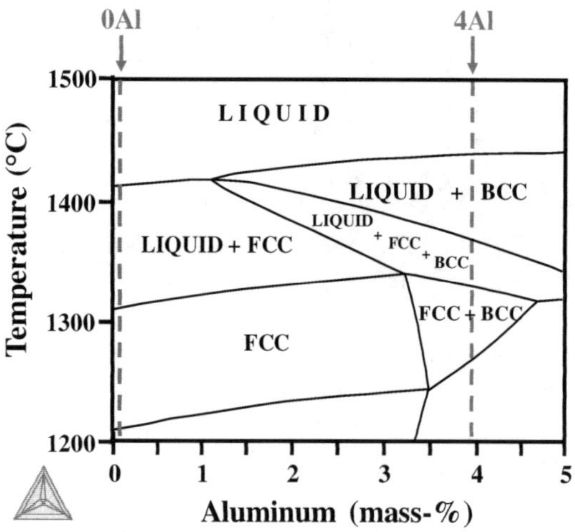


**Figure 4.** Pseudo-binary phase diagram of Fe–16.9Cr–8.7Ni–6.2Mn–0.42C alloy system as a function of Al concentration. The actual position of alloys are indicated by dashed lines in diagram.

For austenitic and ferritic stainless steels with a roughly similar chemical composition, the yield strength of austenite has been reported to exceed that of the ferrite at temperatures above nearly 600 °C [17]. Although the mechanical properties available in Ref. [17] are limited to 900 °C, it is likely that the strength of austenite remains higher than that of ferrite at still higher temperatures, including the solidification temperature range. Therefore, it can be implied that the stresses induced by fluid flow have not been high enough to yield the austenite dendrites in the 0Al alloy but high enough to cause the plastic deformation of the softer ferritic dendrites in the 4Al alloy. The subgrains induced by the plastic deformation of primary dendrites of ferrite in the 4Al alloy are subsequently inherited by austenite and persist down to RT.

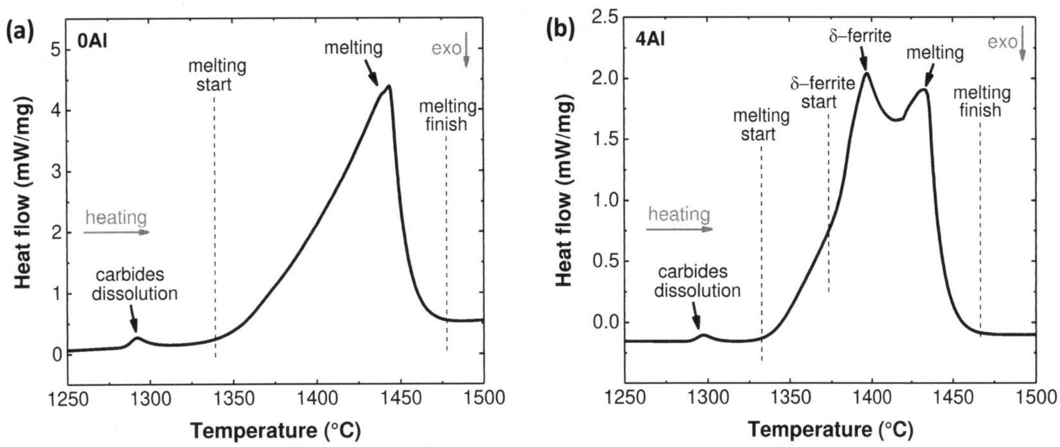


**Figure 5.** DSC curves for (a) 0Al and (b) 4 Al steels in the heating section. The heating rate was 50 K/min.

### *3.3. Phase transformation during cooling*

The effect of solidification and transformation sequence on the final microstructure of welded steels has been extensively studied [18–24]. These investigations have mainly focused on the morphology and volume fraction of ferrite, due to their relevance to the crack susceptibility and in-service properties of welds with duplex microstructure [25]. Although most of the previous studies have focused on welded metals, the principles of solidification are equally applicable to cast metals. Based on the succession of the formation of phases, the solidification mode of steels can be divided into three categories, namely austenitic (A), ferritic (F), and ferritic-austenitic (F-A) modes. In steels with the A-mode of solidification, austenite is the primary phase nucleating from the melt and remains stable during further cooling. The delta ferrite, if any, may only form from the remaining liquid as a secondary phase or as one of the constituents of a eutectic phase transformation. In such cases, the delta ferrite will assume a roundish or vermicular morphology and appears between the cellular dendrites of austenite [26]. The microstructure of F-mode steels, on the other hand, is composed of large grains with a lathy morphology. In some cases, Widmanstätten γ-austenite also forms from the ferrite in the solid state [27]. In steels with the F-A solidification mode, ferrite is formed as the primary phase and austenite nucleates afterwards either in the melt or in the solid state. The morphology of ferrite in the latter case might be circular and/or lathy [28]. The $Cr_{eq}/Ni_{eq}$ ratio is believed to govern the solidification mode and final microstructure of steels, where $Cr_{eq}$ and $Ni_{eq}$ are chromium and nickel equivalents as originally proposed by Schaeffler [29]. For $Cr_{eq}/Ni_{eq}$ values below 1.48 and above 1.95, the solidification modes are expected to be A and F, respectively. For steels with $Cr_{eq}/Ni_{eq}$ values between 1.48 and 1.95, the F-A mode of solidification is expected [26–28].

The alloying elements in the present steels can be classified into γ-stabilizers and α-stabilizers, whose balance controls the final microstructure. The γ-stabilizers are C, Mn, and Ni and the α-stabilizers are Al, Cr, and Si. Since the chemical composition of both steels is almost identical except for the Al content, it is plausible to assume that the 4Al alloy has a higher value of $Cr_{eq}/Ni_{eq}$ and is consequently more susceptible to the F-mode of solidification. The expectations based on the ferrite potential and the predictions based on the thermodynamic equilibrium calculations (Figure 4) were indeed confirmed by the DSC results in Figure 5.

As mentioned before, the subgrain pattern was found to be a distinctive characteristic of the Al-added steel. This dissimilarity may originate from the difference in their mode of solidification. Based on the presented results, the sequence of phase transformations during cooling of the 0Al alloy from the liquid state can be summarized as $L \rightarrow L + \gamma \rightarrow \gamma$. For the 4Al alloy, the transformation sequence changes to $L \rightarrow L + \delta \rightarrow L + \delta + \gamma_I \rightarrow L + \gamma_I \rightarrow \gamma_I + \gamma_{II}$, where $\gamma_I$ denotes the austenite formed in the presence of δ-ferrite and $\gamma_{II}$ denotes the austenite deposited directly from the melt after the transformation of δ-ferrite into $\gamma_I$.

As schematically represented in Figure 6, the subgrains within the coarse austenitic grains of the 4Al alloy can be attributed to the replacement of δ-ferrite by $\gamma_I$. Although the formation of $\gamma_I$ can be via a solid state transformation ($\delta \rightarrow \gamma_I$), it is more likely via a peritectic reaction initiating at interfaces of δ-ferrite and liquid ($L + \delta \rightarrow \gamma_I$). Small misorientations between the independently nucleated regions of austenite will then lead to the formation of LAGBs at the points of impingement. The subgrain-free regions in the 4Al alloy (Figure 2(b)), on the other hand, can be the $\gamma_I$ regions in the schematic of Figure 6, namely the austenite formed directly from the melt after the full exhaustion of δ-ferrite.

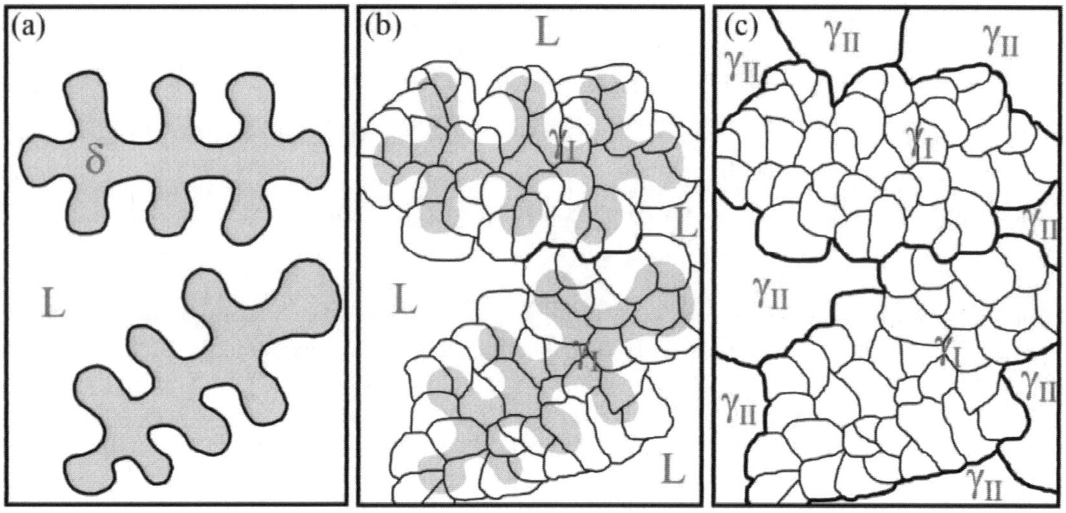


**Figure 6**. Schematic showing the microstructural evolution of the 4Al alloy during solidification: (a) formation of δ-ferrite primary dendrites at the beginning of solidification; (b) nucleation of austenite ($\gamma_I$) with the involvement of δ-ferrite leading to the creation of LAGBs represented by thin lines; (c) transformation of the remaining liquid into austenite ($\gamma_{II}$) free of subgrains. Thick lines in (b) and (c) denote the HAGBs.

## 4. Conclusions

Microstructures of two austenitic stainless steels with Al (4Al) and without Al (0Al) were examined using various characterization methods and the differences were discussed in view of the dissimilar solidification modes. Whereas the 0Al alloy had a uniform distribution of coarse γ grains free of subgrains, the austenite grains in the 4Al alloy contained subgrains separated by low-angle grain boundaries. The occurrence of subgrains in the 4Al alloy was justified by its ferritic mode of solidification as predicted by thermodynamic calculations and confirmed experimentally by dynamic scanning calorimetry measurements. The subgrains can be a consequence of the plastic deformation of soft primary ferrite dendrites by the fluid flow and their subsequent inheritance by the austenite. Alternatively, they can be induced during the austenite formation from δ-ferrite, most likely via a peritectic reaction, in which case the impingement of austenite grains at small misorientation angles would lead to the formation of subgrains. The subgrains were absent in the 0Al alloy with an austenitic mode of solidification.

## 5. References

[1] Doherty R D, Lee H-I and Feest E A 1984 Microstructure of stir-cast metals *Mater. Sci. Eng.* **65** 181–9
[2] Annavarapu S and Doherty R D 1993 Evolution of microstructure in spray casting *Int. J. Powder Metall. Princet. N. J.* **29** 331–43
[3] Flemings M C 1991 Behavior of metal alloys in the semisolid state *Metall. Trans. B* **22** 269–93
[4] Mullis A M, Walker D J, Battersby S E and Cochrane R F 2001 Deformation of dendrites by fluid flow during rapid solidification *Mater. Sci. Eng. A* **304–306** 245–9
[5] Dragnevski K, Mullis A M, Walker D J and Cochrane R F 2002 Mechanical deformation of dendrites by fluid flow during the solidification of undercooled melts *Acta Mater.* **50** 3743–55
[6] Yamaguchi M and Beckermann C 2013 Simulation of solid deformation during solidification: Compression of a single dendrite *Acta Mater.* **61** 4053–65
[7] Doherty R D 2003 Comments on "Mechanical deformation of dendrites by fluid flow during the solidification of undercooled melts" *Scr. Mater.* **49** 1219–22
[8] Samajdar I and Doherty R D 1994 Grain boundary misorientation in DC-cast aluminum alloy *Scr. Metall. Mater.* **31** 527–30
[9] Bhattacharyya A, El-Danaf E, Kalidindi S R and Doherty R D 2001 Evolution of grain-scale microstructure during large strain simple compression of polycrystalline aluminum with quasi-columnar grains: OIM measurements and numerical simulations *Int. J. Plast.* **17** 861–83
[10] Dantzig J A and Rappaz M 2009 *Solidification* (EPFL Press)
[11] Crussard C, Aubertin F, Faoul B and Wyon G 1950 Polygonization in strongly deformed metals *Prog. Met. Phys.* **2** 193–202
[12] Wendler M, Reichel B, Eckner R, Fabrichnaya O, Krüger L, Weiß A and Mola J 2016 Effect of Vanadium Nitride Precipitation on Martensitic Transformation and Mechanical Properties of CrMnNi Cast Austenitic Steels *Metall. Mater. Trans. A* **47** 139–51
[13] Lejcek P 2010 *Grain Boundary Segregation in Metals* (Springer)
[14] Rahimi R, De Cooman B C, Biermann H and Mola J 2014 Microstructure and mechanical properties of Al-alloyed Fe–Cr–Ni–Mn–C stainless steels *Mater. Sci. Eng. A* **618** 46–55
[15] Rahimi R, Ullrich C, Klemm V, Rafaja D, De Cooman B C, Biermann H and Mola J 2016 Influence of Al on the temperature dependence of strain hardening behavior and glide planarity in Fe–Cr–Ni–Mn–C austenitic stainless steels *Mater. Sci. Eng. A* **649** 301–12
[16] Li J R, Zhao J Q, Liu S Z and Han M 2008 Effects of low angle boundaries on the mechanical properties of single crystal superalloy DD6 Proceedings of the International Symposium on Superalloys pp 443–51
[17] Fischer F D, Rammerstorfer F G and Bauer F J 1990 Fatigue and fracture of high-alloyed steel specimens subjected to purely thermal cycling *Metall. Trans. A* **21** 935–48
[18] Fredriksson H 1972 The solidification sequence in an 18-8 stainless steel, investigated by directional solidification *Metall. Trans.* **3** 2989–97
[19] Suutala N, Takalo T and Moisio T 1979 The relationship between solidification and microstructure in austenitic and austenitic-ferritic stainless steel welds *Metall. Trans. A* **10** 512–4
[20] Allen D J 1983 solidification sequences in austenitic steel weld metal deposited on ferritic base material. *Met. Technol.* **10** 24–7
[21] Inoue H, Koseki T, Okita S and Fuji M 1997 Solidification and transformation behaviour of austenitic stainless steel weld metals solidified as primary ferrite: Study of solidification and subsequent transformation of Cr-Ni stainless steel weld metals (2nd Report) *Weld. Int.* **11** 937–49
[22] Rajasekhar K, Harendranath C S, Raman R and Kulkarni S D 1997 Microstructural evolution during solidification of austenitic stainless steel weld metals: A color metallographic and electron microprobe analysis study *Mater. Charact.* **38** 53–65

[23] Ma J C, Yang Y S, Tong W H, Fang Y, Yu Y and Hu Z Q 2007 Microstructural evolution in AISI 304 stainless steel during directional solidification and subsequent solid-state transformation *Mater. Sci. Eng. A* **444** 64–8
[24] Inoue H, Koseki T, Ohkita S and Fuji M 1997 Effect of massive transformation on formation of acicular structure in austenitic stainless steel weld metal solidified as ferritic single phase - Study on solidification and subsequent transformation of Cr-Ni stainless steel weld metals (Report 5) *Yosetsu Gakkai RonbunshuQuarterly J. Jpn. Weld. Soc.* **15** 305–13
[25] Badji R, Bouabdallah M, Bacroix B, Kahloun C, Belkessa B and Maza H 2008 Phase transformation and mechanical behavior in annealed 2205 duplex stainless steel welds *Mater. Charact.* **59** 447–53
[26] Takalo T, Suutala N and Moisio T 1979 Austenitic solidification mode in austenitic stainless steel welds *Metall. Trans. A* **10** 1173–81
[27] Suutala N, Takalo T and Moisio T 1979 Single-phase ferritic solidification mode in austenitic-ferritic stainless steel welds *Metall. Trans. A* **10** 1183–90
[28] Suutala N, Takalo T and Moisio T 1980 Ferritic-austenitic solidification mode in austenitic stainless steel welds *Metall. Trans. A* **11** 717–25
[29] Schaeffler A L 1947 Selection of austenitic electrodes for welding dissimilar metals *Weld. J.* **26** 601–20

**Acknowledgments**
The authors would like to express their appreciations to the German Research Foundation (Deutsche Forschungsgemeinschaft) for the financial support of this work under the grant number MO 2580/2-1. The provision of TEM by B.C. De Cooman, Pohang University of Science and Technology, and the assistance of C. Ullrich from the Institute of Materials Science of TU Bergakademie Freiberg in the SEM examinations are gratefully acknowledged.

# Influence on the weld strength of high-strength fine-grained structural steels by thin-film-coated GMA welding electrodes

**V Wesling, K Treutler and T Gehling**

TU Clausthal, Institut für Schweißtechnik und Trennende Fertigungsverfahren, Agricolastraße 2, D-38678 Clausthal-Zellerfeld, office@isaf.tu-clausthal.de

**Abstract.** The increasing demand for high-strength fine-grain steels for the purpose of reducing the weight of steel structures and increasing payloads, is driving forward the development of fine-grain steels. However, with the currently available welding consumables, it is not possible to achieve the strengths of modern fine-grain steels with a grade of S1300QL in the weld metal. As a result, there is a need for novel welding consumables, which guarantee high-strength joints. The application of thin, functional layers on GMA welding electrodes by a PVD coating process makes it possible to influence the mechanical-technological properties to an extent which cannot be achieved by state-of-the-art technology. The sputtering of different elements on the substrate material of GMA electrodes and the layer thickness has an influence on the welding process and the welding metal. Especially, the influence on the microstructure by a niobium coating of different layer thickness is presented. Furthermore, the changes in the mechanical properties of the weld metal depending on the layer thickness or rather the niobium content are shown. In particular, the fine-grain and carbide-forming effect of niobium and the influence on the mechanical properties are presented. Here niobium contents between 0.5 wt.-% and 1.5 wt.-% are in the focus of investigations.

## 1. Introduction

Due to their high strength and toughness, fine-grain steels are used, among others, in the field of steel structures and in mobile crane construction with the aim of reducing the weight and increasing payloads,[1]. Considering the requirements in these applications, the development of fine grain steels is progressing towards increasing yield strengths while maintaining good toughness properties through the addition of alloying elements such as chromium, nickel or molybdenum and corresponding tempering treatments **Figure 1**, [2].
Steels with a carbon content of ≤ 0.2% and small amounts of alloying elements have a low tendency to harden and are therefore easily weldable without a particularly high risk of cracking, [3].

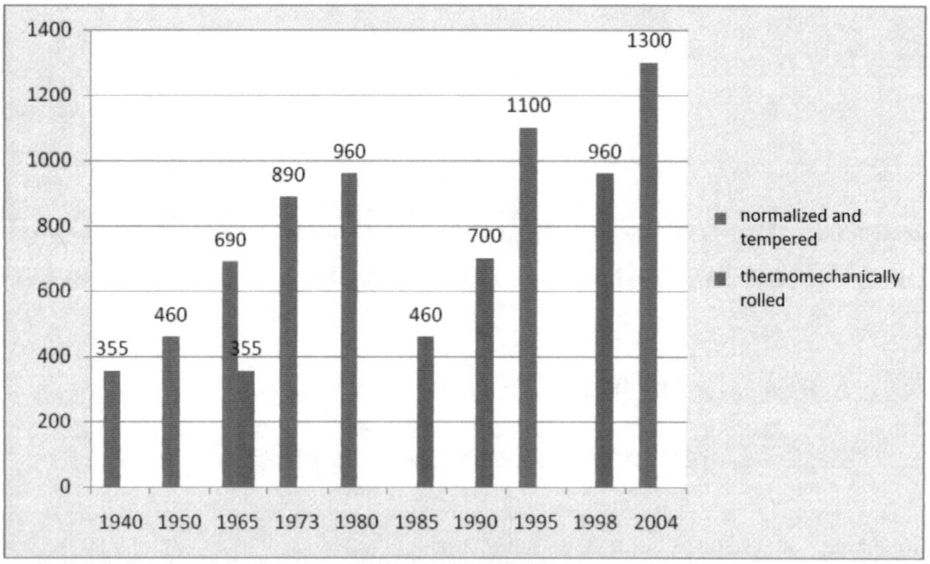


**Figure 1:** Development of the yield strength of fine grain steels based on [2]

During welding processing, a drop in strength occurs in the area of the weld. This can be influenced by the filler material and its alloy. Due to the deformability required in the wire drawing process, the design of the alloy composition of the welding wire and thus also the influence on the weld metal is limited.
The idea of using the physical vapor deposition to coat GMA welding wire electrodes enables the application of functional layers on the surface after the wire drawing process, thereby expanding the possibilities of designing the alloy composition in the weld metal. The influence of low niobium contents in the heat affected zone of welds in fine grain steels was investigated simulatively by [4] whereby a grain refining effect was detected and a positive influence on the yield strength and the tensile strength was determined. The use of niobium as a micro-alloying element in a proportion of less than 0.1% by weight in the production of modern fine-grained steels was also investigated by Hulka, Kern and Schriever. In addition to the wear resistance, the influence of niobium on the toughness properties of welded joints is considered and a positive effect on the notched impact strength is determined, as well as a shift of the transition temperature T27 towards lower temperatures, [5].
With the aid of the coating process described here, alloying elements in previously unachieved mass percentages can be introduced into the weld metal via the filler material and thus examined with regard to the effects on microstructure and mechanical-technological properties. The aim is the development of a welding consumable for the welding processing of modern, high-strength fine-grained steels with yield strengths in the range of 1300 MPa, whereby the strength in the base material is to be achieved or even exceeded to allow an overmatching. Currently, there are no welding consumables that guarantee these strengths in the weld metal, [6].
Initial investigations with coated welding wire electrodes show that they are suitable for increasing the strength in the weld metal. The tensile strength and yield strength of the weld metal can be influenced by the targeted integration of alloying elements over thin-film coated welding wire electrodes.

## 2. Process of welding wire electrode coating

The substrate material used is a Mn4Ni2CrMo alloy GMA welding wire electrode with a diameter of 1.2 mm. Magnetron sputtering in a PVD system is used to apply niobium thin-film layers to the substrate material, **Figure 2**. The procedure has already been described in detail in [7]. An inert gas is decomposed in ions and electrons by applying a high voltage in the evacuated process chamber. Positively charged ions are accelerated and, when hitting the cathode (target), eject atoms and molecules from the coating material by impulse transfer. In the case of the magnetron sputtering used here, permanent magnets are additionally arranged behind the targets. The superposition of magnetic and electric fields directs the electrons to a helical path and increases the ionization probability. This additionally increases the deposition rate, [8].

**Figure 2**: PVD coating system [7]

**Figure 3:** Device for holding the welding wire in the system [7]

In the experiments shown here, the welding wire is fed to the machine and coated batchwise. For this purpose, it is wound onto a device in the coating chamber, **Figure 3**. Among other parameters, in this process, the layer thickness correlates with the process time and can thus be influenced in a targeted manner. The micrographs of the welding wire electrodes in **Figure 4** show thin-film layers of niobium with a layer thickness of 3 µm and 6 µm.

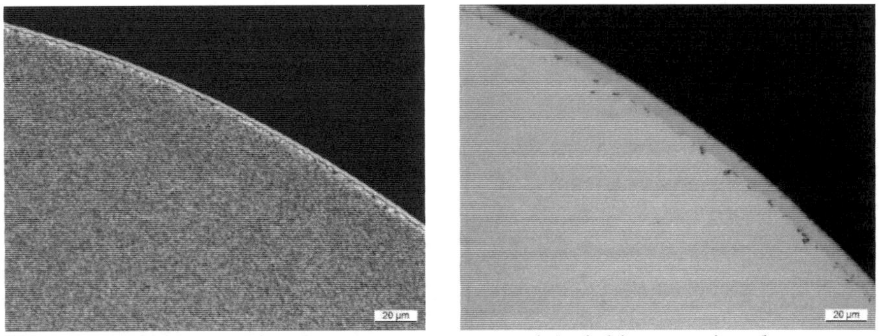


a.) niobium coating 3µm b.) niobium coating 6µm

**Figure 4:** Micrographs of GMA welding wire electrodes with niobium coatings in different layer thicknesses

**3. Material and metallography**

To carry out the welding tests, the high-strength fine-grain structural steel S960QL is used as the base material. The chemical composition of this steel is defined according to DIN EN 10025-6, as shown in **Table 1**. In direct comparison, the composition of the filler metal Mn4Ni2CrMo according to DIN EN ISO 16834 is listed. In order to be able to detect the transition of the elements from the welding consumable into the weld metal, single-layer surface welds were performed, from which samples were taken for spectrographic investigations. The results are also summarized in **Table 1**.

IOP Conf. Series: Materials Science and Engineering **373** (2018) 012006 doi:10.1088/1757-899X/373/1/012006

**Table 1: Chemical composition of the base material, the steel and the corresponding weld metal depending on the filler material**

| | % max. DIN EN 10025-6 | welding wire Mn4Ni2CrMo DIN EN ISO 16834 | weld metal Mn4Ni2CrMo | weld metal Mn4Ni2CrMo + 6µm Niob |
|---|---|---|---|---|
| **C** | 0,20 | 0,12 | 0,1184 | 0,0979 |
| **Si** | 0,80 | 0,60 - 0,90 | 0,4540 | 0,4555 |
| **Mn** | 1,70 | 1,60 - 2,10 | 1,190 | 1,193 |
| **P** | 0,020 | 0,015 | 0,0066 | 0,0045 |
| **Cr** | 1,5 | 0,20 - 0,45 | 0,4434 | 0,4013 |
| **Mo** | 0,70 | 0,45 - 0,70 | 0,4267 | 0,4292 |
| **Nb** | 0,06 | - | 0,0092 | 1,022 |
| **Ni** | 2,0 | 1,8 - 2,3 | 1,619 | 1,752 |
| **Ti** | 0,05 | - | 0,0130 | 0,0078 |

In the first investigations with the filler material of the alloy Mn4Ni2CrMo and a 6µm thick niobium coating (Mn4Ni2CrMo + 6 µm niobium), the transition of the niobium into the weld metal can be detected by optical emission spectroscopy.

The spectroscopic analysis shows that the carbon content in the investigated compounds is about 0.1% and about 1% niobium are included. Up to about 0.15% C, the formation of Fe3Nb2 is observed, [9]. According to the ternary system iron-niobium-carbon the formation of α-ferrite, niobium carbides (Nb4C3) and Laves phases (Fe3Nb2) is to be expected, **Figure 6**. To assess the microstructure an image recording in the cover layer of the weld and one in the layer below (hereafter referred to as "middle-layer") was made, **Figure 5**.

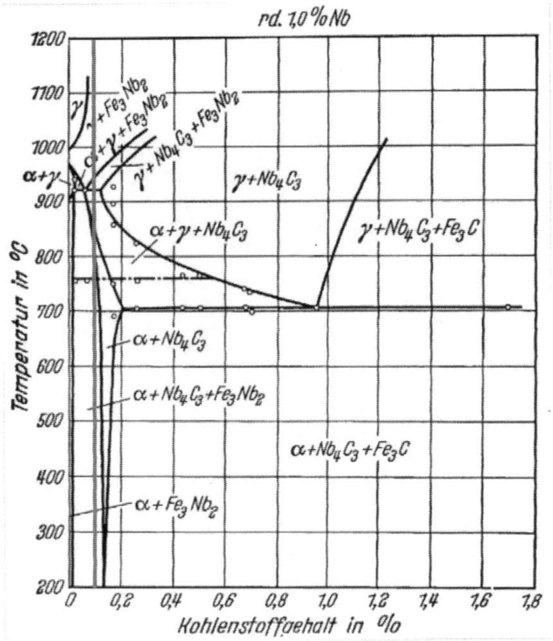


**Figure 5:** Section through the ternary system iron-niobium-carbon parallel to the iron-carbon side [9]

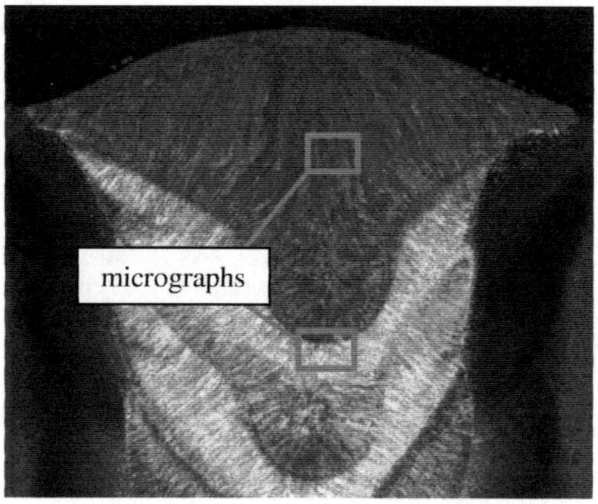


**Figure 6:** Position of micrographs in the welding seam

The influence of the increased niobium content on the microstructure becomes clear from the micrographs in Figure 7a.) bis f.). In each case, micrographs of the top layer (left) and the middle layer (right) are shown.

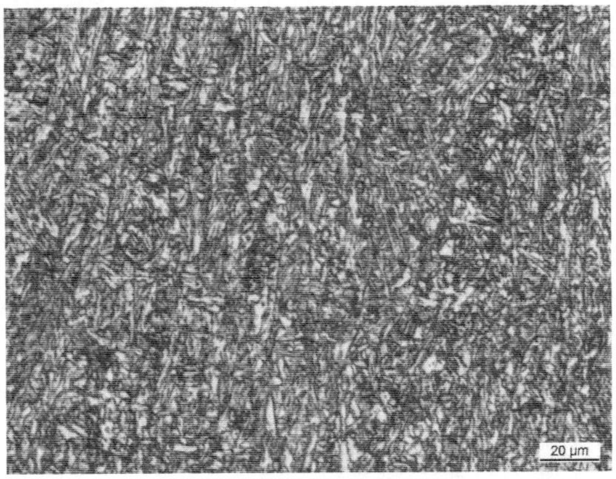


a) Mn4Ni2CrMo top layer

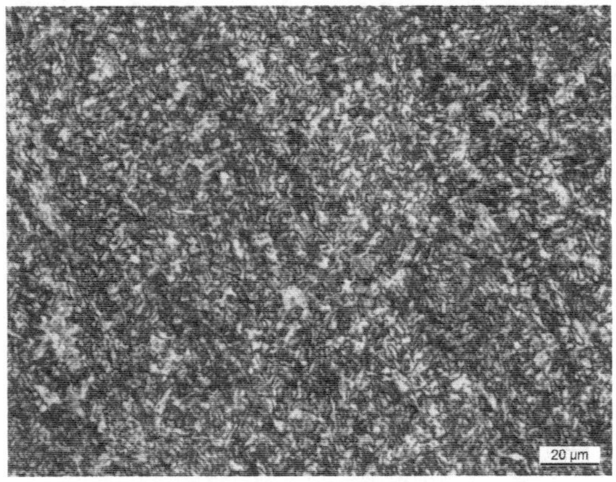


b) Mn4Ni2CrMo middle layer

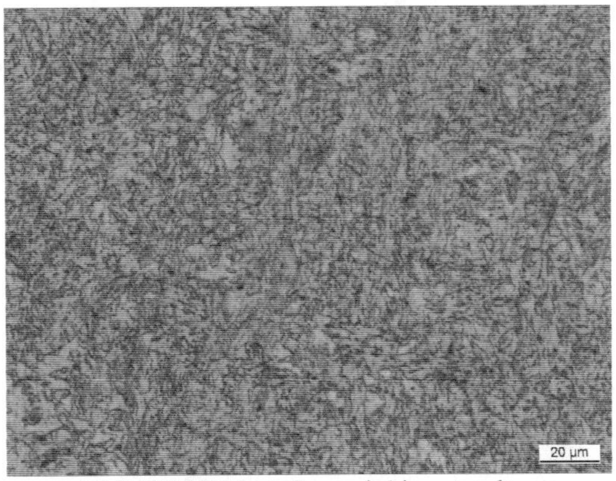


c) Mn4Ni2CrMo + 3µm niobium top layer

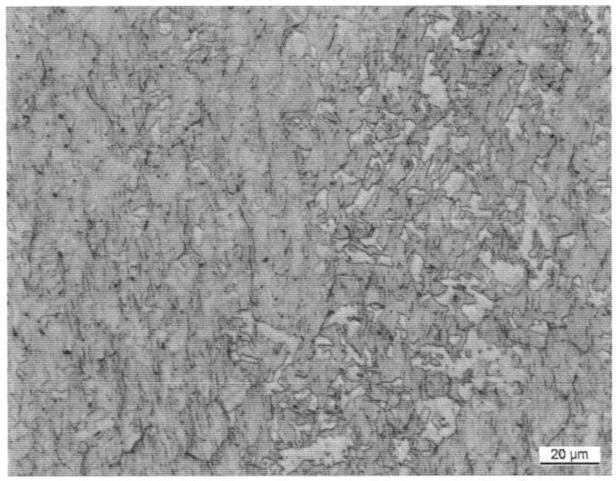


d) Mn4Ni2CrMo + 3µm niobium middle layer

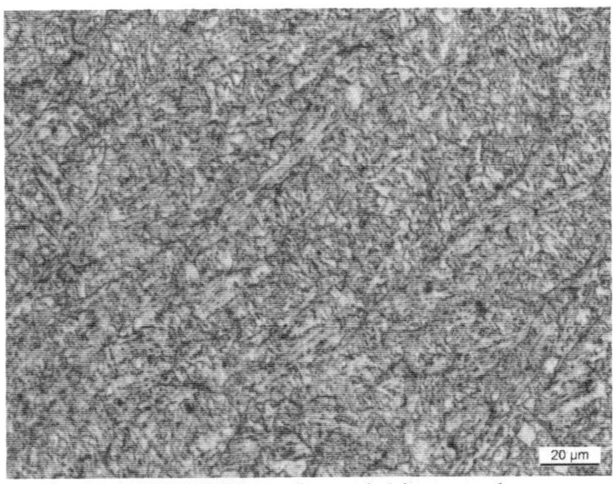


e) Mn4Ni2CrMo + 6µm niobium top layer

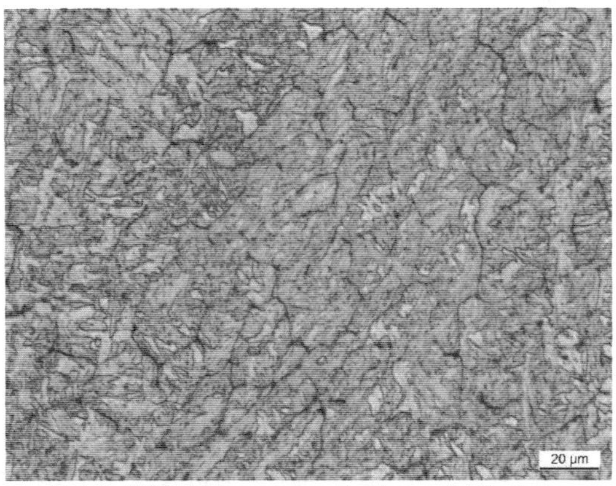


f) Mn4Ni2CrMo + 6µm niobium middle layer

**Figure 7:** Welded products of the filler material Mn4Ni2CrMo without coating and with different thicknesses of niobium coating

In all cases, microstructural changes between the cover layer and the middle layer of the welded joint can be recognized, which are due to diffusion and recrystallization processes by the reheating.

IOP Conf. Series: Materials Science and Engineering **373** (2018) 012006 doi:10.1088/1757-899X/373/1/012006

**Figure** 7**a** und **Figure 7b** show the structure of the uncoated filler material Mn4Ni2CrMo. This mainly produces ferritic-bainitic microstructure. Micro-alloying elements allow the fine-grained structure to be retained in the center of the weld by precipitates that block grain growth and reduce the re-crystallization rate of austenite, [10].

In direct comparison, a much finer grain structure is found in the cover layer of the weld metal, which is produced with Mn4Ni2CrMo + 3 μm niobium. On closer examination, the position of niobium carbides (Nb4C3) and Laves phases (Fe3Nb2) can be assumed. In the middle layer, the fine-grained structure dissolves due to the repeated temperature influence of the cover layer and diffusion processes occur at the grain boundaries, Figure 7c und Figure 7d.

The microstructure, which is produced with Mn4Ni2CrMo + 6 μm niobium, appears coarser in the micrograph, Figure 7e und Figure 7f. The brittle-hard phases (carbide and Laves phases) are more pronounced in the cover layer. The micrograph of the middle layer allows the hypothesis that the diffusion processes are therefore more distinct here and therefore the primary austenite grain boundaries are clearly recognizable after the production of the cover layer.

## 4. Determination of mechanical technological properties

The influence of the microstructures on the mechanical technological properties of the welded joints was investigated by notched bar impact tests according to DIN EN ISO 148-1 and tensile tests according to DIN EN ISO 6892-1,[11] [12].

The focus of the investigations is the mechanical strength of the weld metal, which is why the samples for the analysis of the microstructure as well as for the investigation of the mechanical technological properties are taken according to DIN EN ISO 15792-1,[13]. The location of the samples in the weld metal is shown in **Figure 8** and **Figure 9**.

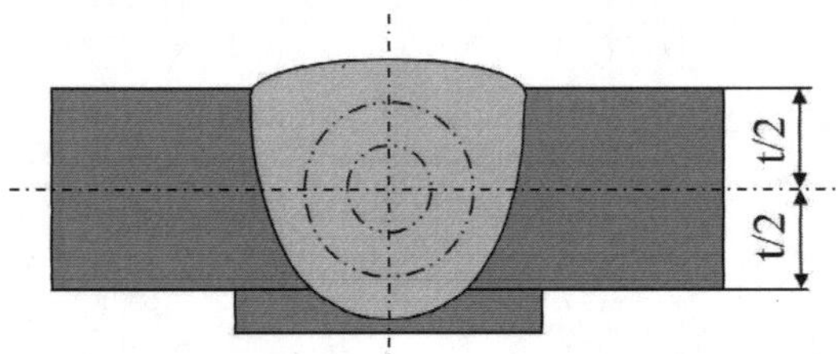


**Figure 8:** Location of the tensile specimens in the weld metal

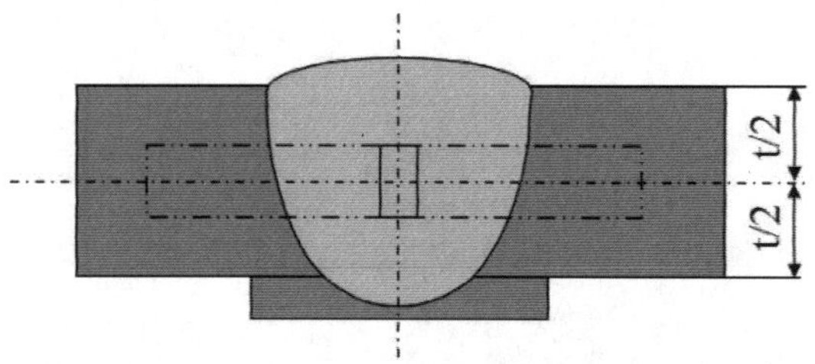


**Figure 9:** Location of notched bar impact tests in weld metal

For initial investigations, compounds with Mn4Ni2CrMo, Mn4Ni2CrMo + 3 μm niobium and with Mn4Ni2CrMo + 6 μm niobium are welded before impact test samples and tensile specimens are taken. The welding parameters are given in **Table 2**.

**Table 2:** welding parameters

| parameter | value |
|---|---|
| wire feed | 9,3 m/min |
| voltage | 30 V |
| current | 290 A |
| welding speed | 300 mm/min |
| energy per section | 1,74 kJ/mm |

In the notched-bar impact tests, the notched-bar impact work falls considerably due to the niobium introduced into the weld metal via the filler metal. **Table 3** compares the average notched impact values of three impact tests each.

**Table 3**: Impact work as a function of the niobium coating

| Filler metal | Impact work as a function of the niobium coating in J |
|---|---|
| Mn4Ni2CrMo | 107,5 |
| Mn4Ni2CrMo + 3μm niobium | 6,67 |
| Mn4Ni2CrMo + 6μm niobium | 5,17 |

The brittle material failure can be traced back to the brittle-hard, intermetallic Laves phases which are not plastically deformable at room temperature and the niobium carbides. An embrittlement of the material was therefore to be expected. The fracture surfaces of the impact test samples are shown in **Figure 10**.

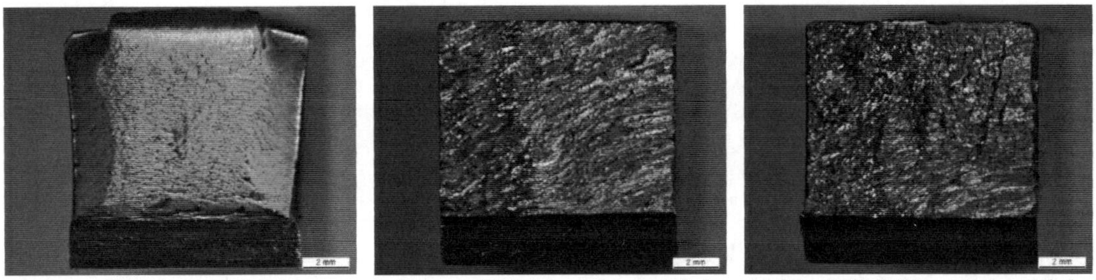

a.) Mn4Ni2CrMo b.) Mn4Ni2CrMo + 3 μm Niob c.) Mn4Ni2CrMo + 6 μm Niob

**Figure 10:** Fracture surfaces of impact test samples with different niobium content

**Figure 10a** shows the fracture surface of an impact test sample from the weld metal, which was welded by the uncoated filler material Mn4Ni2CrMo. The comparatively ductile behavior of the sample is reflected in the illustrated deformation fracture.

**Figure 10b** and **Figure 10c** show the fracture surfaces of the samples using the niobium-coated welding wire. Both samples show macroscopically no plastic deformation at the fracture surface and show brittle material failure. The sample with a lower niobium content indicates a more oriented fracture surface structure, which is significantly less pronounced in the sample with a higher niobium content.

The influence of the niobium contents on the strength of the weld metal becomes clear in the tensile test and the corresponding stress-strain diagrams.

IOP Conf. Series: Materials Science and Engineering **373** (2018) 012006 doi:10.1088/1757-899X/373/1/012006

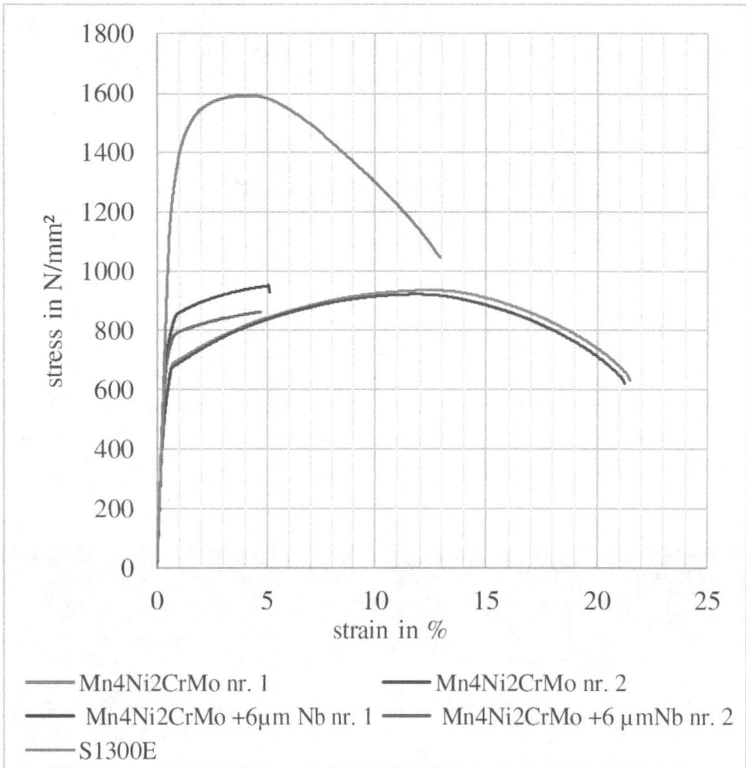


**Figure 11:** Stress-strain diagram for filler metals Mn4Ni2CrMo and Mn4Ni2CrMo + 6 µm niobium

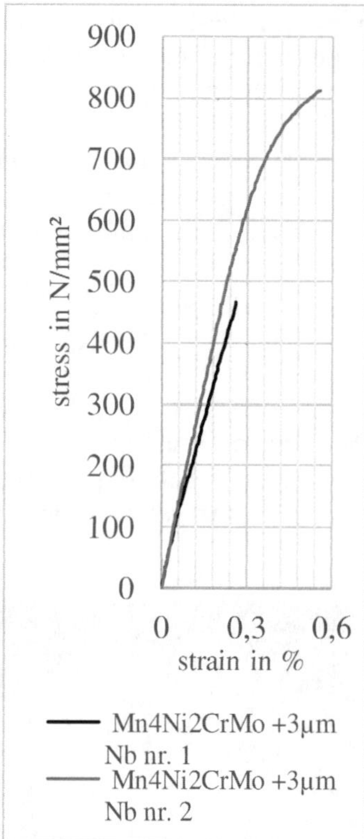


**Figure 12:** Stress-strain diagram for Mn4Ni2CrMo + 3 µm niobium

With a layer thickness of 3 µm niobium on the welding wire, the weld metal has such a brittle behavior that neither data for the yield strength of Rp0.2 nor for the tensile strength can be determined, **Figure 12**. With a further increase in the niobium content (layer thickness 6 µm), the elongation at break is lower in comparison to the uncoated filler metal and the behavior of the material is also more brittle, **Figure 11**. In addition, the stress-strain behavior of the modern fine grain structural steel S1300E was determined. The comparison shows that the yield strength can be positively influenced by the coated filler metals, but further research work is necessary in view of such high strengths. The averaged values are summarized in **Table 4**

**Table 4:** Average characteristics of the tensile test according to DIN EN ISO 6892-1

| | $Rp_{0,2}$ in MPa | tensile strenght in MPa | elongation at break in % |
|---|---|---|---|
| Mn4Ni2CrMo | 605,3 | 927,9 | 21 |
| Mn4Ni2CrMo + 3µm niobium | not evaluated | not evaluated | 0,1 |
| Mn4Ni2CrMo + 6µm niobium | 765 | 873,5 | 4,5 |

The yield strength of the weld metal can be increased by about 26% by the addition of 1 wt .-% niobium. With increasing niobium content, the elongation at break decreases first and then increases again to 4.5%. By comparison, the base material S960QL used according to DIN EN 10025-6 has an elongation at break of at least 10%, [14]

**Figure 13** shows the determined values for the elongation at break and an indicated possible course over the layer thickness of the niobium. For a comparison with the base material, the elongation at break of the S960QL is also shown.

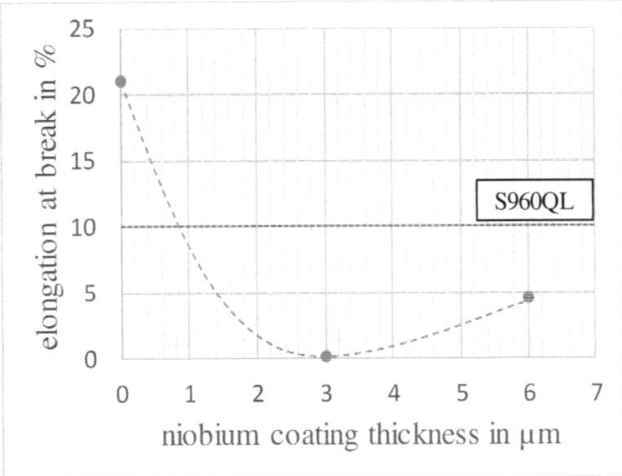


**Figure 13**: elongation at break depending on the niobium coating thickness

Further investigations with a higher niobium content are intended to show whether the illustrated relationship is a local minimum of the elongation at break and to what extent the elongation at break can be raised again with increasing niobium content in order to reach the level of the base material with simultaneously increased yield strength.

## 5. Summary and research needs

In view of the demand for welding consumables for modern fine-grain structural steels, the presented method opens up new opportunities for research. The investigations show that it is possible to develop welding consumables with multifunctional thin-film layers, which not only have a direct influence on the weld metal and its strength, but also on the arc and the processability of the filler material. This allows the investigation of alloying elements in previously unachieved proportions of the weld metal.
The yield strength of the weld metal can be increased by 26% through the filler metal Mn4Ni2CrMo, on which a 6 μm thick layer of niobium was applied. However, the toughness of the weld metal decreases significantly. On the basis of the knowledge acquired, investigations of further niobium content as well as other alloying elements are planned. whereby the influence of the thin film coatings on the mechanical technological characteristics of the weld metal should be considered.
The goal is to achieve or even exceed the strengths of modern fine grain steels (overmatching). The challenge is the combination of increased strength with sufficient toughness properties, which should result directly from the weld structure without any possibilities for aftertreatment.

## 6. References

[1] Ummenhofer, T., Spannaus, M., Steidl, G., Hölbling, W. u. Di Rosa, V.: Die Anwendung hochfester Feinkornbaustähle im konstruktiven Ingenieurbau. Stahlbau 82 (2013) 4, S. 236–245

[2] DVS – DEUTSCHER VERBAND FÜR SCHWEISSEN UND VERWANDTE VERFAHREN E.V.: Metall-Schutzgasschweißen von Feinkornbaustählen. Merkblatt DVS 0916, 2012

[3] Bargel, H.-J. u. Schulze, G.: Werkstoffkunde. Berlin, Heidelberg: Springer Berlin Heidelberg 2012

[4] Wang, Z., Wu, J., Li, J., Wu, X., Huang, Y. u. Li, X.: 2017 Effects of Niobium on the Mechanical Properties and Corrosion Behavior of Simulated Weld HAZ of HSLA Steel. Metallurgical and Materials Transactions A 11, S. 14

[5] Hulka, K., Kern, A. u. Schriever, U.: 2005 Application of Niobium in Quenched and Tempered High-Strength Steels. Materials Science Forum 500-501, S. 519–526

[6] Heuser, H. u. Jochum, C.: 2007 Schweißzusatzwerkstoffe können/müssen nicht immer die Anforderungen der Grundwerkstoffe erfüllen. Materialwissenschaft und Werkstofftechnik 38, 7, S. 515–520
[7] Wesling, V., Schram, A., Müller, T. u. Treutler, K.: 2016 Influencing the arc and the mechanical properties of the weld metal in GMA-welding processes by additive elements on the wire electrode surface. IOP Conference Series: Materials Science and Engineering 118, S. 12006
[8] Klocke, F. u. König, W.: Fertigungsverfahren 1. Drehen, Fräsen, Bohren. VDI-Buch. Berlin, Heidelberg: Springer-Verlag 2008
[9] Houdremont, E.: Handbuch der Sonderstahlkunde. Berlin, Heidelberg: Springer Berlin Heidelberg 1943
[10] Dilthey, U.: Schweißtechnische Fertigungsverfahren 2. Verhalten der Werkstoffe beim Schweißen. VDI-Buch. Berlin, Heidelberg: Springer-Verlag Berlin Heidelberg 2005
[11] DIN Deutsches Institut für Normung e. V.: Metallische Werkstoffe - Kerbschlagbiegeversuch nach Charpy. Teil 1: Prüfverfahren. DIN EN ISO 148-1. Berlin: Beuth Verlag GmbH 2017
[12] DIN Deutsches Institut für Normung e. V.: Metallische Werkstoffe - Zugversuch. Teil 1: Prüfverfahren bei Raumtemperatur. DIN EN ISO 6892-1. Berlin: Beuth Verlag GmbH 2017
[13] DIN Deutsches Institut für Normung e. V.: Schweißzusätze - Prüfverfahren. Teil 1: Prüfverfahren für Prüfstücke zu Entnahme von Schweißgutproben an Stahl, Nickel und nickellegierungen. DIN EN ISO 15792-1. Berlin: Beuth Verlag GmbH 2012
[14] DIN Deutsches Institut für Normung e. V.: Warmgewalzte Erzeugnisse aus Baustählen. Teil 6: Technische Lieferbedingungen für Flacherzeugnisse aus Stählen mit höherer Streckgrenze im vergüteten Zustand. DIN EN 10025-6. Berlin: Beuth Verlag GmbH 2011

**Acknowledgment**
The investigations presented were supported within the framework of the Central Innovation Program for SMEs (ZIM) by the Federal Ministry for Economic Affairs and Energy on the basis of a resolution of the German Bundestag (grant number: ZF4419001FH7).
We thank the Hermann Fliess GmbH & Co KG and the SSAB Swedish Steel GmbH for the support of the investigations.

# Electrodeposition and characterisation of Al-W alloy films from ionic liquid

**D Höhlich, D Wachner, M Müller, I Scharf and T Lampke**

Institute of Materials Science and Engineering, Chemnitz University of Technology, 09125 Chemnitz, Germany

**Abstract.** Al–W alloy films were prepared by electrodeposition using anhydrous 1-ethyl-3-methylimidazolium chloride (EMIMCl) ionic-liquid solution with aluminium chloride in a ratio of 1:1.5. As a commercially available tungsten precursor, tungsten hexachloride was used. The metal is dissolved in the ionic liquid at concentrations up to 0.06 mol/l. The deposition took place outside the glove box with a continuous argon stream over the electrolyte at a temperature of 60 °C with a current density of 1 A/m$^2$. Resulting alloys show a tungsten content higher than 20 wt% (3.5 at%). The phase composition of Al-W alloys was observed by X-ray diffraction (XRD), and the chemical composition was characterised by scanning electron microscopy (SEM) and inductively-coupled plasma optical emission spectroscopy (ICP-OES).

## 1. Introduction

Many components can be made from inexpensive and readily available raw materials if they are functionalised by a suitable metallic coating. This should provide protection against corrosion or chemical degradation, impart abrasion resistance, scratch resistance and hardness to the surface, or improve the electrical conductivity. For this purpose, the electrochemical coating of materials out of aqueous electrolytes is used. However, some technically important metals (e.g. aluminium) require deposition potentials where water decomposes electrolytically. As an alternative to aqueous electrolytes, the use of ionic liquids (IL) is possible. ILs are salts with large unsymmetrical ions, which are liquid at a temperature below 100 °C. They possess a high decomposition voltage and thus a larger usable electrochemical window compared to water. As a result, noble metals [1, 2] and very non-noble metals can be deposited from IL without a decomposition of the electrolyte. Further advantages of ILs are the low vapour pressure, low toxicity, and non-flammability. However, the handling of the ionic liquids is complicated due to their strong hygroscopic properties. Even small amounts of water from the surrounding atmosphere can lead to unwanted side reactions such as the release of HCl in chloride-containing ILs. Therefore, working in closed environments (such as a glove box with an inert gas atmosphere) is required. This results in challenges in the production of technically usable metallic coatings. Anhydrous work on an industrial scale is not state of the art due to high costs and the complex technical realisation.

The principles of galvanic deposition from IL are described in the two standard works of H. Ohno [3] and F. Endres et al. [4]. Several reviews have explained the great potential of the IL for the deposition of base metals and semiconductors [5–7]. Among others, aluminium was deposited for example on high-strength steel screws [8]. Aluminium layers are interesting because the metal shows a high resistance to corrosion due the formation of a natural passive film (alumina) on the surface. Therefore, they are suitable as corrosion protection for metals that are more reactive. But these layers can be also damaged by corrosion. The aluminium is locally corroded by pitting [9]. As an alternative to technically used aluminium alloys, the alloy with tungsten offers good protection against corrosion it by shifting the corrosion potential to a much more positive value [10]. In recent years, attempts have been made to deposit refractory metals from IL. In 2010, Nohira and co-authors deposited tungsten from N-ethyl-N-methylpyrrolidinium chloride (EMPyrCl) - $ZnCl_2$ on Ni at 150 °C [11]. The precursors $WCl_4$ or $WCl_6$ were also used in the presence of KF as additive. In the case of $WCl_6$ precursor, a denser and smoother, but still rough and cracked film was obtained. Al–W alloy films produced by sputtering [12–13], ion implantation [14], laser surface alloying [15], and electrodeposition [16–20] were reported previously. Hirato and co-authors deposited W-Al alloys from IL using a $W_6Cl_{12}$ precursor [21]. In the case of aluminium deposition, the system $AlCl_3$-EMIMCl has often been used. In this system, a general problem is the moisture sensitivity of the IL, especially for the chloride-based one. However, the methods listed here have disadvantages for a later technical implementation. The applied thin-film techniques such as sputtering are e.g. not suitable for obtaining layers > 2 µm. Processes with specially synthesised precursors such as $W_6Cl_{12}$ are unfavourable for a later scale up. Furthermore, a process that only works in the glove box is difficult to translate into a larger technical framework. Many attempts have been made to use less moisture and air sensitive systems, but until now there is no alternative to $AlCl_3$ [22]. In consequence, all relevant tests for aluminium deposition are carried out in a glove box under inert gas atmosphere. Initial trials for technical implementation outside a closed facility were carried out for the first time, a steel strip coating succeeded in an open plant [23]. Nevertheless, it is still far from a technical implementation.

The aim of this work is a description of the electrochemical deposition of Al-W layers from ionic liquids outside a glove box with commercially available electrolyte components. In the future, this process can be further developed to a technical application.

## 2. Materials and Methods

The electrolyte prepared consists of tungsten(VI)-chloride $WCl_6$ (99,9 %, ABCR) and 1-Ethyl-3-methylimidazolium chloride-aluminium chloride $EMIMAlCl_3$ (> 98 %, Iolitec) ionic liquid. The molar ratio of EMIM:$AlCl_3$ was 1:1.5. For the purpose of using the electrolyte outside the glove box, the contact of the electrolyte to water must be avoided completely. All vessels must be thoroughly flushed with argon. Also, all tools must be dry. The substances are highly hygroscopic and/or release HCl on contact with atmospheric moisture. The preparation of 40 ml electrolyte took place in a round bottom flask under argon. Before every experiment, the IL was heated up to 100 ° C and dried for 5 h under vacuum at 4 mbar. A Vacuubrand PC 3004 Vario vacuum pump was used. Thereafter, the flask must be put back under the protective Ar atmosphere. Subsequently, the $WCl_6$ was added, preferably in small portions. It was again dried for 2 days under vacuum (15 mbar) at 80–90 °C until the colour of the electrolyte was clear orange, which indicates the chemical equilibrium. For the galvanostatic deposition, the electrolyte was transferred into an argon-purged vessel, in which a few grams of molecular sieve were submitted. The molecular sieve protects the electrolyte by taking up any traces of water that may be contained in the electrolyte. Figure 1 shows the coating cell for working under non-aqueous conditions.

Steel sheets (5×5 mm) were used as cathode after appropriate pretreatment. As anode, an aluminium sheet with the same size was used. The electrochemical deposition was carried out under constant stirring (100 rpm) with a magnetic stirring bar in the size of 20×5 mm and argon counterflow with a plating temperature of 60 °C. The electrodeposited coating was from a plating bath containing 0.06 mol/l $WCl_6$. The Metrom Autolab PGSTAT204 potentiostat was used as power source in the galvanostatic mode. The applied cathodic current density was 1 $A/dm^2$ with a deposition time of 3600 s. The potential values were not measured. After the deposition, the cathode was immediately removed from the electrolyte and thoroughly cleaned with ethanol and water. Every deposition was carried out in freshly prepared electrolyte. The deposit morphology and composition was determined with a scanning electron microscope (LEO 1455VP, Zeiss), an X-ray diffractometer with a pointfocus, Lynxeye XE detector and Co-Kα radiation (D8 Discover, Bruker AXS) and with the optical emission spectrometry (Optima 8300, Perkin Elmer).

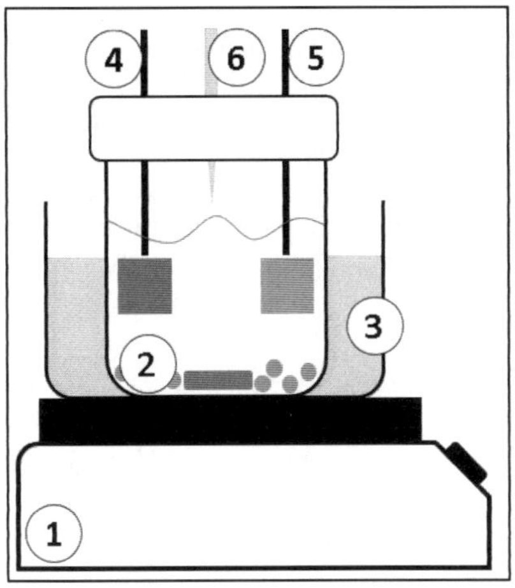


**Figure 1.** Illustration of the coating cell; with Heater and magnetic stirrer (1), beaker with magnetic stirring bar and molecular sieve (2), oil bath for constant tempering (3), anode (4), cathode (5) and the inert gas inlet (6).

## 3. Results and discussion

The obtained Al-W deposits were adherent, compact, light to dark grey and evenly dull. The layer thickness is heavily dependent on the flow rate. Too little bath movement (50 rpm) as well as too fast stirring (250 rpm) leads to less layer growth (< 2µm). An optimum has been shown at 100 rpm. Likewise, the current density had an influence on the layer properties. Variations between 0.05 A/dm² and 1 A/dm² showed that a higher tungsten content could be incorporated into the layer with a current density of 1 A/dm². A lower current density lead to a less uniform distribution of the layer thickness. It was also found that for lower current densities the layer thickness distribution became more uneven. The results were increased layer thicknesses in the edge areas with increased tungsten content.

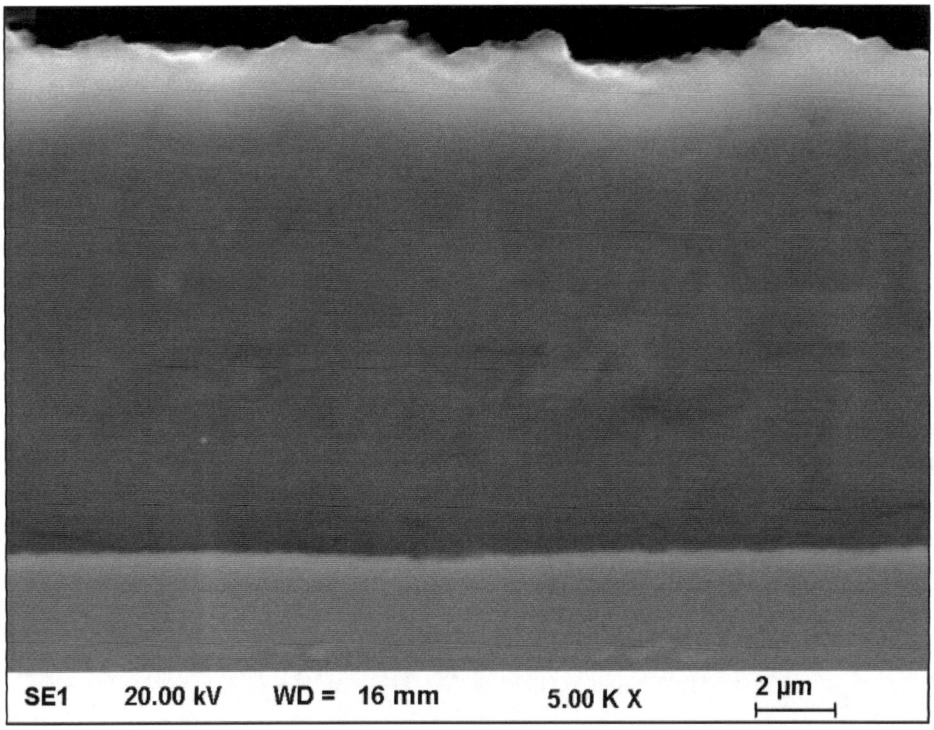


**Figure 2.** SEM image of the Al-W layer (1 A/dm², 60 °C, 100 rpm, 3600 s) with 20 wt% (3.5 at%) tungsten deposited outside the glove box.

Figure 2 shows the image of the cross-section from the Al–W layer. The resultant layer has a thickness of 16 μm and the tungsten content was determined by inductively coupled plasma optical emission spectroscopy (ICP-OES) and is 20 wt%, which corresponds to 3.5 at%. For this purpose, an aqua regia digestion was carried out in order to bring substrate and layer into solution. Thus, the measurement was carried out against corresponding standards. Literature shows that the layer becomes more porous with higher tungsten content. Results from the literature show that a firmly adhering layer can contain up to 12.4 at% Tungsten [22]. This shows that the results obtained can be compared with the systems known from the literature and a deposition outside the glove box is possible with commercially available precursors.

Furthermore, X-ray diffraction (XRD) measurements were carried out. A typical diffractogram measured is shown in figure 3. The layer shows a characteristic diffraction pattern for fcc Al. At $2\theta$ = 45° there is a clear peak, which indicates Al {111} and at $2\theta$ = 52° and 77° the peaks of the substrate material Fe. Elemental tungsten was not found because of a clear formation of an Al–W solid solution which was identified due to the shift of the aluminium lattice parameter, which is also in agreement with the literature (figure 4). In addition, the tungsten content is too low and the crystallite size could be in the order of a few nm. Thus, a possible tungsten phase can be X-ray amorphous. Thus, the resulting amorphous phase cannot be measured, which is consistent with the literature findings shown in figure 3 (right). Determined from the peak positions the lattice parameter shows a value of 0.4034 nm, which fits well into the previously known results.

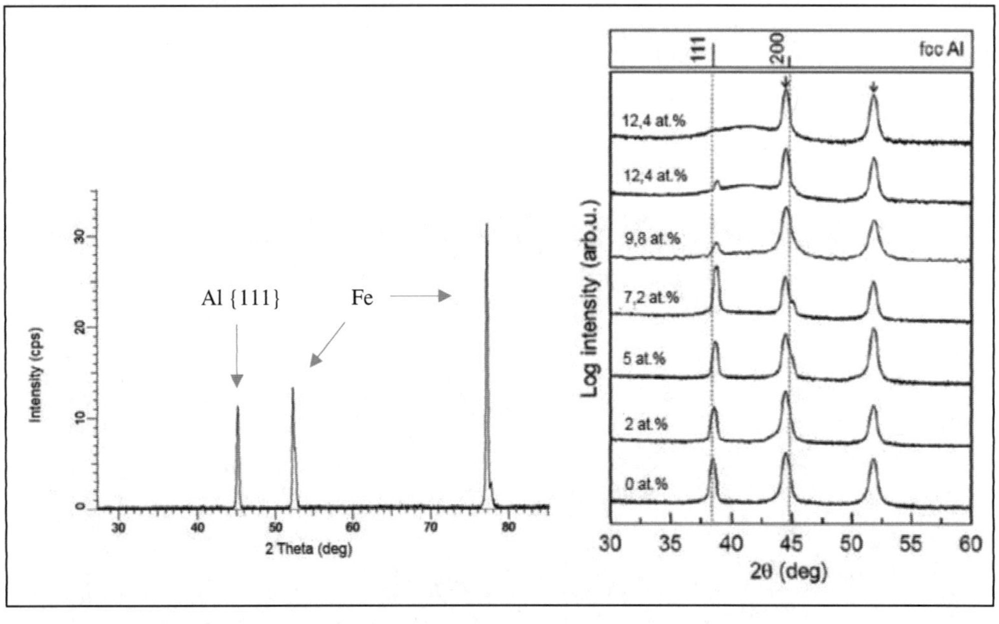


**Figure 3.** XRD measurement of the Al–W film with 20 wt % (3.5 at%) tungsten deposited outside the glove box (left) and XRD diffraction diagrams of aluminium and Al–W alloy films obtained from EMIMCl–$AlCl_3$ baths from reference [22] measured with Cu-Kα radiation (right).

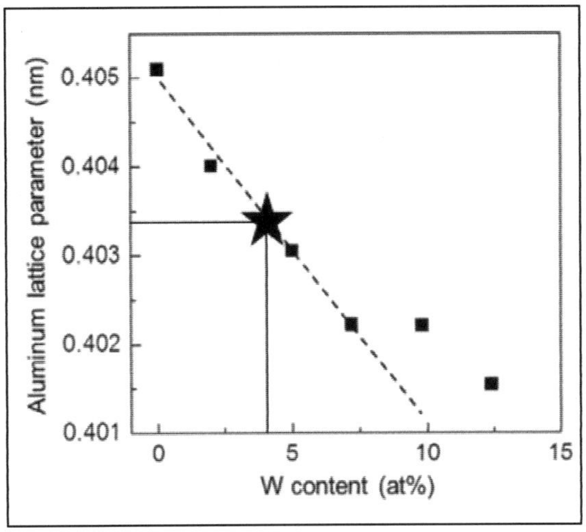


**Figure 4.** XRD measurement from reference [21], lattice parameter of fcc Al phase in aluminium and Al–W films on Nickel. The star corresponds to the lattice parameter from the measurements in this study.

## 4. Summary and conclusion

This work shows the successful preparation of Al–W alloys by electrodeposition from an EMIMCl-$AlCl_3$ electrolytic bath with a concentration of 0.06 mol/l $WCl_6$ outside the glove box at 60 °C from commercially available substances. The resultant layer has a thickness of 16 μm and a tungsten content of 20 wt% (3.5 at%). Layer studies were carried out using SEM, XRD and ICP-OES. This shows, in good agreement with existing literature, that it is possible to successfully electrodeposite Al–W layers outside closed systems like a glove box and obtain a metallic layer with technically relevant properties.

## 5. References

[1] Böck R Lanzinger G Freudenberger R Mehner T Nickel D Scharf I and Lampke T 2013 *J Appl Electrochem* 43 **12** 1207-16
[2] Lanzinger G Böck R Freudenberger R Mehner T Scharf I and Lampke T 2013 *Transactions of the IMF* 91 **3** 133-40
[3] Ohno H 2011 *Electrochemical Aspects of Ionic Liquids* (Hoboken: John Wiley & Sons, Inc.)
[4] Endres F MacFarlane A and Abbott A 2008 *Electrodeposition from Ionic Liquids* (Weinheim: Wiley-VCH)
[5] Abbott A Dalrymple I Endres F and Macfarlane D 2008 *Why use Ionic Liquids for Electrodeposition, in: Electrodeposition from Ionic Liquids* (Weinheim: Wiley-VCH)
[6] Zein El Abedin S and Endres F 2006 Electrodeposition of Metals and Semiconductors in Air- and Water-Stable Ionic Liquids, *Chemphyschem* **7** 58–61
[7] Endres F 2002 *Chemphyschem* **3** 144-54
[8] Zein El Abedin S and Endres F 2013 *J of Solid State Electrochem* **17** 1127–32
[9] Szklarska-Smialowska Z 1999 *Corros Sci* **41** 1743–67
[10] Shaw B Fritz T Davis G and Moshier W 1990 *J Electrochem Soc* **137** 1317–18
[11] Nohira T Kitagawa K Hagiwara R Nitta K Majima M and Inazawa S 2010 *Trans Mat Res Soc Japan* **35** 35–7
[12] Shaw B Fritz T Davis G and Moshier W 1990 *J Electrochem Soc* **137** 1317–18
[13] Shaw B Davis G Fritz T Rees B and Moshier W 1991 *J Electrochem Soc* **138** 3288–95
[14] Rangel C Travassos M and Chevallier J 1997 *Surf Coat Technol* **89** 101–7
[15] Rajamure R Vora H Srinivasan S and Dahotre N 2015 *Appl Surf Sci* **328** 205–14
[16] Tsuda T Hussey C Stafford G 2007 *ECS Trans* **3** 217–31
[17] Tsuda T Ikeda YArimura T Imanishi A Kuwabata S Hussey C and Stafford G 2012 *ECS Trans* **50** 239–50

[18] Tsuda T Ikeda Y Arimura T Hirogaki M Imanishi A Kuwabata S Stafford G and Hussey C 2014, *J Electrochem Soc* **161** D405–12
[19] Sato K Matsushima H Ueda M 2016 *ECS Trans.* **75** 305–12
[20] Higashino S Miyake M Fujii H Takahashi A and Hirato T 2017 *J Electrochem Soc* **164** D120–25
[21] Higashino S Miyake M Takahashi A Matamura Y Fujii H Kasada R andHirato T 2017 *Surf. Coat Tech* **325** 346–51
[22] Zein El Abedin S Moustafa E Hempelmann R Natter H and Endres F 2006 *Chemphyschem* **7** 1535–43
[23] Dietz A 2012 *Technische Entwicklungen bei der Al-Abscheidung in Proc. 34. Ulmer Gespräch der DGO* (Bad Saulgau: Leuze-Verlag)

**Acknowledgements**
The authors acknowledge the financial support by by the German "Bundeministerium für Bildung und Forschung" in the joint project "GALACTIF" 13XP5017D. We would also like to acknowledge the VDI Technologiezentrum and all project partners for supporting the research. We would also like to thank Mr Marc Pügner for the XRD measurements and Mr Steffen Clauß for the SEM measurements.

IOP Conf. Series: Materials Science and Engineering **373** (2018) 012008 doi:10.1088/1757-899X/373/1/012008

# Electroplating of aluminium microparticles with nickel to synthesise reactive core-shell structures for thermal joining applications

**S Schreiber and M F Zaeh**

Institute for Machine Tools and Industrial Management (*iwb*), Technical University of Munich, Boltzmannstrasse 15, 85748 Garching, Germany

**Abstract.** Reactive particles represent a promising alternative for effectively joining components with freeform surfaces and different material properties. While the primary application of reactive systems is combustion synthesis for the production of high-performance alloys, the highly exothermic reaction can also be used to firmly bond thermosensitive joining partners. Core-shell structures are of special interest, since they function as separate microreactors. In this paper, a method to synthesise reactive nickel-aluminium core-shell structures via a two-step plating process is described. Based on an electroless process, the natural oxide layer of the aluminium particles is removed and substituted with a thin layer of nickel. Subsequently, the pre-treated particles are electroplated with nickel. The high reactivity of aluminium and the oxide layer play a significant role in adjusting the process parameters of the Watts bath. Additionally, the developed experimental set-up is introduced and the importance of process control is shown. In order to achieve reproducible results, the electroplating process was automated. Ignition tests with electromagnetic waves demonstrated that the particles undergo an exothermic reaction. Therefore, they can be used as a heat source in thermal joining applications.

## 1. Introduction

A mere invention does not automatically lead to innovation, as the latter depends on whether an idea is successfully put into industrial practice or not [1]. Combustion synthesis is nowadays considered as an established production technology to efficiently synthesise high-performance materials [2; 3]. Apart from the main objective of alloying, the occurring highly exothermic reaction has also attracted interest in academia and industries. The released energy can be used as a heat source and therefore represents a promising approach for thermal joining applications. Multi-material designs and thus hybrid joints are gaining importance in the field of production engineering due to their lightweight potential. Consequently, inventive and innovative joining strategies are required.

### *1.1. Reactive systems as innovative heat source*

The combinations of elements (e. g. Ni + Al, Ti + 2B, Hf + C, etc.), which are used in combustion synthesis, are denoted as reactive systems. In a specific stoichiometric ratio, the reactants are able to undergo a self-sustaining, exothermic reaction. During this reaction, the reactants can either be in a solid, liquid or gaseous state depending on the respective adiabatic temperatures [4]. The ignition of a reactive system is of major importance, since it significantly influences the combustion mode. Generally, it can be distinguished between two combustion modes: the self-propagating or propagation mode [5] and the thermal explosion [6] or volume combustion synthesis mode [7]. Together with the fact that the term self-propagation high temperature synthesis (SHS) is often synonymously used with the term combustion synthesis as well as the self-propagating mode, it becomes evident that a vast variety of terms is present in scientific publications. In case that the reaction is initiated by uniform heating (e. g. in a conventional oven, with electromagnetic waves [8] or in an induction furnace [9]), the thermal explosion mode occurs. If the energy input is not homogeneous, but locally confined to a certain point or area (e. g. induced by a mechanical impact [10], a heated wire [11] or a laser impulse [12]), a planar, self-sustaining reaction wave propagates through the reactive system which is referred to as self-propagating mode. Aside from the initiation, the reaction behaviour is also influenced by the morphologic structure of the reactants. Most common in combustion synthesis are powders or pressed pellets [2]. In addition to that, reactive foils, also denoted as multilayer systems, have attracted increased attention over the past decades. An overview over the different structures is depicted in figure 1.

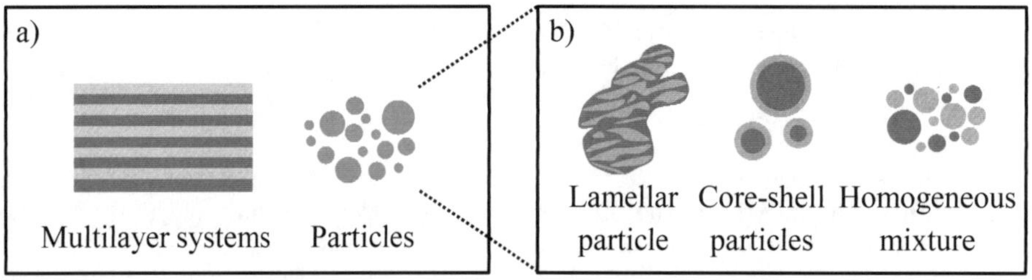


**Figure 1.** Schematic representation of distinct a) structures of reactive systems and b) morphologic forms of reactive powders.

As illustrated in figure 1 a), multilayer foils and particles can be considered as the two basic forms of reactive systems. Moreover, the category of particles can be further subdivided into homogeneous mixtures of particles, lamellar composites and particles with a core-shell structure (see figure 1 b)). The structure of the reactive system as well as the morphologic form significantly influence the reaction characteristics like the reaction rate, the ignition temperature or the maximum combustion temperature. Taking into consideration that multilayer systems, which consist of alternating layers of two metals in the range of nanometres, differ from particles with dimensions in the millimetre range by several orders of magnitude, the previous statement becomes most obvious.

*1.2. Reactive joining with nickel and aluminium*

While a great variety of reactive systems exists, the combination of nickel and aluminium has attracted particular attention. The ordered crystal structures in nickel aluminides result in low densities, high melting points, high mechanical strengths as well as comparatively good oxidation and corrosion resistances [13]. Due to these excellent material properties, nickel aluminides are not only used as lightweight construction materials, but also in order to influence the combustion characteristics of propellants [14] or as elaborated coatings [15]. Although experimentally determined values for the reaction enthalpy vary, a value of $-62 \pm 2$ kJ/mol corresponds to most publications [16; 17]. Furthermore, maximum reaction temperatures of up to 2083 K [18] are reached within milliseconds or seconds, which forms a promising basis for thermal joining applications.

The idea of using the released heat of reactive systems as a joining technology has led to sophisticated research efforts [19]. To the present day, reactive multilayer systems consisting of nickel and aluminium represent the only system that is commercially available (trade name NanoFoil®). Different intrinsic structures and characteristics, namely the bilayer or the overall thickness, strongly influence the reaction characteristics [20] and thus can be an effective means to adapt the desired reaction enthalpy to a specific joining task. However, this valuable joining concept features certain disadvantages. Due to the brittle nature of the multilayer systems, the geometries of the joining partners are limited to even surfaces. In addition to that, the comparatively low degree of automation [21] combined with high manufacturing costs because of the physical vapour deposition process impede the industrial application.

Reactive particles, which provide the advantageous characteristics of reactive systems as well as an increased flexibility for joining geometries, represent a promising alternative to multilayer systems.

Pellets of homogeneous mixtures of nickel and aluminium powders have been used to join nickel base superalloys [22; 23] or refractory ceramics [24]. However, comparatively little attention is being paid to reactive core-shell structures. Research efforts in the field of core-shell particles, also denoted as clad particles, are limited and mainly focussed on the reaction behaviour [25–27] for the use in propellants or combustion synthesis. Every single particle represents a reactive system and thereby contributes to the reaction during the joining process. Thus, the synthesis of core-shell particles, which forms the basis for the resulting combustion characteristics, must be further studied.

### *1.3. Electrochemical synthesis of reactive particles*
Reactive core-shell particles are commonly synthesised via wet-chemical coating processes, with aluminium as the substrate and nickel as the deposited material. In contrast to the electroless deposition [28], the electroplating of aluminium powders with nickel is not properly described. Generally, aluminium particles are suspended in a nickel ions containing electrolyte and electroplated with nickel as soon as they come into contact with the cathode. In a continuous process, the suspension is circulated so that the particles repeatedly collide with the cathode [29; 30]. A discontinuous process consists of three steps: sedimentation, electroplating and stirring [31]. First, particles settle onto the cathode. As soon as the circuit is closed, particles are electroplated. Finally, vigorous stirring separates agglomerates, which formed during the electroplating step.
For aluminium particles with a natural oxide layer, a specific pre-treatment is required, since the particles have to be electrically conductive. In addition to that, numerous parameters influence the deposition of nickel on aluminium particles and therefore have to be taken into consideration. Figure 2 shows essential parameters for obtaining reactive core-shell particles.

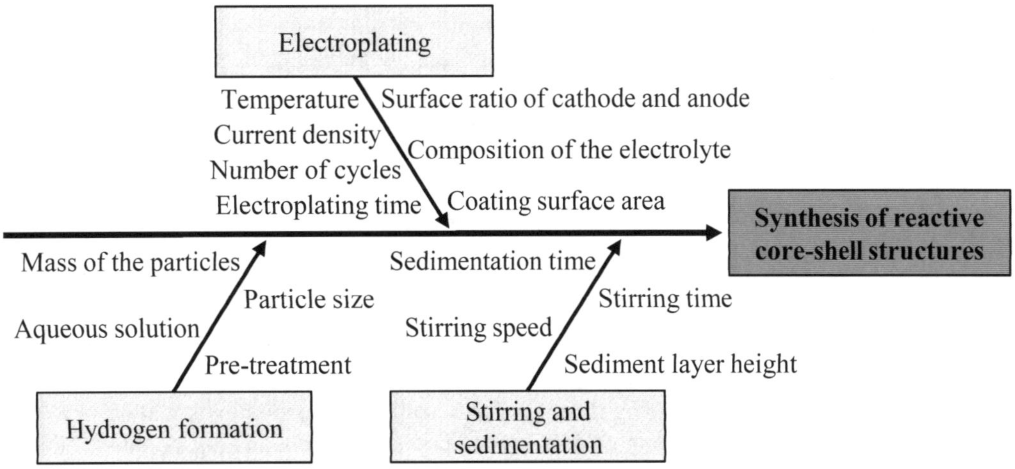


**Figure 2.** Essential parameters for the electroplating of aluminium particles with nickel.

TAKESHIMA et al. [29] present a continuous process to obtain coated aluminium particles with diameters ranging from 0.1 to 10 μm. On the contrary, YIH [31] describes a discontinuous process without any restrictions regarding the particles size, but does not explicitly mention aluminium particles to be suitable for electroplating. Therefore, this research focusses on the electrochemical synthesis of reactive core-shell particles, which are composed of aluminium as a substrate and nickel as coating. A two-step wet-chemical process is studied, which consists of an electroless pre-treatment for the removal of the oxide layer and a discontinuous electroplating process.

## 2. Experimental procedure
The pre-treatment of the aluminium particles (New Materials Development GmbH, 6.9 μm and 23.5 μm, 99.7 % purity) requires the following chemicals: nickel(II) chloride hexahydrate (Alfa Aesar, 99.3 %),

ammonium chloride (Sigma-Aldrich, 99.998 %), sodium citrate dihydrate (Sigma-Aldrich, ≥ 99 %) and sodium fluoride (Fluka, ≥ 99 %). The electrolyte is a common Watts bath and consists of the following chemicals: nickel(II) chloride hexahydrate (see above), boric acid (Sigma-Aldrich, 99.7 %) and nickel(II) sulphate hexahydrate (Sigma-Aldrich, ≥ 99.95 %). All chemicals were used as received.

*2.1. Pre-treatment and electroplating of nickel*

Prior to the electroplating, it is essential to remove the natural oxide layer of the aluminium particles. The pre-treatment solution is based on RAMASESHAN et al. [32] and serves to remove possible contaminations as well as to prevent further oxidation by substituting the oxide with a layer of nickel. The adapted composition for the electroless pre-treatment at room temperature is shown in table 1.

**Table 1.** Composition of the pre-treatment solution.

| Chemical | Formula | Pre-treatment solution |
|---|---|---|
| Nickel(II) chloride hexahydrate | $NiCl_2 \cdot 6H_2O$ | 30 g/l |
| Ammonium chloride | $NH_4Cl$ | 7 g/l |
| Sodium citrate dihydrate | $C_6H_5Na_3O_7 \cdot 2H_2O$ | 20 g/l |
| Sodium fluoride | NaF | 0.5 g/l |

Subsequently, aluminium particles are electroplated with nickel until a stoichiometric ratio of 1:1 is reached. Table 2 describes the composition of the electrolyte solution, which is a Watts bath.

**Table 2.** Composition of the Watts electroplating solution.

| Chemical | Formula | Electroplating solution |
|---|---|---|
| Nickel(II) sulphate hexahydrate | $NiSO_4 \cdot 6H_2O$ | 300 g/l |
| Nickel(II) chloride hexahydrate | $NiCl_2 \cdot 6H_2O$ | 20 g/l |
| Boric acid | $H_3BO_3$ | 20 g/l |

*2.2. Experimental set-up*

The electroplating of the aluminium particles with nickel requires a specific experimental set-up, which guarantees that the aluminium particles are in contact with the cathode. In this study, a discontinuous three-step process patented by YIH [31] was used. A schematic representation of the experimental set-up is depicted in figure 3. An automated process control allows to adapt the time-related parameters sedimentation, electroplating, and stirring time. Uniform heating is ensured by a heating jacket which encloses the glass cylinder with the electrolyte solution. The cathode is made out of aluminium so that the oxide layer, in this case, can prevent nickel deposition [31].

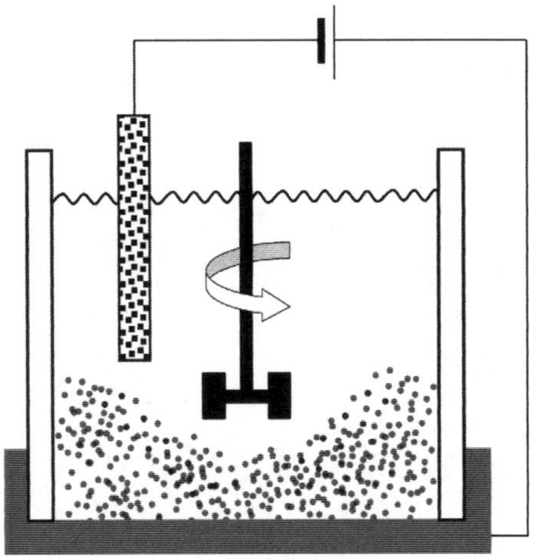

**Figure 3.** Scheme of the experimental set-up based on [31].

## 3. Results and discussion

Before discussing the electroplating process, the oxidation of aluminium, which can be found in common literature, but is not mentioned in the patents cited above, is studied. This phenomenon is of crucial importance for the laboratory- and industrial-scale synthesis of reactive core-shell particles.

### *3.1. Reaction of aluminium particles*

Aluminium particles are also under investigation due to the oxidation of aluminium in aqueous solutions and the resulting formation of hydrogen. Starting with the overall reaction equation, the occurring reactions can be described as follows [33]:

$$2Al + 6H_2O \rightarrow 2Al(OH)_3 + 3H_2\uparrow \quad (1)$$

$$Al_2O_3 + H_2O \rightarrow 2AlOOH \quad (2)$$

$$AlOOH + H_2O \leftrightarrow Al(OH)_3 \quad (3)$$

Since the reaction shown in equation (1) is both a source for hydrogen and highly exothermic with a heat release of 16.95 kJ/g [34], aluminium particles are suitable for energy and hydrogen storage. The oxide layer on the aluminium particles and its reaction with water (see equation (2) and (3)) as well as low temperatures inhibit further oxidation and are hence considered disadvantageous [33; 35].
As it can be consequently anticipated that the suspension of the aluminium particles in the aqueous Watts electrolyte solution will result in the formation of hydrogen, preliminary studies with only deionised water and aluminium particles (6.9 µm and 23.5 µm) were conducted. It was investigated how the ratio of the mass of aluminium particles to the volume of water, the temperature of the suspension, a pre-treatment, the particle size and stirring influence the subsequent electroplating process.

It was found that a high ratio of the mass of the aluminium particles to the volume of the water, stirring and a smaller particle size promote the formation of hydrogen. In contrast to that, lower temperatures ranging from 318 to 323 K and a pre-treatment of the aluminium particles impede the exothermic reaction. Furthermore, the formation of hydrogen in the electrolyte solution is weaker.

### *3.2. Electroplated core-shell structures*

In order to obtain reactive core-shell particles, the pre-treated aluminium particles are electroplated with nickel until a stoichiometric ratio of 1:1 is reached. The pre-treatment took place at room temperature and no significant formation of hydrogen was observed. A scanning electron micrograph of the pre-treated aluminium particles is shown in figure 4. In order to avoid further or a new formation of $Al_2O_3$, the particles were directly added into the electrolyte solution after washing.

**Figure 4.** Scanning electron micrograph of a pre-treated aluminium particle.

Based on the preliminary studies, on experiments with a Hull cell and on the information given in the cited literature, essential parameters were identified and varied within a specific range as listed in table 3. It is recommended to adjust a current density over 5 A/dm$^2$ and to vary the height of the sediment layer from 1 to 30 mm depending on the particle size [31].

**Table 3.** Parameters for the electroplating process.

| Parameter | Value or range |
|---|---|
| Stirring time | 10 s |
| Sedimentation time | 120 s |
| Current density of the cathode plate | 5 A/dm$^2$ |
| Electroplating time | 150 s |
| Sediment layer height | 1 – 10 mm |
| Temperature | 323 K |
| Stirring speed | 500 rpm |
| Number of cycles | 50 |

At first, particles with a diameter of 23.5 µm were studied at a temperature of 323 K and a sediment layer height of 10 mm. The results of the electroplated aluminium particles are illustrated in figure 5.

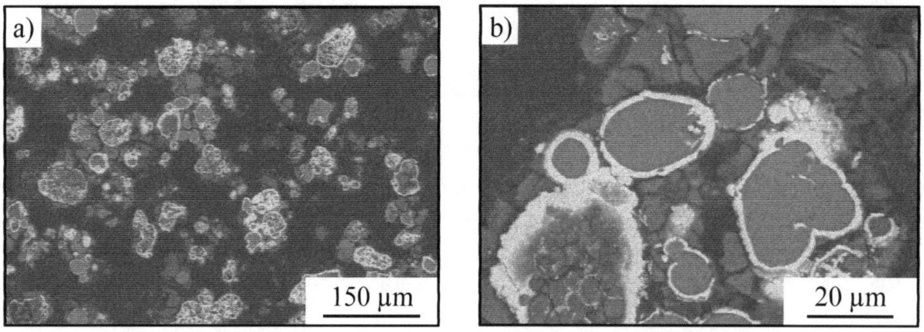


**Figure 5.** Scanning electron micrographs of the metallographic section of the electroplated aluminium particles (dark grey) with nickel (light grey).

Even though the aluminium particles are electroplated with nickel, the coating only results in some core-shell particles (see figure 1 a)). The thickness of the nickel layer as well as the uniformity of the nickel shell differ (see figure 1 b)). As significant hydrogen formation was observed during the electroplating process, no proper sediment layer could be formed, which led to the circumstance that particles were not constantly in contact with each other during the electroplating step.

Moreover, particles with a diameter of 6.9 μm were investigated. In order to reduce the expected increase in the hydrogen formation due to the smaller particle size, the height of the sediment layer was reduced to 3 mm. The result is shown in figure 6.

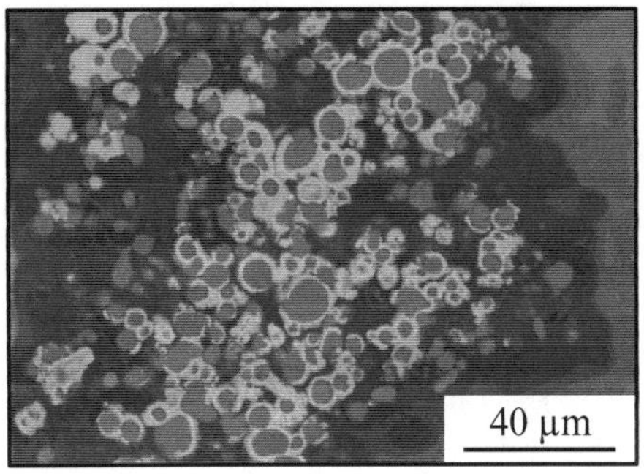


**Figure 6.** Scanning electron micrograph of the metallographic section of the electroplated aluminium particles (dark grey) with nickel (light grey).

The formation of hydrogen and thus the movement of the suspension could be reduced. As a result, a more uniform layer of nickel was deposited onto the aluminium particles. However, there are still particles without any nickel shell and agglomeration was enhanced. After the electroplating process, a significant amount of nickel could be found on the cathode as well as adhering aluminium particles. Ignition tests with microwaves at a frequency of 2.45 GHz were used to demonstrate the reactivity. However, the described inhomogeneities led to an irregular combustion.
Consequently, the electroplating process must be further investigated to determine the cause-effect relationships between electroplating and hydrogen formation to obtain reactive core-shell particles.

## 4. Conclusions

Reactive systems are not only successfully used in combustion synthesis for the production of high-performance alloys, but do also represent a promising approach for joining different materials like metals, ceramics and plastics. The combustion characteristics and thus the released energy of a reactive system strongly depend on the morphologic structure of the reactants. In contrast to multilayer systems, particles provide an increased flexibility for joining geometries. In addition to that, reactive particles, with each single particle being reactive, are of great interest as a tailored heat source in thermal joining applications. Thus, a two-step wet-chemical process for the synthesis of reactive core-shell particles has been developed. The occurring oxidation of aluminium in aqueous solutions and the resulting formation of hydrogen significantly hamper electroplating. On the basis of experimental studies it could be demonstrated that the production of reactive core-shell structures is possible. However, further studies to investigate the cause-effect relationships must be performed to obtain a stable and safe process.

## 5. References

[1] Schumpeter J A 1947 The Creative Response in Economic History *J. Econ. Hist.* 7 **2** 149–59
[2] Merzhanov A G 1995 History and recent developments in SHS *Ceram. Int.* 21 **5** 371–9
[3] Rogachev A S and Mukasyan A S 2014 *Combustion for Material Synthesis* (Hoboken: Taylor and Francis)
[4] Subrahmanyam J and Vijayakumar M 1992 Self-propagating high-temperature synthesis *J. Mat. Sci.* 27 **23** 6249–73
[5] Moore J J and Feng H J 1995 Combustion synthesis of advanced materials *Prog. Mater. Sci.* **39** 243–273
[6] Yi H C and Moore J J 1990 Self-propagating high-temperature (combustion) synthesis (SHS) of powder-compacted materials *J. Mater. Sci.* **25** 1159–68
[7] Varma A, Rogachev A S, Mukasyan A S and Hwang S 1998 Combustion Synthesis of Advanced Materials: Principles and Applications *Adv. Chem. Eng.* **24** 79-226
[8] Karimbeigi A, Zakeri A, Aboutalebi M R and Sharifi P 2013 Study on microwave synthesis and characterisation of bulk nickel aluminides *Mater. Sci. Technol.* 28 **1** 86–91
[9] Lee H-Y, Ikenaga A, Kim S-H and Kim K-B 2007 The effects of induction heating rate on properties of Ni–Al based intermetallic compound layer coated on ductile cast iron by combustion synthesis *Intermet.* 15 **8**, 1050–56
[10] Reeves R V, Mukasyan A S and Son S F 2010 Thermal and Impact Reaction Initiation in Ni/Al Heterogeneous Reactive Systems *J. Phys. Chem. C* 114 **35** 14772–80
[11] Hardt A P and Holsinger R W 1973 Propagation of gasless reactions in solids—II. Experimental study of exothermic intermetallic reaction rates *Combust. Flame* 21 **1** 91–7
[12] Biffi C A, Bassani P, Sajedi Z, Giuliani P and Tuissi A 2017 Laser ignition in Self-propagating High temperature Synthesis of porous NiTinol Shape Memory Alloy. *Mater. Lett.* **193** 54–7
[13] Deevi S C and Sikka V K 1995 Nickel and iron aluminides: an overview on properties, processing, and applications *Intermet.* **4** 357–75
[14] Shafirovich E, Bocanegra P E, Chauveau C, Gökalp I, Goldshleger U, Rosenband V and Gany A 2005 Ignition of single nickel-coated aluminum particles. *Proc. Combust. Inst.* 30 **2** 2055–62
[15] Rasmussen A J, Agüero A, Gutierrez M and Østergård M J L 2008 Microstructures of thin and thick slurry aluminide coatings on Inconel 690 *Surf. Coat. Technol.* 202 **8** 1479–85
[16] Hu R and Nash P 2005 The enthalpy of formation of NiAl. *J. Mater. Sci.* 40 **5** 1067–69
[17] Nash P and Kleppa O 2001 Composition dependence of the enthalpies of formation of NiAl *J. Alloys Compd.* 321 **2** 228–31
[18] Zhu P, Li J C M and Liu C T 2002 Reaction mechanism of combustion synthesis of NiAl *Mater. Sci. Eng. A* **329-331** 57–68
[19] Mukasyan A S and White J D 2007 Combustion joining of refractory materials *Int. J. Self-Propag. High-Temp Synth.* 16 **3** 154–68
[20] Gunduz I E, Fadenberger K, Kokonou M, Rebholz C and Doumanidis C C 2008 Investigations on the self propagating reactions of nickel and aluminum multilayered foils *Appl. Phys. Lett.* 93 **13** 134101
[21] Boettge B, Braeuer J, Wiemer M, Petzold M, Bagdahn J and Gessner T 2010 Fabrication and characterization of reactive nanoscale multilayer systems for low-temperature bonding in microsystem technology *J. Micromech. Microeng.* 20 **6** 64018
[22] Pascal C, Marin-Ayral R M and Tédenac JC 2002 Joining of nickel monoaluminide to a superalloy substrate by high pressure self-propagating high-temperature synthesis *J. Alloys Compd.* **337** 221–5
[23] Rosa R, Colombini E, Veronesi P, Poli G and Leonelli C 2012 Microwave Ignited Combustion Synthesis as a Joining Technique for Dissimilar Materials. *J. Mater. Eng. Perform.* 21 **5** 725–32
[24] Wu W W, Gubarevich A V, Wada H and Odawara O 2011 Volume combustion synthesis of NiAl as applied to ceramics joining *Int. J. Self-Propag. High-Temp Synth.* 20 **2** 94–9
[25] Andrzejak T, Shafirovich E and Varma A 2007 Ignition mechanism of nickel-coated aluminum particles *Combust. Flame* **150** 60–70

[26] Thiers L, Mukasyan A S and Varma A 2002 Thermal explosion in Ni-Al system *Combust. Flame* **131** 198–209
[27] White J D, Reeves R V, Son S F and Mukasyan A S 2009 Thermal explosion in Al-Ni system *J. Phys. Chem. A* 113 **48** 13541–7
[28] Schreiber S, Theodossiadis G D and Zaeh M F 2017 Combustion synthesis of reactive nickel-aluminum particles as an innovative approach for thermal joining applications *Mater. Sci. Eng.* **181** 12008
[29] Takeshima E, Takatsu K, Kojima Y and Fujii T *Electroplating of fine particles with metal* US Patent US4908106 A
[30] Semrau W *Coated metal powder and method for producing the same* Patent WO1999061183 A3
[31] Yih P *Method for electroplating metal coating(s) particulates at high coating speed with high current density* US Patent US006010610 A
[32] Ramaseshan R, Seshadri S K and Nair N G 2001 Electroless nickel-phosphorus coating on Ti and Al elemental powders *Scripta Mater.* **45** 183–9
[33] Shmelev V, Nikolaev V, Lee J H and Yim C 2016 Hydrogen production by reaction of aluminum with water *Int. J. Hydrogen Energy* 41 **38** 16664–73
[34] Rosenband V and Gany A Application of activated aluminum powder for generation of hydrogen from water *Int. J. Hydrogen Energy* 35 **20** 10898–904
[35] Dudoladov A O, Buryakovskaya O A, Vlaskin M S, Zhuk A Z and Shkolnikov E I 2016 Generation of hydrogen by aluminium oxidation in aqueous solutions at low temperatures *Int. J. Hydrogen Energy* 41 **4** 2230–37

**Acknowledgement**
Funded by the Deutsche Forschungsgemeinschaft (DFG, German Research Foundation) – 3376550608

IOP Conf. Series: Materials Science and Engineering **373** (2018) 012009 doi:10.1088/1757-899X/373/1/012009

# Corrosion protection mechanisms of TiMgN hard coatings on steel

**A Heyn[1], T Mueller[2], M Balzer[3], H Kappl[3] and M Fenker[3]**

[1] Otto-von-Guericke-University Magdeburg, Institute of Materials and Joining Technology, Magdeburg, Germany
[2] Bundesanstalt für Materialforschung und –prüfung, Division 7.6 Corrosion and Corrosion Protection, Berlin, Germany
[3] fem, Forschungsinstitut Edelmetalle + Metallchemie, Schwäbisch Gmünd, Germany

**Abstract.** Hard coated steel components are used in a wide application range, mostly for protective, wear resistant and decorative purposes. Despite of these coatings being regarded as relatively dense, there is always a high risk of localized corrosion when a coated low alloyed steel component encounters a surrounding high humidity atmosphere or even an aqueous medium. An approach to enhance the corrosion properties is the addition of magnesium to physical vapor deposited hard coatings, like TiN. It has been found that there is a remarkable increase in corrosion resistance in dependence of magnesium content of the TiMgN coating and its surface properties. In this work the authors will explain the underlying corrosion protection mechanisms by means of electrochemical and analytical studies. The positive impact of magnesium in the coating relates on its preferred dissolution vs. steel. This causes the potential to shift to more negative direction with respect to the steel substrate and additionally leads to a temporarily passivation of the steel due to alkalization of the surrounding electrolyte by formation of magnesium hydroxide.

## 1. Introduction

By using hard coatings on steel wear resistant and protective functional surfaces can be created. The most popular hard coating type "TiN" is almost anodically inert and usually gets deposited with only a few micrometers of thickness. These coatings are only corrosion protective versus the underlying steel due to their barrier functionality. Unavoidable defects in the coating, often originated during their production or in operation, can lead to localized corrosion phenomena if aqueous electrolyte (water / water vapor) reaches the unprotected steel.

An alternative approach to change the electrochemical, mechanical and visual properties of the coating is the incorporation of magnesium into a PVD hard coating such as TiN via a magnetron sputtering process. For example, hardness or oxidation resistance at higher temperatures can be changed as well as color from gold-yellowish to blue-greyish [1]. A first indication of a change of their corrosion protection properties were revealed by potential measurements in 0.8 M NaCl solution. They showed that the open circuit potentials (OCP) of TiMgN coatings deposited on glass are temporarily lower than those of steel and that the steel can be temporarily protected cathodically. The more Mg is deposited, the lower the potentials and the longer this effect will last. In [2], Fenker et al. found an improvement in the corrosion behavior of TiMgN hard coatings on high-speed steel (HSS) from about 17 at-% Mg content based on results in the salt spray test.

IOP Conf. Series: Materials Science and Engineering **373** (2018) 012009 doi:10.1088/1757-899X/373/1/012009

Based on the observations that co-deposited magnesium has a positive effect on corrosion protection properties of the TiMgN hard coatings (see Fig. 1), further research was carried out starting in 2012. The work was financially supported by the German Research Foundation (DFG) and aimed to elucidate the underlying mechanisms and further improve the layer properties [3]. In this project, TiMgN coatings with 0, 10, 20 and 35 at% Mg, respectively on polished and blasted substrates made of glass and HSS substrates, were considered.

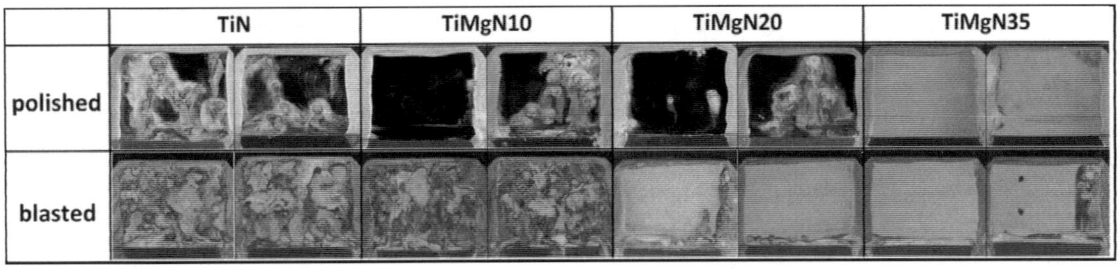


Fig. 1. Appearance of TiN and TiMgN hard coatings on HSS-substrates after 24 h NSS test

Based on OCP measurements with subsequent Mg content measurements of the NaCl-solutions, galvanic current measurements between TiMgN and steel, polarization measurements and camera-based experiments with gel pads, several underlying hypotheses for the improved corrosion protection have been established [4], e.g. the effect of the cathodic protection of the steel by preferably dissolving magnesium. However, since the TiMgN coating's OCP was found to be only temporarily more negative than steel, at least one additional mechanism had to be effective, to which the improved protection against local corrosion of the steel substrates could be attributed. A further focus was on the released Mg ions, which alkalinize the electrolyte at the interface to the exposed steel (Fe) by formation of hydroxides ($Mg(OH)_2$ and/or FeOOH), whereby the steel gets in a passive state. To demonstrate this effect, the release rates of Mg from the TiMgN hard coatings and the protection of steel in different $Mg(OH)_2$ solutions with different NaCl contents were investigated.

## 2. Specimen properties and experimental procedures

### *2.1. TiMgN fabrication and properties*

Semi-circular samples of hardened low alloyed steel 100Cr6 with a diameter of 30 mm were used as substrate material for the various coatings. For corrosion tests the substrate materials were used in two surface states: corundum-blasted and polished for triggering very high and very low growth defect concentrations, respectively. These extremely opposite types of surfaces should help to understand the influence of growth defects on the corrosion behavior of the coating-substrate systems. Studies on $Ti_xMg_yN$ systems [2] have shown that there were specific ranges of Mg content resulting in different corrosion behavior of the coatings. In scope of this work two different Mg contents (20 and 35 at-% Mg) were chosen to characterize the electrochemical behavior of TiMgN coating systems. The true Mg-concentration, measured by energy dispersive X-ray spectroscopy (EDX), was 19.9 at-% and 33.8 at-% respectively.

All coatings were deposited by reactive DC magnetron sputtering with a thickness between 2.5 and 3.0 µm in a vacuum plant Leybold L560 UV using a special magnetron arrangement. It consists of three circular 2" magnetrons (US Inc.) arranged in a triangle all facing a one-fold rotating flat substrate holder at 4.5 cm distance. Two cathodes were used for each deposition, one equipped with a Ti- (99.9 at-% purity) and one with a Mg-target (99.95 at-% purity). The desired Mg-contents of the TiMgN films were realized by adjusting the power fed to the Mg-target while keeping a constant target power for Ti. All coatings were deposited at a bias voltage $U_{Bias}$ = -50 V and a total pressure of p = 0.4 Pa. The basic $N_2$ partial pressure was $p_{N_2}$ = 0.05 Pa and the substrate temperature was 200 °C. Prior to the depositions the substrates were plasma etched at $U_{Bias}$ = -1000 V at $p_{Ar}$ = 2.6 Pa with additional cathode support.

The surface of the hard coatings on polished substrate showed a columnar structure, but also common growth defects, which were characterized using "large area high resolution" (LAHR) mapping, as described by Balzer in [5]. Typical coating defects are so called hillocks of different sizes, which may have pores or crevices down to steel substrate, or be broken out, exposing the substrate to the environment at open holes of different size. On the other hand, hard coatings on blasted substrate showed a very high amount of disorder, a porous structure and a much rougher surface.

### *2.2. Electrochemical and analytical investigations*

Samples of TiMgN with 20 and 35 at-% Mg deposited on polished and blasted 1.3505 steel were exposed to aqueous solutions with different levels of NaCl for up to 180 min and the OCP and dissolution kinetics were recorded. For this purpose, small areas of the sample surface were prepared by means of a perforated adhesive film. Then, a 3 mm in diameter measuring cell was attached thereto. The OCP was measured versus a saturated Ag/AgCl reference electrode, with a reference potential of 199 $mV_{NHE}$. In order to continuously determine the kinetic characteristic of the currently occurring corrosion reactions without significantly influencing the sample surface, the method of electrochemical frequency modulation (EFM) was used. In this case, a sinusoidal polarization (0.1 Hz base frequency, 0.2 and 0.5 Hz secondary frequencies) with an amplitude of 5 mV was impressed on the OCP at time-defined intervals (here every 5 min). As a result, the current polarization resistance (RP) and the simultaneously determined a/b factors (anodic and cathodic slopes) are used to obtain the instantaneous corrosion current. Details of this electrochemical methodology are described by Bosch et al. [6, 7]. The investigations were carried out with the measuring system Interface 1000 and the associated software Framework Data Acquisition from Gamry Instruments.

To investigate the release rate of magnesium, support from Max-Planck-Institute for Iron Research GmbH Düsseldorf (working group of M. Rohwerder) was accepted thankfully. There is an analytical method available, which uses a miniaturized scanning flow cell (SFC, cell diameter approx. 200 µm), that transfers the measuring electrolyte from the sample surface continuously for analysis by inductively coupled plasma mass spectrometry (ICP-MS). Thus, metal ions currently in solution (here Mg, Fe, Ti) can be analyzed with a short time shift (about 30 s). TiMgN samples with 20 and 35 at-% Mg were investigated, respectively on polished and blasted steel substrates and on glass. A 0.05 M NaCl solution was used as test electrolyte. Details of this methodology can be found in [8].

Furthermore, investigations on steel samples of 1.0535 carbon steel (C55, wet ground with P320 sandpaper) were carried out in differently concentrated $Mg(OH)_2$ solutions with different contents of NaCl. The aim was to determine the mode of action and the interaction of these electrolyte constituents on the passivity and corrosion of unalloyed steel. From literature [9] it is known, that the saturation concentration of $Mg(OH)_2$ is about 8.9 mg/L (at 18 °C) and that can set a maximum pH of 10.2. From about pH 9, iron can be present in the passive state, resulting in a narrow examination range of pH 9 to about pH 10. This was accomplished by dissolving quantities of 1, 3, 5 and 8 mg/L of solid $Mg(OH)_2$ in distilled water. Due to the electrolyte's susceptibility to lose alkalinity by reacting with $CO_2$ from ambient air to form $MgCO_3$, the electrolytes were kept under exclusion of air until the measurement was started and the current pH was determined directly before each measurement. Furthermore, defined amounts of NaCl were added to the electrolyte, initially 150, 367 and 584 mg/L. The background for the interaction of chlorides and pH values is a relationship that is well known in the field of corrosion protection of steel in concrete. Here, a [$Cl^-$ to $OH^-$]-ratio is often used, up to a level where the steel is still expected to be passive. There is a review article from Angst et al. [10] on this subject, in which the authors have found limiting values from 0.09 to 45. Since in the present case the pH is limited at about 10, additionally even lower NaCl contents of 75, 50 and 25 mg/L were used. Fig. 2 shows the resulting experimental design.

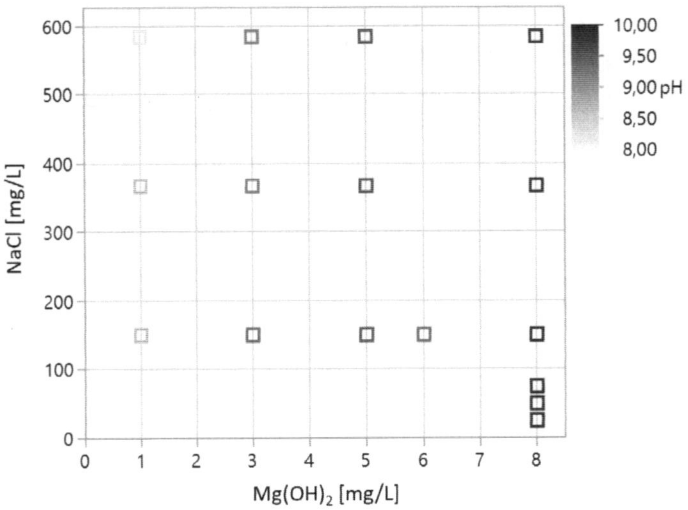


Fig. 2. Experimental design for the determination of the influence of $Mg(OH)_2$ concentration (pH-value) and NaCl concentration on the corrosion behavior of unalloyed steel

To demonstrate the influence of pH (or $Mg(OH)_2$ concentration) in combination with NaCl on the passivity and corrosion of steel in the context of the problem of TiMgN coatings, a special experimental procedure was devised, which takes account to the ongoing mechanisms on this system. For a certain time after contact with the electrolyte, the steel is known to be cathodically protected by the preferred Mg dissolution. To verify this, a pre-polarization of -500 $mV_{NHE}$ of the steel for 10 min was realized immediately after immersing the samples in the test electrolyte. This was followed by another 10-minute period during which the OCP (without external polarization) of the sample was measured. This was followed by an electrochemical impedance spectroscopy measurement (EIS) in the range of 50 kHz to 0.05 Hz, with an amplitude of 10 $mV_{RMS}$ at the last measured OCP. From the EIS measurement, the RP and the electrolyte resistance can be determined (and separated from each other) and thus the corrosion current can be calculated. Further results can be determined from the OCP measurement, e.g. a critical pitting potential ($E_{crit}$) and the time until it is reached (time to $E_{crit}$) if localized corrosion or activation has already been initiated within the 10-minute OCP measurement.

## 3. Results

### *3.1. OCP's and corrosion kinetic of TiMgN20 and TiMgN35 on polished and blasted HSS substrates*

Fig. 3 shows the curves of measurements on TiMgN with 20 and 35 at-% Mg, respectively on polished and blasted steel. The OCP's mostly start quite negative, but with a tendency to more positive values. The differences between 20 and 35 at% Mg and between polished and corundum blasted substrate are very clear. The same applies to the RP curves, whereby it should be noted here that falling RPs at 20 at-% Mg are due to localized corrosion after about 15 min. For low RPs of <1,000 $\Omega cm^2$, e.g. at 35 at-% Mg, an enhanced active dissolution of Mg from the TiMgN hard coatings is probably responsible. Higher RPs in the range of 10,000 to > 50,000 $\Omega cm^2$ indicate almost anodically inert surfaces, from which no or only relatively small amounts of Mg dissolve. For a better summary, the values at 10 minutes (boxed area in Fig. 3) from each of three repeated measurements were used for further statistical evaluation by analysis of variance (ANOVA).

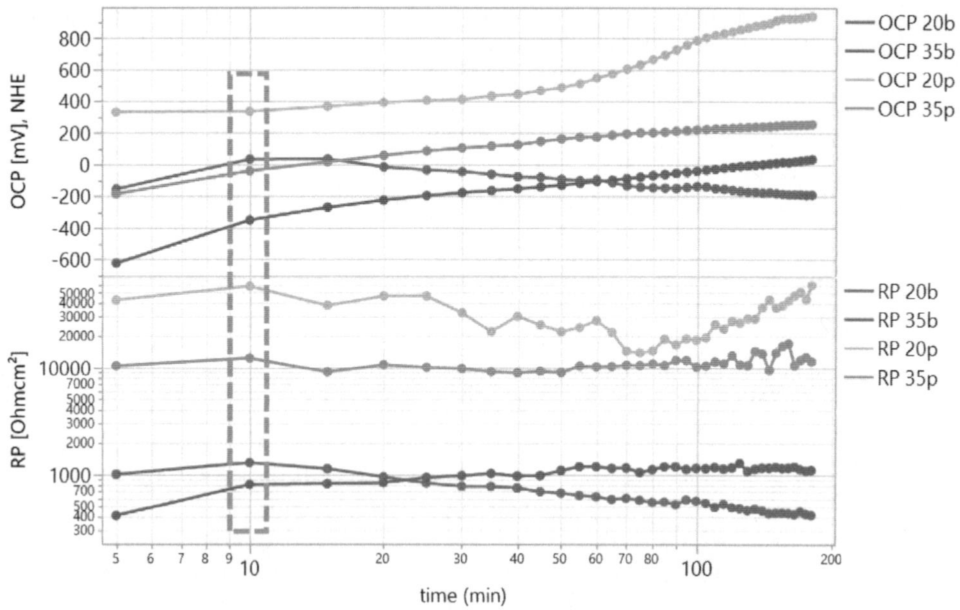


Fig. 3. Example of OCP´s and RP´s over 180 min in 0,8 M NaCl solution, TiMgN with 20 and 35 at-% Mg on corundum blasted (b) and polished (p) HSS, boxed area with values for statistical evaluation

The statistical evaluation (ANOVA) of the values after 10 min indicates a significant influence of the OCP ($r^2 = 0,95$) on the Mg content and surface state and the RP ($r^2 = 0,94$) on the Mg content, surface state and an interaction effect between them. The relationships of the factors are summarized in Fig. 4. For a better assessment of the OCP's, the potential range for actively corroding steel (approx. -350 $mV_{NHE}$) is shown. After 10 minutes, only samples with 35 at-% Mg on blasted substrate are still well below the potential of active steel. However, since improvements in the corrosion behavior were also observed in other states, it can already be assumed here, that the potential shift cannot be solely responsible for the corrosion protection. In addition, the leaching of magnesium from the TiMgN lattice and the associated changes in the electrolyte at the phase boundary appear to be significant. The strong dependency of the RP's on the Mg content and surface condition already gives an important hint here.

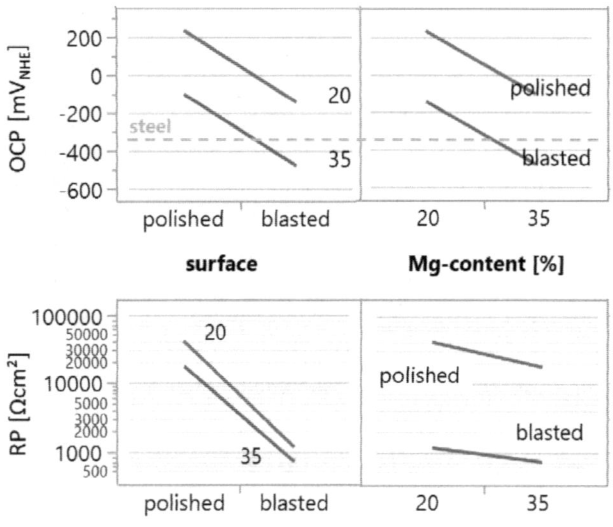


Fig. 4. Factor effects of Mg-content and surface conditions on OCP and RP after 10 min of immersion of TiMgN hard coatings in neutral solutions with different chloride contents, evaluated electrochemical data from Fig. 3 (boxed area)

### *3.2. Mg dissolution kinetic of TiMgN20 and TiMgN35 on polished and blasted HSS substrates*

TiMgN samples with Mg contents of 20 and 35 at-% Mg, both on polished and blasted substrates, were investigated for the released amount of Mg ions by means of a flow cell and ion analysis. Fig. 5 shows the trend of the Mg-ion concentration over an investigation period of 6000 s in a 0.05 M NaCl solution.

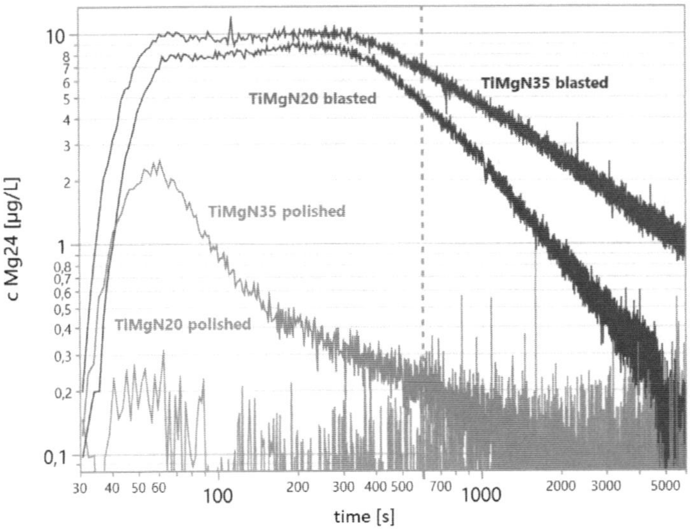


Fig. 5. Instantaneous released Mg ions from TiMgN hard coatings with 20 and 35 at-% Mg content on polished and blasted HSS substrates in 0.05 M NaCl solution using a flow cell with ICP-MS

Already after a very short time of the electrolyte contact, the Mg release increases rapidly. The differences between the Mg contents in combination with the surface state are very clear. TiMgN on corundum blasted substrates and the resulting larger surface release significantly more Mg ions than the polished states. At 35 at-% Mg (polished), a rapid release can be seen at first, but it soon stops again. From the samples with 20 at-% Mg (polished), almost no release of Mg ions is detectable (background noise at 0.2 - 0.3 µg/L). In the corundum blasted state, on the other hand, there is a very long-lasting and strong release. Considering the results from 3.1, this results in a coherent relationship regarding the situation of the OCP's and the RP's. If the release rate of Mg ions is higher, the RP is lower (or the corrosion current is correspondingly higher) and the OCP's tend to be more negative.

### *3.3. Corrosion behavior of steel in $Mg(OH)_2$ solutions with different NaCl content*

The experimental design (Fig. 2) was carried out according to the procedure described in 2.2. After the fixed cathodic polarization at -500 $mV_{NHE}$ for 10 min, the polarization was ceased and the progress of the OCP was followed for another 10 min. As an example, Fig. 6 shows potential curves for three different chloride contents in the case of nearly saturated $Mg(OH)_2$ solution (pH 9.9). In comparison, OCP curves are shown with the same chloride content, but in a pH 7 solution without $Mg(OH)_2$. It can be clearly seen, that in neutral solution with pH 7, corrosion occurs very rapidly within 2-3 min, characterized by the reversal of the potential in the negative direction after an initial rise in the positive direction. The behavior in the $Mg(OH)_2$ solution is different from that. Here, the OCPs increase longer in the positive direction and corrosion occurs only at more positive potentials, whereby the potential drops in the negative direction. The position of the critical potential ($E_{crit}$) and the time until reaching the critical potential (time to $E_{crit}$) are used for further evaluation. Following the OCP measurement, an EIS measurement is carried out to determine the RP. Fig. 7 shows the curves for the same three chloride contents in the case of nearly saturated $Mg(OH)_2$ solution (pH 9.9) and in a solution without $Mg(OH)_2$ at pH 7.

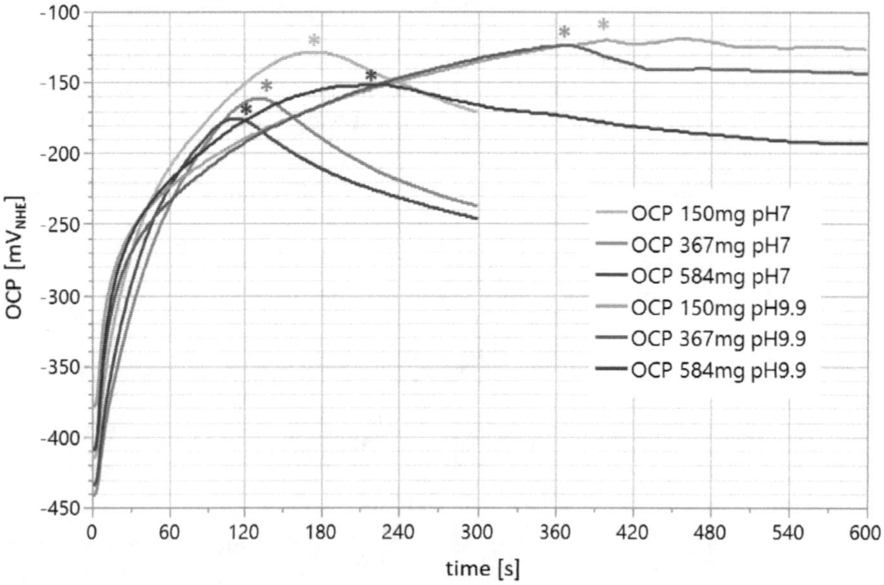


Fig. 6. OCP vs. time plot of C55 unalloyed steel in solutions of $Mg(OH)_2$ (pH 9.9) and neutral water (pH 7) with variation of the NaCl content, "*" indicates $E_{crit}$

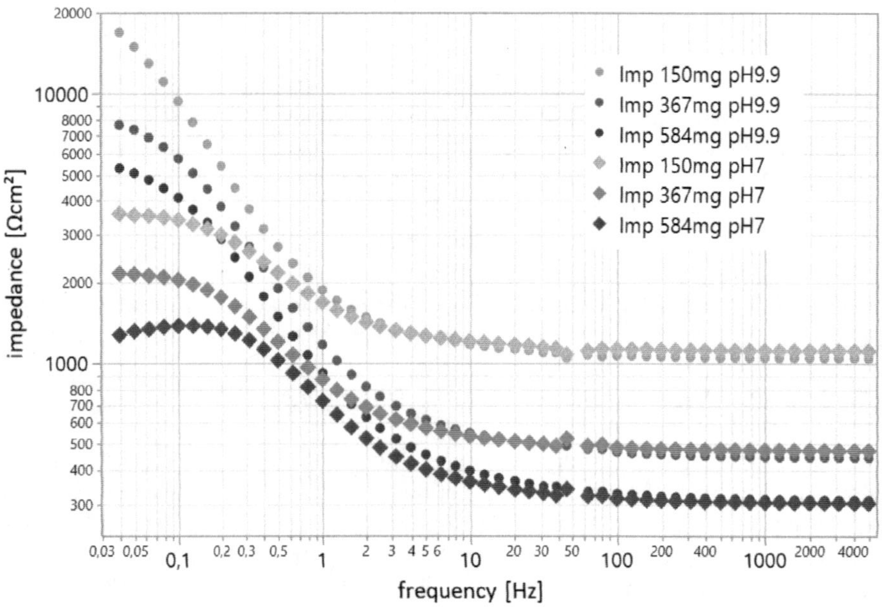


Fig. 7. Impedance vs. frequency plot of C55 unalloyed steel in solutions of $Mg(OH)_2$ (pH 9.9) and neutral water (pH 7) with variation of the NaCl content

For a further statistical evaluation (ANOVA) of the conducted test plan, the corrosion current ($i_{corr}$) was calculated from the RP. The required B value ($i_{corr}$ = B / RP) was determined by preliminary investigations in analogy to [11]. For steel it is between 0.06 and 0.025 V, depending on the RP. Furthermore, the critical potential and the time until its onset were used for further analysis. For the pH range of 9 to 10, the statistical evaluation showed that $E_{crit}$ primarily depends on the NaCl content ($r^2$ = 0,74), without significant influence of the pH-value. In contrast, the corrosion current depends on both the pH and the NaCl content ($r^2$ = 0,82) and the time to $E_{crit}$ depend on pH, pH*pH and NaCl content ($r^2$ = 0,80).

Furthermore, the $OH^-$-ion concentration was calculated from the measured pH value and, together with the given chloride ion concentration, the [$Cl^-$ to $OH^-$] ratio was determined. The results of all investigated electrolyte variants are shown in Fig. 8 as a function of this value, as well as the results of the statistical evaluation of the effects of pH value and NaCl content on $i_{corr}$ and the time to $E_{crit}$. One can clearly see the effect of alkalization on the relevant electrochemical characteristics. From a [$Cl^-$ to $OH^-$] ratio of 100 a positive effect starts, below 30 it becomes significant. The corrosion current already decreases by more than a decade at less than a ratio of 30.

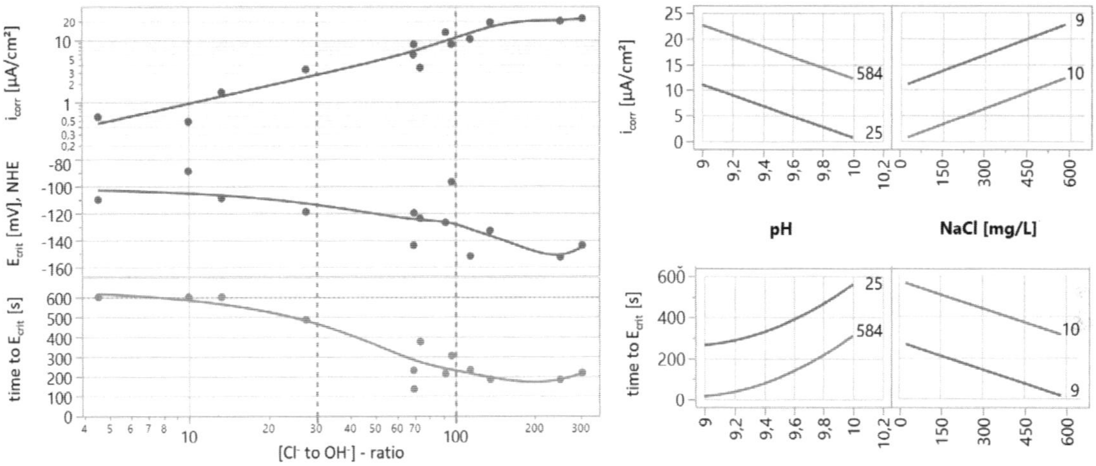


Fig. 8. Influences of pH (due to $Mg(OH)_2$) and NaCl on the corrosion behavior of unalloyed C55 steel, left: electrochemical values vs. [$Cl^-$ to $OH^-$]-ratio, right: factor effects from ANOVA model

## 4. Discussion

Electrochemical and analytical data can be used to give a more accurate picture of the corrosion protection mechanisms involved, leading to a significant improvement in the corrosion behavior of TiMgN coatings on steel substrates. The release rate of Mg ions proved to be crucial. With sufficiently high kinetics the potential can be lowered, so that the steel exposed by coating defects is protected cathodically for a short time. The more important and longer-lasting effect is the alkalization of the adjacent electrolyte due to the dissolved Mg ions, which form $Mg(OH)_2$ and allow passivation of steel.

If, for example, the theoretical release of Mg from the first atomic layer (about 0.3 nm) of a model defect of 1 µm diameter at 2.5 µm layer thickness of the TiMgN is calculated, the electrolyte volume of the model defect is enriched by 100-1,000 times that of the solubility for $Mg(OH)_2$, depending on the Mg content of the coatings. As a result, $Mg(OH)_2$ should precipitate out of the electrolyte and strongly buffer it, which in turn prevents the loss of alkalinity, e.g. by reactions of the electrolyte with $CO_2$ from the ambient air or by formation of magnesium chlorides.

In addition, $Mg(OH)_2$ ensures that no significant hydrolysis reactions occur in conjunction with chlorides, which can also lower the pH, as reported by Lamaka et al. in [12]. Passive steel in alkaline electrolytes in the presence of chloride ions, however, is subject to the risk of localized corrosion, which occurs at a critical pitting potential. However, this potential is much more positive (around 200 mV) than the OCP for general active corrosion, so that the temporary cathodic protection can extend further in time. The studies on unalloyed steel in the pH range from 9 to 10 showed, that critical chloride contents are to be considered. Significant positive effects are already evident with a [$Cl^-$ to $OH^-$]-ratio of less than 30. However, since the maximum achievable pH is limited to about 10, only relatively low levels of chloride can thus be effectively tolerated. At a ratio of 30 at pH10 it is 3 mmol/L $Cl^-$, respectively 289 mg/L NaCl.

However, positive protection effects were also observed under the conditions of the neutral salt spray test (50 g/L NaCl, see Fig. 1). Therefore, another protective effect should be considered which reduces the access of chlorides to the steel via small defects. Since the migration of chlorides always takes place in the direction of the anodic areas, they are temporarily transported to the location of the anodic Mg dissolution and thus away from the steel, which is temporarily cathodically polarized. It can therefore also be assumed that the effective amount of chloride ions at small defects (diameter << coating thickness) is less than the nominal amount in the bulk solution.

## 5. Conclusions

From the investigations shown here, several responsible mechanisms of corrosion protection in TiMgN hard coatings on steel can be concluded.

- The first mode of action can be explained as a temporary cathodic protection of the low-alloyed steel by the preferred anodic Mg dissolution.
- Related to this is the alkalization of the adjacent electrolyte by formation of $Mg(OH)_2$, which causes the steel to passivate and achieve some tolerance to chloride ions. The relatively large amount of soluble magnesium is likely to cause a buffering effect, which extends the protective effect in time.
- It is also plausible that chloride ions get attracted to the places of the anodic dissolution of Mg from the TiMgN, thus reducing their concentration on the steel substrate.

For different defect types (aspect ratio of diameter to depth or coating thickness), however, a different effective protection can be expected. In large open hole defects, where steel is exposed over a wider area, both the polarization effect and the amount of available $Mg(OH)_2$ are limited, which in addition can lose its alkalinity more quickly by reacting with ambient air ($CO_2$). By contrast, smaller defects with a high aspect ratio can build up a stable buffered alkaline electrolyte and thus maintain the protection longer.

## 6. References

[1] M. Fenker, M. Balzer, H. Kappl, O. Banakh: Some properties of (Ti,Mg)N thin films deposited by reactive dc magnetron sputtering, Surf. Coat. Technol. 200 (2005) 227-231

[2] M. Fenker, M. Balzer, H. Kappl, Corrosion behavior of decorative and wear resistant coatings on steel deposited by reactive magnetron sputtering – Tests and improvements, Thin Solid Films 515 (2006) 27-32

[3] Untersuchung des Einflusses von Mg auf das Korrosionsverhalten hartstoffbeschichteter Stahlsubstrate, DFG-funded research project FE 613/3-2 and HE 6160/5-2, 2012 - 2017

[4] Müller, Thoralf; Heyn, Andreas; Balzer, Martin; Fenker, Martin: Untersuchungen zum Einfluss von Magnesium auf das Korrosionsverhalten von TiMgN-Hartstoffschichten auf Stahlsubstraten, Kongress: 15. Sommerkurs Werkstoffe und Fügen (2014), ISBN: 978-3-940961-85-3, S. 187-196 (this article is also available on researchgate.com via authors profiles)

[5] M. Balzer, Identification of the growth defects responsible for pitting corrosion on sputter-coated steel samples by Large Area High Resolution mapping, Thin Solid Films 581 (2015) 99–106
[6] R.W. Bosch, W.F. Bogaerts: Instantaneous Corrosion Rate Measurement with Small-Amplitude Potential Intermodulation Techniques, Corrosion-Vol. 52, No. 3, 204-212, 1996
[7] R.W. Bosch, J. Hubrecht, W.F. Bogaerts, B.C. Syrett: Electrochemical Frequency Modulation: A New Electrochemical Technique for Online Corrosion Monitoring, Corrosion-Vol. 57, No. 1, 60-70, 2001
[8] S.O. Klemm, A.A. Topalov, C.A. Laska, K.J.J. Mayrhofer: Coupling of a high throughput microelectrochemical cell with online multielemental trace analysis by ICP-MS, Electrochemistry Communications 13 (2011), pp. 1533–1535
[9] Weast, R.C. (ed.). Handbook of Chemistry and Physics, 60th ed. Boca Raton, Florida: CRC Press Inc., 1979, p. B-94
[10] U. Angst et al., Critical chloride content in reinforced concrete - A review, Cement and Concrete Research 39 (2009), 1122
[11] M. Babutzka, A. Heyn: Dynamic tafel factor adaption for the evaluation of instantaneous corrosion rates on zinc by using gel-type electrolytes, IOP conference series / Materials science and engineering, Vol. 181.2017, Art. 012021, 12 pages
[12] S. V. Lamaka, R. M. Souto, M. G. S. Ferreira: In-situ visualization of local corrosion by Scanning Ion-selective Electrode Technique (SIET), Microscopy: Science, Technology, Applications and Education, Vol. 3, Formatex Research Center, Badajoz, Spain (2010), p. 2162, available on http://www.formatex.info/microscopy4/2162-2173.pdf.

## Acknowledgements

The funding of this work by the German Research Foundation (DFG) under grant no. HE 6160/5-1(2) and FE 613/3-1(2) is gratefully acknowledged!

# Oxidation behavior of thermal barrier coating systems with Al interlayer under isothermal loading

**I Ali[1], P Sokołowski[2, 3], T Grund[1], L Pawłowski[2] and T Lampke[1]**

[1] Institute of Materials Science and Engineering, Chemnitz University of Technology, Chemnitz, Germany
[2] Institut de recherche sur les céramiques IRCER, University of Limoges, Limoges, France
[3] Faculty of Mechanical Engineering, Wrocław University of Technology, Wrocław, Poland

**Abstract.** In the present study, the phenomena related to the Thermally Grown Oxides (TGO) in atmospheric plasma sprayed Thermal Barrier Coatings (TBCs) are discussed. CoNiCrAlY bond coatings were sprayed on Inconel 600 substrates. Subsequently, thin Al layers were deposited by DC-Magnetron sputtering. Finally, yttria-stabilized zirconia (YSZ) top coatings were deposited to form a three-layered TBC system. The thus produced aluminum interlayer containing thermal barrier coatings (Al-TBC) were subjected to isothermal exposure with different holding times at 1150 °C and compared with reference TBCs of the same kind, but without Al interlayers (R-TBC). The oxide film formation in the interface between bond coating (BC) and top coating (TC) was investigated by scanning electron microscope (SEM) after 100 and 300 h of high temperature isothermal exposure. The growth of this oxide film as a function of the isothermal exposure time was studied. As a result, the designed Al-TBC system exhibited better oxidation resistance in the BC/TC interface than the two-layered R-TBC system. This was lead back to the Al enrichment, which slows down the formation rate of transition metal oxides during thermal loading.

## 1. Introduction

Progress in gas turbines technology is largely associated with developments in the research area of Thermal Barrier Coatings (TBCs). Plasma spraying and electron beam physical vapor deposition methods have been used to produce TBCs since the 1960s and 1980s, respectively [1-5]. Recently, suspension plasma spraying (SPS), liquid solution precursor plasma spraying (SPPS), suspension high velocity oxy-fuel spraying and plasma-spray physical vapor deposition (PS-PVD) were introduced to modify the microstructure of the ceramic top coatings of the TBC systems. The recent generation of TBCs thus shows a smaller thermal conductivity and better performance under thermal loads in comparison to older generations [5-8].

The current design of a TBC system in turbine applications consists of two main coatings: (i) a metallic bond coating (BC), which is sprayed directly onto the base material and (ii) a ceramic top coating (TC), which insulates the coated component from high-temperature gas atmosphere [9-10]. The BC is composed of MCrAlY alloys (M=Ni, Co or NiCo alloy) and plays two important roles. Firstly, it provides an effective protection of the substrate against high-temperature oxidation. And secondly, it reduces thermal stresses that affect the ceramic TC material in thermal cycling loads due to CTE mismatches between the metallic and ceramic partners in the TBC system [11-12]. The TCs are most frequently composed of phase-stabilized zirconia (rare-earth-doped zirconia, mainly with yttrium) [13-14]. The so-called TGO layer in TBC systems summarizes thermally grown oxides which develop and grow during high-temperature service in the BC/TC interface. Since TGO introduce additional stress into the TBC system and hence cause TBC failure, one possible way to increase the lifetime of TBCs is to control the formation of the TGO and minimize their growth rate [15-16].

In the present work, thin Al interlayers were applied between BC and TC obtained by atmospheric plasma spraying in order to improve the oxidation resistance of the BC by direct and fast formation of alumina films on top of this interlayer. TGO development and growth during isothermal loading was then analyzed and quantified. The obtained results were compared to those obtained with conventional reference TBC systems (R-TBC).

## 2. Experimental work

The single coating layers included in TBCs were prepared by air plasma spraying (APS) on Inconel 600 (DIN 2.4816, UNS N06600) substrates with rectangular dimensions of 30 x 10 x 3 mm. They included a CoNiCrAlY bond coating (CoNi32Cr21Al8Y0.5, 60.95.1, GTV GmbH) and an yttria-stabilized zirconia (YSZ) top coating ($ZrO_2$ + 8 wt.-% of $Y_2O_3$, Oerlikon Metco 204NS). Substrate sheets were grit blasted and cleaned before spray coating application. BCs were sprayed with a thickness of 150 µm using a F6 plasma torch (GTV GmbH). Thin Al films of 2 µm thickness were sputtered on the as-sprayed BC surfaces using a non-commercial DC-Magnetron PVD coating unit B0 TSP. (working pressure: 8 x $10^{-3}$ mbar, Al target: Al 99.999 %, EVOCHEM Advanced Materials GmbH, sputtering gas: Ar 99.9999 %. Roughness measurements by contact stylus instrumentation and basing on statistical determination according to ISO 11562 were conducted on the raw as well as the Al-covered BC surfaces. Afterwards, YSZ TCs were sprayed using plasma torch SG-100 (Praxair S.T. ) to a thickness of 150 µm. BCs and TCs were sprayed according to the given parameters in Table 1. The designs of TBCs used in the present study are shown in Figure 1. The microstructure of the as-sprayed TBC systems was investigated using laser scanning microscopy. The porosity content in the coatings was determined using digital image analysis.

The TBC systems were submitted to isothermal exposure tests. Samples were heated up in air from room temperature (RT = 20 °C) up to 1150 °C with a heating rate of 6 K/min, then held for 100 or 300 h at 1150 °C and finally cooled down to RT with cooling rate of 0.5 K/min. After thermal treatment, all specimens were metallographically polished and analyzed using scanning electron microscopy (SEM) and energy dispersive X-ray analysis (EDXS).

Table 1: Spraying parameters of BC and TC using atmospheric plasma spraying

| **Spray powder** | **Co32Ni21Cr8Al0.5Y** | **YSZ** |
|---|---|---|
| **Ar [slpm]** | 65 | 45 |
| **$H_2$ [slpm]** | 14 | 5 |
| **Plasma current [A]** | 600 | - |
| **Plasma power [kW]** | - | 40 |
| **Spray distance [mm]** | 140 | 100 |

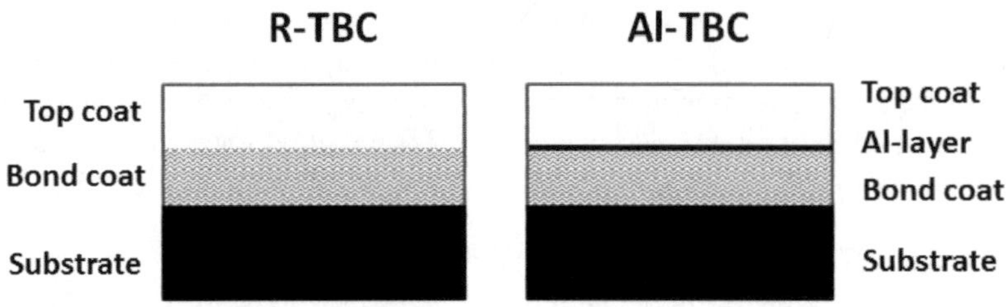


Figure 1: Schemes of the two TBC designs analyzed in the present study

## 3. Results

### *3.1. Microstructure of as-sprayed TBC*

The microstructures of R-TBC and Al-TBC samples are shown in Figure 2. The BC in both systems reveals the typical microstructure of APS sprayed metal coatings showing oxide scales, pores, flakes and inter-lamellar delamination between molten splats. Porosity and oxide content are about 2 % and 22 %, respectively. The porosity of the YSZ TC is about 7 %. The TBCs show generally good contact between TC and BC by mechanical interlocking, anticipating a good adhesion. In the Al-TBC design, the Al interlayer wis not influenced by application of the TC. The interlayer is, however, not homogeneously and continuously formed, that results in zones of direct contact between the BC and TC (Figure 2, right, regions marked by circles). This is either due to shadowing effects resulting from the roughness of the as-sprayed BC surface or due to inhomogeneous grain growth of the DC-sputtered aluminum on the rough and chemically inhomogeneous BC surface. Consequently, the sprayed YSZ particles could have removed some parts of the Al interlayers. In comparison to the uncoated BC, the surface roughness of the Al-covered BC is not altered by the 2 µm thik Al layer . Measurements revealed $R_a = 7$ µm and $R_z = 49$ µm for the uncoated BC surface, but $R_a = 7$ µm and $R_z = 50$ µm for the Al-coated BC surface.

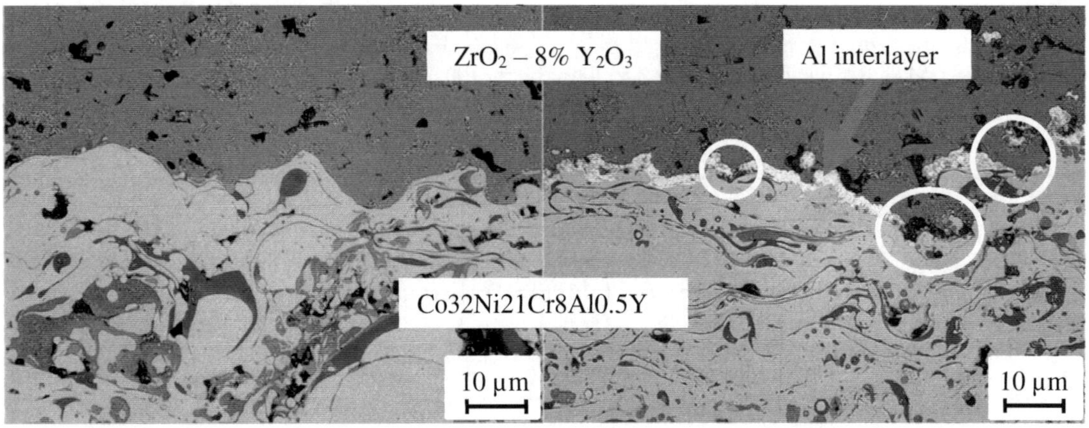


Figure 2: Laser intensity images show the as-sprayed microstructures of the TBC systems, left: without Al interlayer, right: with Al interlayer; the marked zones highlight limited BC surface areas that have not been fully covered by the DC-PVD Al interlayer and reveal direct contact between BC and TC material.

IOP Conf. Series: Materials Science and Engineering **373** (2018) 012010 doi:10.1088/1757-899X/373/1/012010

### *3.2. Microstructure of as-sprayed TBC*

The cross sections of the TBC after both 100 h and 300 h of isothermal exposure are shown in Figure 3. The Al-TBC system (Figure 3, right) exhibits a thinner TGO layer (dark grey region between BC and TC) in comparison to the R-TBC system. The difference is already visible after 100 h, but much more pronounced after 300 h, wherein the optically estimated reduction is about factor 3.

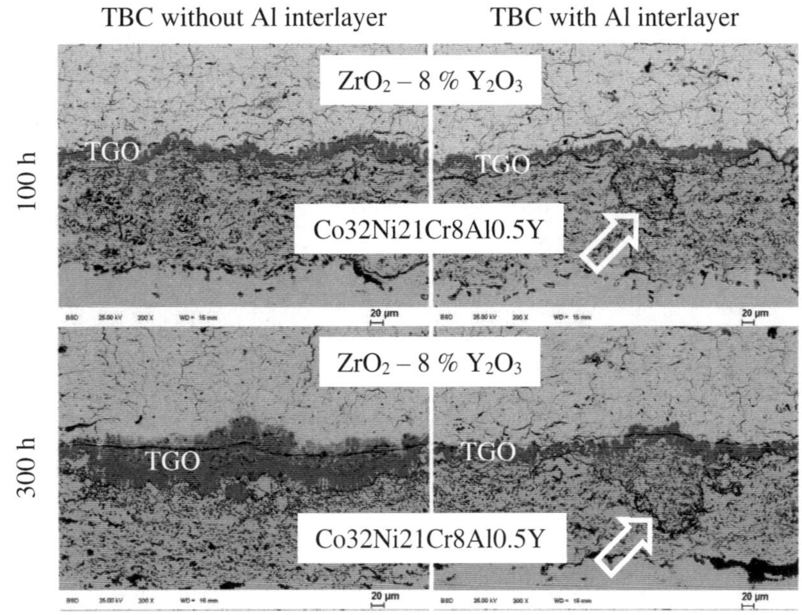


Figure 3: Back-scattered electron SEM micrographs of the cross-sections of Al- and R-TBCs after isothermal exposure at 1150 °C, left: R-TBC without Al interlayer, right: Al-TBC with Al interlayer, top: after 100 h of exposure, bottom: after 300 h of exposure; arrows point to an area/diffusion front with heightened contents of "Al" and "O" on the BC facing side and in comparison to the BC material composition.

EDXS mapping on the Al-TBC samples show, the semi-continuous dark zone into the BC material starting from the BC/TC interface is rich in aluminum and oxygen (Figure 4). This observation suggests that part of the Al diffuses during isothermal exposure into the bond coating material, and the non-continuous dark line hence represents the diffusion boarder . The diffusion is probably supported by interlamellar surface diffusion and hinders diffusion of oxygen and thus, consequently, improves the BC's oxidation resistance.

In both systems - Al-TBC and R-TBC -, the investigated microstructures during isothermal exposure reveal the so-called "isothermal failure mode". Crack paths run across the TGO as described in [17-19]. Since the R-TBC samples show thicker TGO in comparison to Al-TBC ones, crack paths initiate and follow the TGO layer in higher density within these standard TBC systems. This is in accordance with the state-of-the-art. It has been reported that the growth of TGO is correlated with long delamination cracks along the YSZ TC and near the BC/TC interface [20-23]. Consequently, the use of Al interlayer in TBC systems can effectively slow down TGO growth and is hence a useful approach to increase the lifetime of TBC systems.

## 4. Conclusion

The Al-TBC design, showing DC-PVD Al interlayers between BC and TC, form a TGO less intensively at isothermal exposure than R-TBCs. The reason is the presence of Al at BC/TC interface, which results in the diffusion of Al into the BC material and enable a steady TGO growth, probably by reducing oxygen diffusion. This may may lead to an increased service time of Al-TBC in comparison to R-TBC systems. The investigated microstructures of both TBC systems revealed the so-called "isothermal-failure mode", however in different intensities.

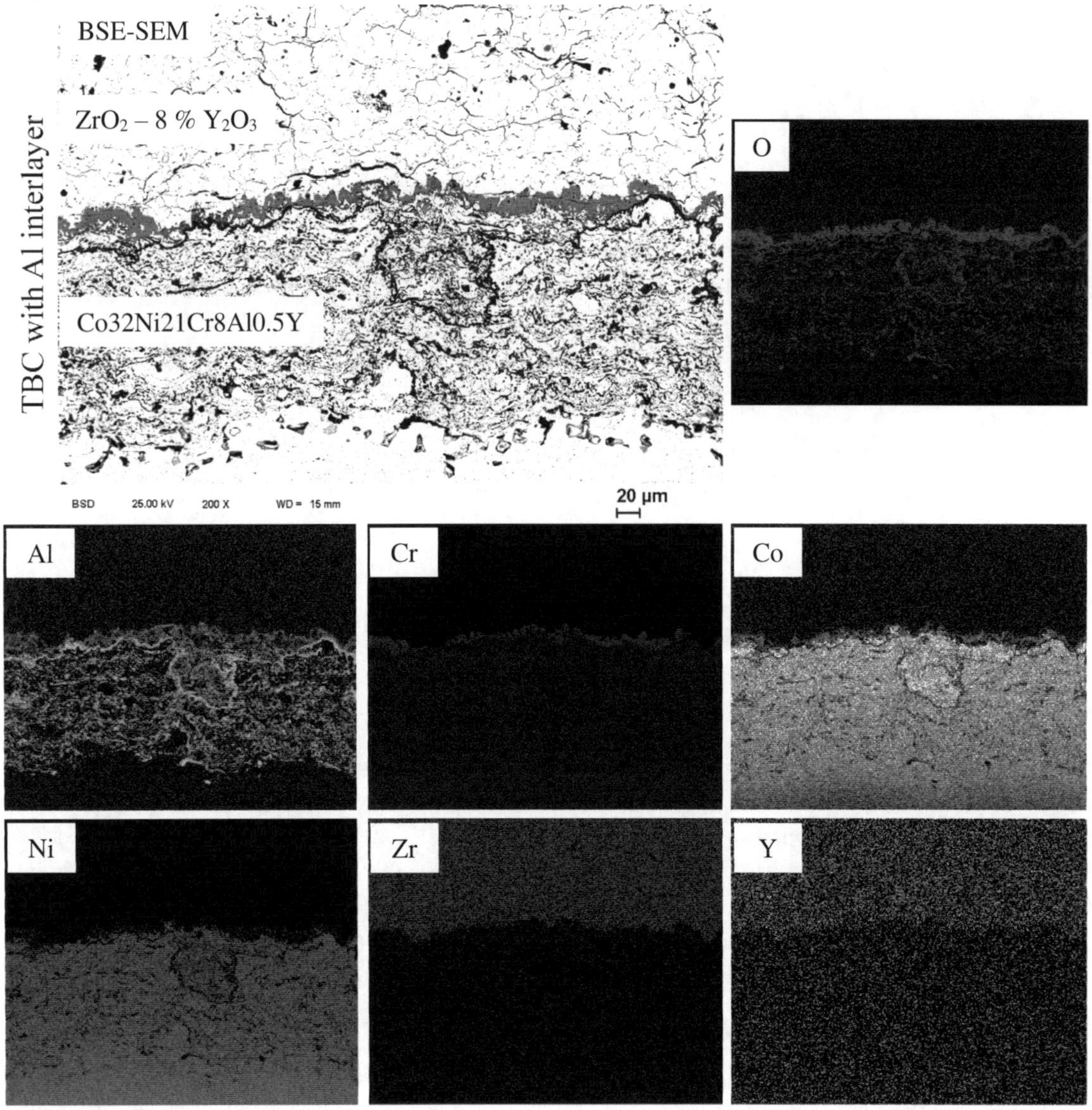


Figure 4: Back-scatteredelectron SEM micrograph and respective EDXS mapping of Al-TBC cross-section showing BC/TC interface with former Al interlayer and oxidation status after 100 h of isothermal exposure at 1150 °C.

## 5. References

[1] Peters M, Leyens C, Schulz U, Kaysser W 2001 Adv. Eng. Mater., 3/4 pp 193–204
[2] Gleeson B, Sordelet D 2014 US Patent 8,821,654 B2
[3] Trunova O, Beck T, Herzog R, Steinbrech R, Singheiser L 2008 Surf. Coat. Technol., 202 pp 5027–32
[4] Jones R 1997 J. Therm. Spray Technol., 6/1 (1997) pp. 77–84
[5] Darolia R 2013 Int. Mater. Rev., 58/6 pp 315–48
[6] Bakan E, Vassen R 2017 J. Therm. Spray Technol., 26 pp 992–1010
[7] Sokolowski P, Pawlowski L, Dietrich D, Lampke T, Jech D 2015 J. Therm. Spray Technol., 25 (1-2) pp 94–104
[8] Podlesak H, Pawlowski L, Haese R, Laureyns J, Lampke T, Bellayer S 2010 J. Therm. Spray Technol., 19/3 pp 657-64
[9] Miller R 1987 Surf. Coat. Technol. 30/1 pp 1-11
[10] Padture N, Gell M, Jordan E 2002 Science, 296/5566 pp 280-84
[11] Jarvis E, Carter E 2002 Comput Sci Eng, 4/2 pp 34-41
[12] Spitsberg I, Mumm D, Evans A 2005 J. Mater. Sci. Eng. A., 394/1–2 pp 176-91
[13] Zhu D, Miller R 2008 Proc., of the 35th Int., Conf., on Metallurgical Coatings and Thin Films (ICMCTF), [Editors: Pauleau Y and Mayrhofer P et al], American Vacuum Society, (San Diego, CA)
[14] Sokołowski P, Łatka L, Pawłowski L, Ambroziak A, Kozerski S, Nait-Ali B 2015 Surf. Coat. Technol. 268 pp 147–52
[15] Ali I, Grund T, Lampke T, Wett D, Nestler D and Wagner G 2015 Proc., of the Int., thermal Spray Conf., ASM Int., [Editors: McDonald A and Agarwal A et al], (Long Beach, CA)
[16] Evans A, Mumm D, Hutchinson J, Meier G, Pettit F 2001 Prog. Mater. Sci. 46 pp 505-53
[17] Herzog R, Trunova O, Steinbrech R, Singheiser L 2006 First Japanese-German workshop on properties and performance of thermal barrier coating systems and factors affecting it., Darmstadt, Presentation
[18] Trunova O, Beck T, Herzog R, Steinbrech R, Singheiser L 2008 Surf. Coat. Technol. 202, pp 5027-32
[19] Herzog R, Warnken N, Steinbach I, Hallstedt B, Walter C, Mueller J, Hajas D, Muenstermann E, Schneider J, Nickel R, Parkot D, Bobzin K, Lugscheider E, Bednarz P, Trunova O, Singheiser L 2006 Adv. Eng. Mater. 8/6 pp 535-62
[20] Karaoglanli A, Altuncu E, Ozdemir I, Turk A, Ustel F 2011 Surf. Coat. Technol. 205 pp 369-73
[21] Kwon J, Lee J, Gil Y 2006 Surf. Coat. Technol. 201 pp 3483-90
[22] Munawar A, Schulz U, Cerri G, Lau H 2014 Surf. Coat. Technol. 245 pp 92-101 (doi: 10. 1016/j. Surfcoat 2014, 02.047)
[23] Tsai P, Tseng C, Yang C, Kuo L, Chou Y, Lee J 2013 Surf. Coat. Technol. 228 pp 11-14

# Electrical Resistivity of Wire Arc Sprayed Zn and Cu Coatings for In-Mold-Metal-Spraying

**K Bobzin[1], M Öte[1], M A Knoch[1], X Liao[1,*], Ch Hopmann[2] and P Ochotta[2]**

[1] Surface Engineering Institute, RWTH Aachen University, Kackertstr. 15, 52072 Aachen, Germany
[2] Institute of Plastics Processing, RWTH Aachen University, Seffenter Weg 201, 52074 Aachen, Germany

[*] e-mail: liao@iot.rwth-aachen.de

**Abstract.** Electrical functionalities can be integrated into plastic parts by integrating thermally sprayed metal coatings into the non-conductive base material. Thermally sprayed conducting tracks for power and signal transmission are one example. In this case, the electrical resistance or resistivity of the coatings should be investigated. Therefore, the electrical resistivity of wire arc sprayed Zn and Cu coatings has been investigated. In case of Zn coatings, spray distance, gas pressure and wire diameter could be identified as significant influencing parameters on the electrical resistivity. In contrast, process gas, gas pressure and voltage do have a significant influence on the electrical resistivity of Cu coatings. Through the use of the In-Mold-Metal-Spraying method (IMMS), thermal degradation can be avoided by transferring thermally sprayed coating from a mold insert onto the plastic part. Therefore, the influence of the transfer process on the electrical resistance of the coatings has also been investigated.

## 1. Introduction

Due to the various advantageous properties of plastics, e.g. low density and high chemical resistance, this type of materials has been widely used in different fields. Although these properties may be desired for certain cases, sometimes the opposite properties are required for certain applications. For example, plastics are usually good electrical insulators, which is disadvantageous for applications where the electrical conductivity is required, such as power and signal transmission. In this case, the concept of combining plastic and metal in a single part can be applied, in order to fulfill the requirement of the application and benefit from other properties of the plastics at the same time. There are different approaches for fabricating such hybrid parts, including the application of metal coatings onto the part surface by means of different coating technologies. Due to the high efficiency and the ability of processing different materials, thermal spraying has attracted lots of interest regarding coating deposition, including metal coatings, onto plastic parts.

The results in the literature show, on the one hand, the possibility of fabricating metal coatings on plastic substrates by means of thermal spraying. On the other hand, the thermal and kinetic energy of the particles and the process gas needs to be adjusted to avoid damage or extensive softening of the substrate [1-8]. This leads to difficulties for coating deposition on substrates made of widely used engineering plastics, which exhibit relatively lower thermal resistance in comparison to the more expensive high performance plastics. Therefore, a new approach, the In-Mold-Metal-Spraying

(IMMS), has been developed at Surface Engineering Institute (IOT) and Institute of Plastic Processing (IKV) of RWTH Aachen University. To avoid the direct contact between the plastic substrate and the high temperature particles and gas, the metal coatings are first applied onto the surface of the mold cavity or the mold insert for injection molding. By using a suitable pre-treatment of the surface, e.g. glass bead blasting, a relatively low bond strength of the coating to the mold cavity or the mold insert surface can be achieved. Therefore, the coating can be transferred to the plastic part after the injection molding process [9-12]. The hybrid metal/plastic parts fabricated by IMMS can be used in different applications, such as conducting tracks for signal and power transmission. To the knowledge of the authors, the requirement on the electrical resistance R or the electrical resistivity $\rho_{el.}$ of such conducting tracks depends on the specific application, and there is no unique standard for all applications. Generally, a low electrical resistivity of the conducting track is desired, in order to ensure high signal quality. In the current study, the process parameters that might have significant influence on $\rho_{el.}$ and their effects were identified and confirmed using two staged design of experiments (DoE) in case of Zn and Cu coatings.

## 2. Experimental setup

In the first stage of DoE, the process parameters of WAS that might have significant effect were identified using fractional factorial design of experiments. The effects of chosen parameters were further confirmed in the second stage of DoE using full factorial design of experiments. Furthermore, the influence of the transfer process on the coating was investigated by the comparison of the electrical resistance of transferred and detached coatings.

### *2.1. Experimental methods for the determination of the coating's electrical resistivity*

In the current study, steel sheets made of 1.0037 (AISI 1015) have been used as mold inserts which are the carrier body for the coatings at the same time. The coated carrier bodies can be placed into the mold cavity for coating transfer. The coatings were deposited by means of WAS using the spraying unit OSU G30/4SF-Push-LD/U2 (Oerlikon Metco AG, Winterthur, Switzerland) with a closed, nonadjustable nozzle system. After blasting the carrier body with glass beads using a blasting unit Superbaby II (BHS, Aachen, Germany), five coatings in the shape of conducting tracks with defined width b and length l can be fabricated by using a mask per experiment run, see figure 1.

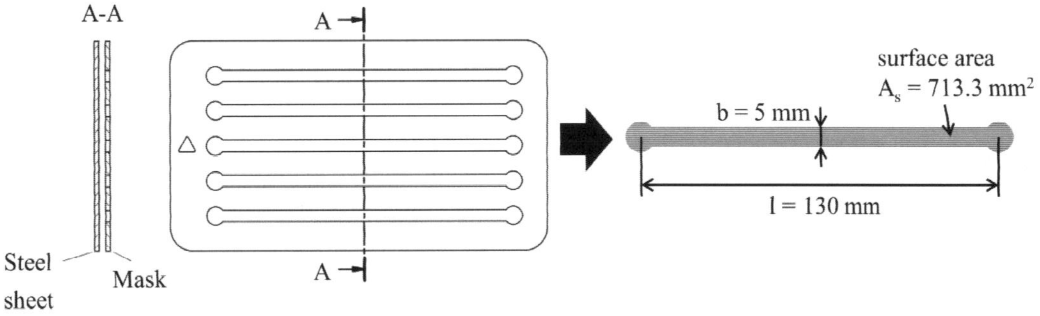


**Figure 1.** Fabrication of coatings in the shape of conducting tracks using a mask.

The influence of the process parameters of WAS on the coating's electrical resistivity was investigated in the first step of the current study. To avoid an overlap of the effects from WAS and injection molding, the deposited coatings were not transferred onto plastic parts, but were directly detached from the carrier bodies. The electrical resistance R is measured by means of 4-wire sensing using an ohmmeter ILOM-508A (RS components GmbH, Mörfelden-Walldorf, Germany), which is necessary to eliminate the influence of lead and contact resistance for a precise determination of small electrical resistance [13]. The electrical resistivity $\rho_{el.}$ can be calculated as shown in equation 1 [14]:

$$\rho_{el.} = R \cdot \frac{A_c}{l} \quad (1)$$

The length l is defined by the mask. The cross-sectional area $A_c$ can be determined by multiplication of the width b and the thickness t of the coating. Because wire arc sprayed coatings exhibit usually a rough surface and inhomogeneity in thickness, the conventional method of determining coating thickness by examining the cross sections of the coating requires great measurement effort to obtain a representative coating thickness. In the current study, two alternative methods were used to determine the cross-sectional area $A_c$. In the first stage of the DoE, the weight m of the coatings was measured. The volume V of the coating can be roughly estimated by dividing m by the density ρ of the feedstock material. Dividing V by the surface area of the coating $A_s$, which is shown in figure 1, yields the average thickness of the coating. Therefore, a rough estimation $\rho_{el.,w}$ of the coating' electrical resistivity is:

$$\rho_{el.,w}=R\cdot\frac{m\cdot b}{\rho\cdot A_s\cdot l} \quad (2)$$

Although there is usually a difference between the densities of the feedstock material and the coating, this method enables an approximation of the coating's electrical resistivity. Furthermore, m can be measured more quickly and precisely. To avoid the influence of the density difference between the coating and the feedstock materials on the result, V can be directly determined by pycnometer measurement. However, the measurement duration with pycnometer is much longer than the weight measurement. Hence, this method has only been used in the second stage of the DoE experiment, after the amount of to be investigated parameters is reduced. In order to distinguish the electrical resistivity obtained by the two methods, the coating's resistivity estimated by the second method is referred to as $\rho_{el.,v}$ which is calculated by equation 3:

$$\rho_{el.,v}=R\cdot\frac{V\cdot b}{A_s\cdot l} \quad (3)$$

*2.2. The procedure of DoE*

The feedstock materials, Zn and Cu, were investigated. For each material, a fractional factorial design of experiments with two levels for each process parameter was used for the first stage of DoE. Each experiment was carried out once. In case of Zn coatings, the process parameters indicated in Table 1 were investigated in the first stage. To ensure a stable process, the voltage can only be adjusted in a very small range for depositing Zn coatings with the used arc spraying unit Therefore, the voltage U has not been included and was set at U = 22 V. The two levels of the other parameters are labeled with "-" and "+" respectively, and their corresponding settings are listed in Table 1. The traverse velocity of the spray gun v was set at v = 200 and 450 m/s for the feed rate F of F = 5.8 and 13.5 kg/h, respectively, in order to achieve a similar coating thickness in all experiments. The spraying gun travels once across the carrier body surface to deposit the coatings in the experiments. To avoid the variation of the process condition caused by changing the wires, the experiments in the first stage were divided in two groups regarding the wire diameter, when the experiments were conducted. In each group, the wire diameter is kept constant, while the settings of the other parameters were varied randomly.

**Table 1.** Levels of the parameters in the screening experiment for Zn coating.

| Parameter | - | + |
|---|---|---|
| **A (wire diameter)** | 1.6 mm | 2.5 mm |
| **B (gas)** | $N_2$ | Pressured air |
| **C (feed rate)** | 5.8 kg/h | 13.5 kg/h |
| **D (pressure)** | 2.5 bar | 4.5 bar |
| **E (stand-off distance)** | 50 mm | 150 mm |

In case of Cu coatings, Cu wire with the diameter Ø of Ø = 2.5 mm was not available in the current study. Furthermore, in the previous study $N_2$ instead of pressured air was used as atomizing gas leading to a much lower electrical resistivity using the same spraying unit with a different nozzle system [15]. Therefore, the two parameters, the wire diameter Ø and the atomizing gas G, are excluded in the first stage of DoE for Cu coatings. They were set to Ø = 1.6 mm and G = $N_2$ respectively. The voltage can be adjusted in a greater range in case of depositing Cu coatings Therefore, the voltage was included in the screening experiment. The parameters and their levels are listed in Table 2. The traverse velocity v of the spraying gun was set to v = 337.5 and 675 m/s for the feed rate F of F = 4.8 and 10.5 kg/h, respectively to achieve a similar coating thickness in all experiments. The spraying gun traveled two times across the carrier body surface to deposit the coatings in the experiments.

**Table 2.** Levels of the parameters in the screening experiment for Cu coating.

| Parameter | - | + |
|---|---|---|
| **A (pressure)** | 3 bar | 6 bar |
| **B (stand-off distance)** | 50 mm | 100 mm |
| **C (voltage)** | 31 V | 40 V |
| **D (feed rate)** | 4.8 kg/h | 10.5 kg/h |

After the first stage of DoE, the effect caused by changing a parameter from the low level to the high level was determined based on the obtained results. Two parameters were chosen for the second stage of DoE with full factorial design of experiments. The level of the parameter's effect determined in the screening experiment is not the sole criterion for choosing the parameters for the second stage of DoE. Other aspects, such as the possible influence on the transferability of the coating, should also be considered. The approach of the second stage of DoE for Zn and Cu coatings will be introduced in section 3 after that the results of the first stage are presented and discussed.

### *2.3. Investigation of the influence of the transfer process on the coating's electrical resistance*

Besides the parameters of WAS processes, the coating's electrical properties might also be influenced by the transfer process. Because the coating adheres stronger to the plastic parts than to the carrier body, the detachment of the coatings from the plastic part without damaging them is difficult. The measurement of the coating's cross-sectional area or thickness with the assistance of pycnometer is no longer possible for transferred coatings. Therefore, the comparison of coating's electrical resistivity before and after transfer process is not a suitable method to evaluate the effect of the transfer process due to the different measurement methods for detached and transferred coatings. In the current study, the coating's electrical resistance instead of the electrical resistivity is used to determine the effect of the transfer process. Ten carrier bodies were coated for each material using the same WAS parameter. The coatings on five of them are transferred to plastic parts by means of IMMS using a polyamide 66 SCHULAMID 66 GF35 H4 with 35 vol. % of glass fibers (A. Schulman GmbH, Kerpen, Germany), while the coatings on the other five carrier bodies were directly detached for the measurement of the electrical resistance. The electrical resistance of the transferred coatings was also measured and compared to that of the detached coatings.

## 3. Results and discussion

The experiments have been conducted as described in section 2. Their results are presented and discussed in this section.

### *3.1. Influence of WAS process on the electrical resistivity of Zn coatings*

A $2^{5-1}$ fractional factorial design of experiments is constructed for Zn coatings as shown in Table 3. As mentioned above, the coating's electrical resistivity was obtained by measuring the weight of the coatings at this stage of DoE.

**Table 3.** The screening experiments in case of Zn coating.

| No. | A (wire diameter) | B (gas) | C (feed rate) | D (pressure) | E (stand-off distance) | $\rho_{el.,w}$/ ($\Omega \cdot mm^2/m$) |
|---|---|---|---|---|---|---|
| **Zn-S-1** | - | + | - | - | + | 0.1194 |
| **Zn-S-2** | + | + | - | - | - | 0.1130 |
| **Zn-S-3** | - | - | - | - | - | 0.1056 |
| **Zn-S-4** | + | - | - | - | + | 0.1334 |
| **Zn-S-5** | - | + | + | - | - | 0.1095 |
| **Zn-S-6** | + | + | + | - | + | 0.1333 |
| **Zn-S-7** | - | - | + | - | + | 0.1111 |
| **Zn-S-8** | + | - | + | - | - | 0.1056 |
| **Zn-S-9** | - | + | - | + | - | 0.1047 |
| **Zn-S-10** | + | + | - | + | + | 0.1337 |
| **Zn-S-11** | - | - | - | + | + | 0.1013 |
| **Zn-S-12** | + | - | - | + | - | 0.0962 |
| **Zn-S-13** | - | + | + | + | + | 0.1089 |
| **Zn-S-14** | + | + | + | + | - | 0.1020 |
| **Zn-S-15** | - | - | + | + | - | 0.0924 |
| **Zn-S-16** | + | - | + | + | + | 0.1161 |

The effect of an investigated parameter can be calculated by subtracting the average $\rho_{el.,w}$ of the experiments with the parameter set at level "+" by the average $\rho_{el.,w}$ of the experiments with the parameter set at level "-". The effects of the parameters on $\rho_{el.,w}$ are compared in figure 3. The stand-off distance has shown the highest effect and was chosen as one of the two parameters to be investigated in the second stage. The wire diameter and the pressure have shown similar effects. As mentioned above, the change of the wires might causes the variation of the process condition, so that the effects of other parameters would be overlapped. Therefore, the pressure was chosen as the second parameter for the full factorial design in the second stage. To confirm the effect of the wire diameter, a separate experiment series, which involves only the wire diameter, was also conducted.

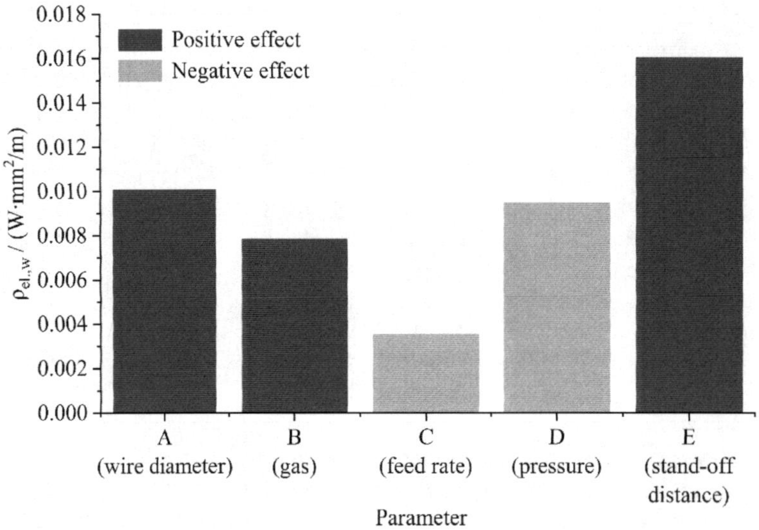


**Figure 2.** Comparison of the effects of the parameters in the screening experiment for the Zn coating.

IOP Conf. Series: Materials Science and Engineering **373** (2018) 012011 doi:10.1088/1757-899X/373/1/012011

A $2^2$ full factorial design of experiments has been conducted to investigate the effects of the parameters, stand-off distance and pressure, on the coating's electrical resistivity. It means that there are $2^2$ different combinations of the settings of the two investigated parameters, (-, -), (+, -), (-, +) and (+, +). The experiments with one of these parameter combinations are referred as factorial points. One additional level labeled "0", which is the average of level "-" and "+", was used to define an additional experiment point, the center point, with the parameter combination (0, 0) in the experimental matrix. By this means, it can be examined if quadratic correlation between the coating's electrical resistivity and the two parameters exist. The levels of the two parameters are given in Table 4. Higher inlet pressure is possible, when $N_2$ is used as atomizing gas. Furthermore, the results of the screening experiment show that higher gas pressure might be beneficial for achieving lower electrical resistivity. Therefore, the "-" and "+" levels of the pressure p were chosen as p = 3 bar and 6 bar respectively. The settings of the other parameters, wire diameter Ø, gas G, feed rate F, voltage U and traverse velocity v, were chosen as Ø = 1.6 mm, G = $N_2$, F = 13.5 kg/h, U = 22 V and v = 450 m/s, respectively.

**Table 4.** Levels of the stand-off distance and the pressure in the characterization experiment for Zn coating.

| Parameter | - | 0 | + |
|---|---|---|---|
| **A (stand-off distance)** | 50 mm | 100 mm | 150 mm |
| **B (pressure)** | 3 bar | 4.5 bar | 6 bar |

At this stage, $\rho_{el.,v}$ is measured and calculated as mentioned in section 2. The results are shown in Table 5. The experiments numbered Zn-C-1 to 4 in Table 5 are factorial points with two replications of each parameter combination. The center point, Zn-C-5, has been replicated four times. To confirm the significance of the effect of the wire diameter, a further parameter set, Zn-C-6, was used in addition to the factorial points and the center point. The feedstock material was Zn wires with the diameter Ø of Ø = 2.5 mm in this parameter set. The settings of the other parameters of the center point, Zn-C-5, were also used for Zn-C-6.

**Table 5.** Results of the second stage of DoE for Zn coating.

| No. | A (stand-off distance) | B (pressure) | Wire diameter /mm | $\rho_{el.,v}/(\Omega\cdot mm^2/m)$ Replication 1 | Replication 2 | Replication 3 | Replication 4 |
|---|---|---|---|---|---|---|---|
| **Zn-C-1** | - | - | 1.6 | 0.0958 | 0.0976 | - | - |
| **Zn-C-2** | - | + | 1.6 | 0.0880 | 0.0884 | - | - |
| **Zn-C-3** | + | - | 1.6 | 0.1072 | 0.1098 | - | - |
| **Zn-C-4** | + | + | 1.6 | 0.1013 | 0.1006 | - | - |
| **Zn-C-5** | 0 | 0 | 1.6 | 0.0975 | 0.0992 | 0.0974 | 0.0991 |
| **Zn-C-6** | 0 | 0 | 2.5 | 0.1027 | 0.1020 | 0.1016 | 0.1018 |

In order to examine the significance of the effects of the parameters, the F-test has been applied on the results. The F-values, which are the ratio between the mean square caused by an effect and the mean square of the experimental error, of the two parameters and their interaction are determined. Due to the integration of a center point in the $2^2$ design, the F-value of the quadratic effect can also be determined and used to examine the significance of a possible quadratic correlation. Details of the calculation of the F-value can be found in [16]. The F-values should be compared with a critical F-value, which corresponds to a chosen confidence level. In the current study, the confidence level is chosen as 95%. The results are shown in the Figure 4. The effects of the two investigated parameters are significant, and their interaction effect and the quadratic effect are insignificant.

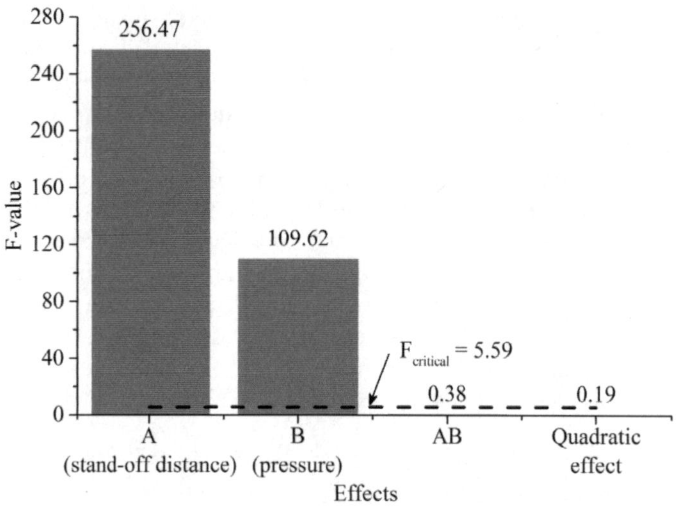


**Figure 3.** The F-values of the effects of stand-off distance and pressure, their interaction effect and the curvature compared with the critical F-value.

Based on the results, a linear model describing the correlation between $\rho_{el.,v}$ and the two parameters can be constructed. Since AB is insignificant, a model containing only A and B was constructed, see equation 4.

$$\rho_v/(\Omega\cdot mm^2/m) = 0.0983 + 1.2775\cdot10^{-4}\cdot A/mm - 2.675\cdot10^{-3}\cdot B/bar \quad (4)$$

The correlation between $\rho_{el.,v}$ and the two parameters based on equation 4 is shown in the contour plot in figure 4. It shows that $\rho_{el.,v}$ decreases with decreasing stand-off distance and increasing pressure. The comparison of the coating's electrical resistivity of the experiments Zn-C-5 and Zn-C-6 in figure 4b shows that the wire diameter exhibit an influence on the coating's resistivity. However, the difference of the coating's electrical resistivity caused by the different wire diameters is small.

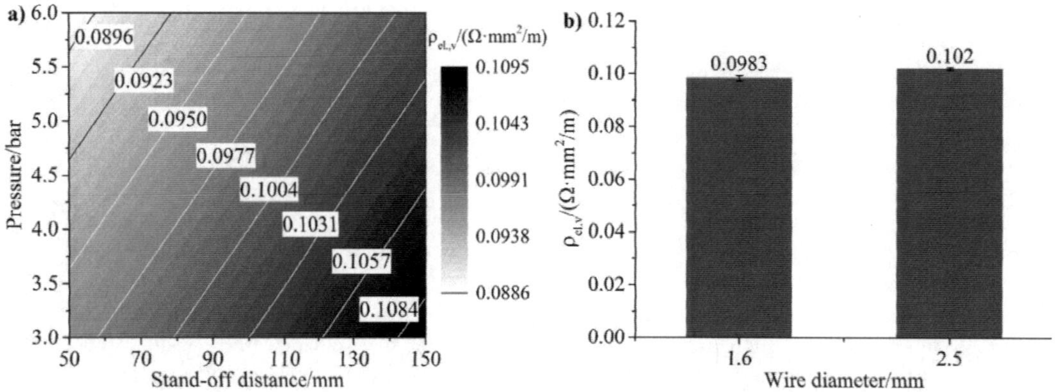


**Figure 4.** Contour plot of the correlation between $\rho_{el.,v}$, the pressure and the voltage in case of Zn coating (a) and the effect of the wire diameter (b).

### *3.2. Influence of WAS process on the electrical resistivity of Cu coatings*

The same procedure has been carried out for the investigation of Cu coatings as in case of Zn coatings. The parameter combinations of the $2^{4-1}$ experiment matrix and the results are listed in Table 6.

**Table 6.** The screening experiments in case of Cu coating.

| No. | A (pressure) | B (stand-off distance) | C (voltage) | D (feed rate) | $\rho_{el.,w}$ /(Ω·mm²/m) |
|---|---|---|---|---|---|
| **Cu-S-1** | - | - | - | - | 0.0348 |
| **Cu-S-2** | + | - | - | + | 0.0302 |
| **Cu-S-3** | - | + | - | + | 0.0416 |
| **Cu-S-4** | + | + | - | - | 0.0375 |
| **Cu-S-5** | - | - | + | + | 0.0398 |
| **Cu-S-6** | + | - | + | - | 0.0361 |
| **Cu-S-7** | - | + | + | - | 0.0525 |
| **Cu-S-8** | + | + | + | + | 0.0461 |

The effects of each parameter are calculated with the same procedure for Zn coatings described in section 3.1 and shown in figure 5. Although the stand-off distance has shown the highest effect among all parameters, the coating adhesion to the carrier body in the experiments with high stand-off distance is low. Lots of coatings detached already during the deposition process and were destroyed by the gas stream in the experiments with high stand-off distance. A higher coating adhesion to the carrier body can be achieved by using carrier bodies blasted with coarser glass beads. On the one hand, the use of the coarsest glass beads available in the current study with the size distribution of 200 to 300 μm could not guarantee a sufficient adhesion. On the other hand, blasting with coarser blasting medium would result in a carrier body surface with more undercuts, which promotes the mechanical interlocking between the coating and the carrier body. This is disadvantageous for the coating transfer in IMMS, because the coating might be damaged at positions, where strong mechanical interlocking is present, during the transfer from the carrier body to the plastic part. Since this parameter would not be adjustable for the aimed application, the stand-off distance was not further investigated in the characterization experiment. The pressure and the voltage have been chosen instead as to be investigated parameters in the second stage. The feed rate has shown a negligible effect in comparison to the other three parameters.

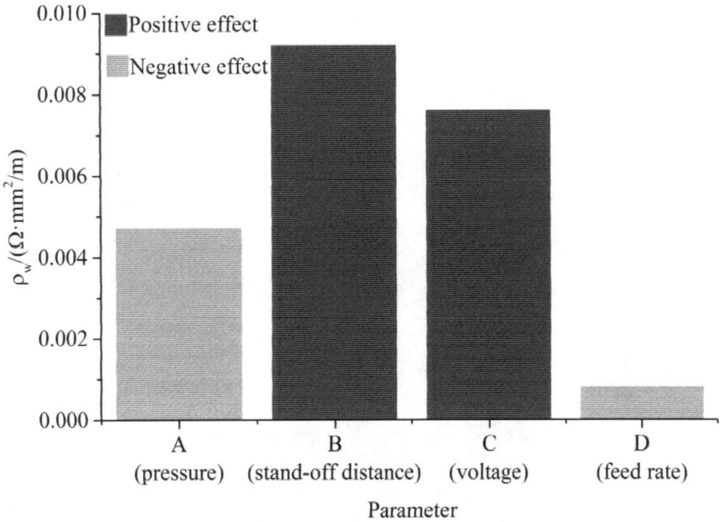


**Figure 5.** Comparison of the effects of the parameters in the screening experiment for the Cu coating.

A $2^2$ experiment design with center point was also used in the second stage of DoE for Cu coatings as in case of Zn coatings. The levels "-", "+" and "0" of the pressure and the voltage are listed in Table 7. The level "0" of the voltage U should be U = 35.5 V, which is the average of level "-" and "+". Due to the stepwise adjustment of the voltage of the spraying unit with relays, the voltage cannot be set at exactly U = 35.5 V. The closest possible value is U = 36 V and was used as level "0" in the experiments. The settings of the other parameters, wire diameter Ø, gas G, feed rate F, the stand-off distance s and the traverse velocity v of the spraying gun, were chosen as Ø = 1.6 mm, G = $N_2$, F = 10.5 kg/h, s = 50 mm and v = 675 m/s, respectively.

**Table 7.** Levels of the pressure and the voltage in the characterization experiment for Cu coating.

| Parameter | - | 0 | + |
|---|---|---|---|
| **A (pressure)** | 3 bar | 4.5 bar | 6 bar |
| **B (voltage)** | 31 V | 36 V | 40 V |

The parameter combinations of the experiments and the measured $\rho_{el.,v}$ were listed in Table 8. As in case of the Zn coatings, the experiments at the factorial points have been replicated twice, while the experiment at the center point has been replicated four times.

**Table 8.** Results of the second stage of DoE for Cu coatings.

| No. | A (pressure) | B (voltage) | $\rho_{el.,v}/(\Omega\cdot mm^2/m)$ Replication 1 | Replication 2 | Replication 3 | Replication 4 |
|---|---|---|---|---|---|---|
| **Cu-C-1** | - | - | 0.0347 | 0.0355 | - | - |
| **Cu-C-2** | + | - | 0.0292 | 0.0288 | - | - |
| **Cu-C-3** | - | + | 0.0490 | 0.0467 | - | - |
| **Cu-C-4** | + | + | 0.0348 | 0.0344 | - | - |
| **Cu-C-5** | 0 | 0 | 0.0359 | 0.0360 | 0.0358 | 0.0355 |

The F-values of the effects of the two parameters, their interaction effects and the quadratic effect were calculated and shown in figure 6. Both effects of the parameters and their interaction effect can be considered as significant based on the F-test. In contrast, the quadratic effect is not significant.

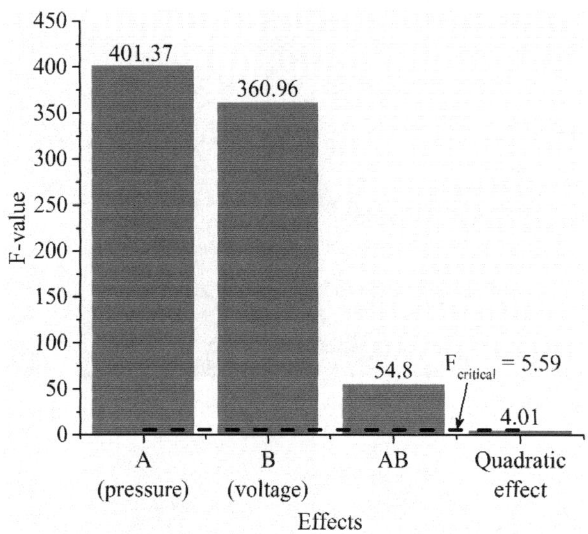


**Figure 6.** The F-values of the effects in the second stage of DoE for Cu coatings.

The correlation between the coating's electrical resistivity and the parameters can be described by equation 5, which is derived by the experimental results shown in Table 8. Because the interaction effect of the two parameters has a F-value higher than $F_{critical}$, the term A·B is included in the equation.

$$\rho_{el.,v}/(\Omega\cdot mm^2/m) = -0.0276 + 6.1759\cdot10^{-3}\cdot A/bar + 2.1111\cdot10^{-3}\cdot B/V - 2.6482\cdot10^{-3}\cdot A\cdot B/(bar\cdot V) \quad (5)$$

The contour plot based on equation 5 is shown in figure 7a. The coating's electrical resistivity increases with decreasing pressure and increasing voltage. The contour lines are curved, which is caused by the interaction effect of the two parameters. The effect of the voltage is slightly higher, when the pressure p is set at p = 3 bar, see figure 7b.

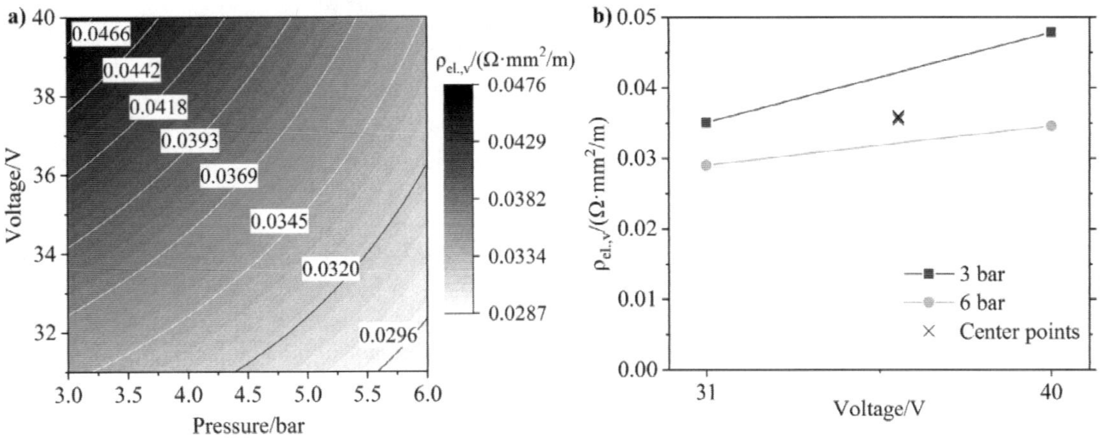


**Figure 7.** Contour plot of the correlation between $\rho_v$, the pressure and the voltage in case of Cu coating (a) and the interaction effect of the parameters (b).

### *3.3. Influence of the transfer process on the coating's electrical resistance*

Cu and Zn coatings were deposited with the parameter sets shown in Table 9. The two parameter sets have been determined as the best parameter sets for achieving low electrical resistivity of the coating in the characterization experiments for the two materials respectively.

**Table 9.** Process parameters used for the deposition of Zn and Cu coatings for the investigation of the influence of the transfer process on the coating's electrical resistance.

| Material | Wire diameter | Voltage | Feed rate | Atomizing gas | Gas pressure | Stand-off distance | Traverse velocity |
|---|---|---|---|---|---|---|---|
| **Zn** | 1.6 mm | 22 V | 13.5 kg/h | $N_2$ | 6 bar | 50 mm | 450 m/s |
| **Cu** | 1.6 mm | 31 V | 10.5 kg/h | $N_2$ | 6 bar | 50 mm | 675 m/s |

The results shown in figure 8 confirms that there is no significant difference between the detached and transferred coatings regarding the electrical resistance. Therefore, it can be concluded that the transfer process has no considerable influence on the coating's electrical resistance.

IOP Conf. Series: Materials Science and Engineering **373** (2018) 012011 doi:10.1088/1757-899X/373/1/012011

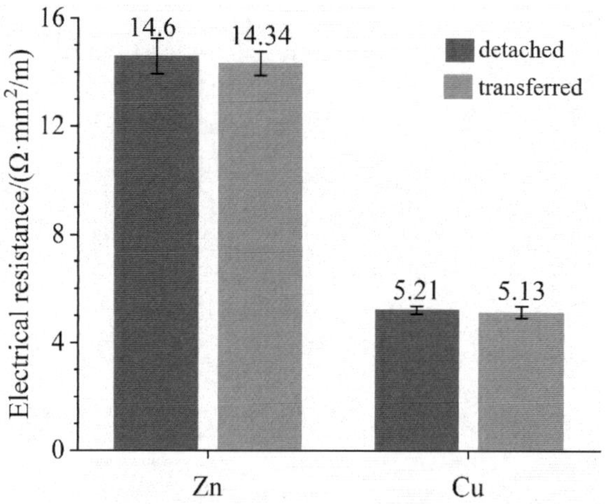


**Figure 8.** Comparison of the electrical resistance of the detached and transferred coatings.

## 4. Conclusion

In case of Zn coatings, the stand-off distance shows the highest effect on the coating's electrical resistivity in the first stage of DoE. The wire diameter and the pressure show similar effects in this stage. To avoid the influence of the variation of the process condition on the investigation caused by changing the wires, only the stand-off and the pressure were included in the $2^2$ full factorial design of experiment in the second stage of DoE. The effect of the wire diameter was investigated by a separate experiment series. The significance of the effects of the three parameters on the electrical resistivity of Zn coatings, which decreases with decreasing stand-off distance and wire diameter as well as increasing pressure, could be confirmed. Likewise, the stand-off distance shows the highest effect on the coating's electrical resistivity in the first stage of DoE for Cu coatings. However, the stand-off distance is not free adjustable for the aimed application. Therefore, the pressure and the voltage were investigated in the second stage. The decrease of the coating's electrical resistivity with increasing pressure and decreasing voltage could be confirmed.

The lowest electrical resistivity of the Zn and Cu coatings achieved in the current study was compared with the electrical resistivity of bulk Zn and Cu given in [24]. The electrical resistivity of the best Zn and Cu coatings was about 1.5 and 1.7 times that of the corresponding bulk materials, respectively, see figure 9.

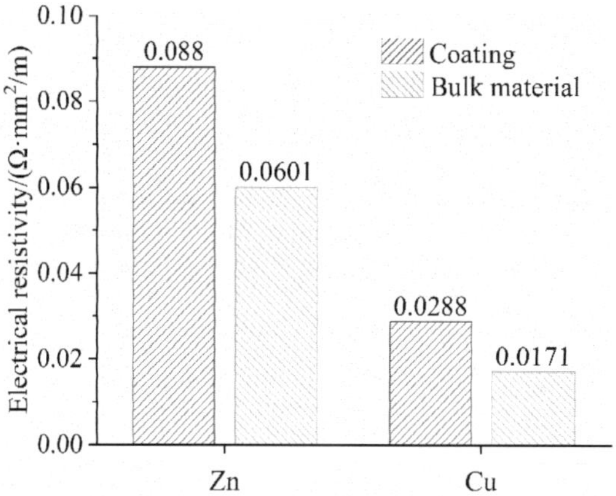


**Figure 9.** Comparison of the electrical resistivity of the coatings and the bulk material.

The influence of the transfer process on the coating's electrical resistance was investigated by the comparison of the electrical resistance of detached and transferred coatings. No noteworthy effect regarding this aspect could be determined in the investigation.

## 5. References

[1] Huang W, Zhao Y, Fan X, Meng X, Wang Y, Cai X and Wang X 2013 *J. Therm. Spray Technol.* **22** 918–25
[2] Liu A, Guo M, Zhao M and Hu M 2006 *Surf. Coat. Technol.* **200** 3073–7
[3] Liu A, Guo M, Gao J and Zhao M 2006 *Surf. Coat. Technol.* **201** 2996–700
[4] Ganesan A, Affi J, Yamada M and Fukumoto M 2012 *Surf. Coat. Technol.* **207** 262–9
[5] Ganesan A, Yamada M and Fukumoto M 2013 *J. Therm. Spray Technol.* **22** 1275–82
[6] Zhou X, Chen A, Liu J, Wu X and Zhang J 2011 *Surf. Coat. Technol.* **206** 132–6
[7] Che H, Vo P and Yue S 2017 *Proc. Int. Thermal Spray Conf. (Düsseldorf)* (Düsseldorff/DVS Media GmbH) pp 98–103
[8] Lupoi R and O'Neill W 2010 *Surf. Coat. Technol.* **205** 2167–73
[9] Bobzin K, Hopmann C, Öte M, Linke T F, Liao X and Ochotta P 2017 Plastics-metal hybrid parts for electrical application *Integrative Production Technology - Theory and Applications* ed Brecher C and Özdemir D (Cham, Springer International Publishing) chapter 7 pp 519–44
[10] Bobzin K, Öte M, Linke T F, Liao X, Hopmann C and Ochotta P 2015 *Therm. Spray Bull.* **67** 28–31
[11] Hopmann C, Bobzin K, Schoeldgen R, Wunderle J, Linke T F and Ochotta P 2016 *J. Polym. Eng.* **36** 549–56
[12] Bobzin K, Öte M, Knoch M A, Liao X, Hopmann C and Ochotta P 2016 *Proc. Int. Thermal Spray Conf. (Shanghai)* (Düsseldorff/DVS Media GmbH) pp 24–9
[13] Bucci D 2017 Fundamentals of sensing and signal conditioning *Analog Electronics for Measuring Systems* ed Bucci D (London, John Wiley & Sons) chapter 1 p 30
[14] Heaney M B 2004 Electrical conductivity and resistivity *Electrical Measurement, Signal Processing and Displays* ed Webster J G (Boca Raton, CRC Press) chapter 7 pp 7-4
[15] Bobzin K, Öte M, Knoch M A, Liao X, Hopmann Ch and Ochotta P 2018 *J. Therm. Spray Technol.* **27** 119–34
[16] Montgomery D C 2012 The $2^k$ Factorial Design *Design and Analysis of Experiments* (Hoboken, John Wiley & Son) chapter 6 pp 233-303
[17] Newberry A P, Grand P S and Neiser R A 2005 *Surf. Coat. Technol.* **195** 91–101
[18] Szulc M, Zierhut J, Atzberger A, Schaup J, Hartz-Behrend K, Zimmermann S, Schein J, Krömmer W and Lang F 2016 *Proc. Int. Thermal Spray Conf. (Shanghai)* (Düsseldorff/DVS Media GmbH) pp 44–8
[19] Oberg K E, Friedman L M, Boorstein W M and Rapp R A 1973 *Metall. Trans.* **4** 61-7
[20] Sano K and Sakao H 1955 *J. Jap. Inst. Mater.* **19** 655-9
[21] Gerlach J, Osterwald J and Stichel W 1968 *Z. Metallk.* **59** 576-9
[22] Wilder T C 1966 *Trans. TMS-AIME* **236** 1035-40
[23] Fischer W A and Ackermann W 1966 *Arch. Eisenhüttenw.* **37** 43-7
[24] Haynes W M, Lide D R and Bruno T J 2017 Properties of Solids *CRC Handbook of Chemistry and Physics* (Boca Raton, CRC Press) chapter 12 pp 12-42-3

**Acknowledgments**
The authors would like to thank the German Research Foundation DFG for the support of the depicted research within the Cluster of Excellence ''Integrative Production Technology for High-Wage Countries''.

# Influence of powder feed rate on corrosion and wear properties of Fe-based HVAF coatings

**K Bobzin, M Öte, M A Knoch and J Sommer***

Institut für Oberflächentechnik, RWTH Aachen University, Germany

* e-mail: sommer@iot.rwth-aachen.de

**Abstract.** Large-area applications, such as Yankee cylinders for paper machines, are often exposed to corrosion/wear, and therefore require adequate surface protection. The goal of this study is to develop a cost-efficient coating system that offers comparable protection to the industrially established Fe-based coating systems produced by wire arc spraying (WAS). Cost efficiency is to be achieved by using an economically priced, novel Fe-based feedstock material, the high velocity air-fuel (HVAF) process as well as a coating thickness reduction. In this study, the feedstock material FeCrB/WC-Co with the grain size fraction -32 +11 µm was investigated. Two coating thicknesses $d_{ct} \approx 120$ µm and $d_{ct} \approx 240$ µm and two different powder feed rates p = 40 g/min and p = 200 g/min were considered. An industrially established WAS FeCrBSiMnC coating system was used as reference. To examine the microstructure, cross sections of the coatings were prepared and investigated with a light microscope. Electro chemical polarization and pin-on-disc- (POD) tests were performed to investigate the corrosion and wear properties of these coating systems. Furthermore, XRD measurements were used to investigate the phase composition. The results show that a dense and crack-free coating can be produced with a powder feed rate of p = 200 g/min and that the powder feed rate p has a significant influence on the corrosion behaviour.

## 1. Introduction

In recent years, economical factors such as raw material and post-processing costs are playing an increasingly important role in the research of novel thermally sprayed wear and corrosion protection coatings. For this reason, cost efficient Fe-based materials have become the focus of research; especially for large area applications. Good corrosion and wear properties of these materials can be achieved by alloying high amounts of Cr and hard phase forming elements such as B or C [1, 2]. A further research trend in the field of thermal spraying technology is the HVAF process. This process is a modification of the high velocity oxygen-fuel (HVOF) process [3] with which higher deposition rates [4, 5] and lower particle in-flight temperatures, compared to standard HVOF processes, can be achieved [4]. The lower process temperature is achieved by using air instead of oxygen as oxidant [6]. Wang et al. showed that dense coatings with a low content of oxides can be produced with the HVAF process and explains this with the lower particle temperatures and the higher particle velocities compared to the HVOF process [4]. These results are confirmed by a number of studies [5 – 10]. However, the variety of different HVOF torch designs must not be neglected, which means that no general statement can be made about the oxygen content in the coatings applied by means of HVOF and HVAF spraying [13].

A typical example for a large area application, which requires high wear and corrosion resistance, is cylinders of paper machines, like Yankee dryers. Usually, these cylinders are coated using WAS with a Fe-based feedstock material, e.g. a FeCrBSiMnC cored wire. Typically, as-sprayed coating thicknesses range from $d_{ct}$ = 700 to 1,200 µm in this application [3, 11]. Experience has shown that the sufficient corrosion resistance of porous coatings can only be achieved with a high coating thickness. The corrosion resistance of the coating can be further increased by using sealers. However, this requires an additional post-production step. Due to the high roughness of WAS coatings, a time-consuming grinding process is necessary to achieve low surface roughness values required for application. Although suitable coatings can be produced by WAS, the process-inherent economic advantages such as low investment costs and high deposition rates are usually countered by the high post production costs. Furthermore, thick and porous coatings exhibit low thermal conductivities, which leads to higher energy expenditure during paper manufacture [11, 12]. Thicker coatings usually result in higher thermal insulation. Coatings exhibiting sufficient corrosion and wear resistance can also be realised in smaller thicknesses using HVOF and HVAF spraying with the utilisation of cemented carbide based feedstock materials. Due to the higher production costs, however, these coating systems are only used for cylinders which are exposed to very high loading conditions [13, 14]. In order to produce good and cost-effective wear and corrosion protection coatings for this application, one approach can be the use of the HVAF process in combination with a highly alloyed Fe-based feedstock material. With the HVAF process, higher powder feed rates can be achieved compared to the HVOF process. This can significantly shorten the process time. Compared to the WAS process, dense, near-net-shape and thinner coatings can be produced, which can have a positive effect on post-processing costs and on the energy consumption during paper production.

In a previous study, the authors have already shown that a novel FeCrB/WC-Co coating exhibits corrosion and wear properties, which are comparable to a HVOF sprayed $Cr_3C_2$/NiCr 75/25 coating system [15], although the grain size fraction G of the used feedstock material was not optimized for the HVAF process. However, the coating system was prone to cracking in case of high powder feed rates ($p >> 40$ g/min). Cracking behaviour of the coating system was attributed to a non-uniform heating of the particles and a high local heat input in the coating due to the high powder feed rate and a low surface velocity of the gas and particle jet. Both effects result in high residual stresses in the coating. The study presented in this paper is a continuation of [15] with the goal of producing dense and crack-free FeCrB/WC-Co coatings with powder feed rates upto p = 200 g/min. This goal is to be achieved among others by using a grain size fraction G = -32 +11 µm. This finer grain size fraction can result in more uniformly molten particles, which can decrease the residual stress and avoid cracks. Moreover, near-net-shape and thinner coatings can be realised with a fine powder fraction. Furthermore, the influence of a coating thickness reduction below $d_{ct}$ = 150 µm on the corrosion and wear behaviour is also investigated. Thinner coatings can save feedstock material, decrease the production time and lead to energy savings due to the reduction of the thermal insulation effect. The focus of this study is, therefore, set to the development of a novel FeCrB/WC-Co coating system with a powder fraction of G = -32 +11 µm and a powder feed rate of p = 200 g/min by means of HVAF spraying. The influence of such a high powder feed rate on the wear and corrosion properties is investigated and compared to properties of a WAS sprayed FeCrBSiMnC reference coating system.

## 2. Experimental setup

All HVAF and WAS coatings were applied on 1.0038 (S235JR) steel substrates with the dimensions of 40 x 50 x 8 $mm^3$. 1.0038 is a widely used construction steel, which has poor wear and corrosion performance in paper manufacturing. However, construction steel and cast iron are widely used as structural material in Yankee dryers. Sufficient wear and corrosion resistance of such parts is achieved through the use of coatings. Prior to the coating process, the substrates were roughened using an injector blasting system with corundum.

As already mentioned, cost-efficient Fe-based coating systems with high depositions rates can be realised by means of WAS. Therefore, a FeCrBSiMnC coating applied by WAS (G30/4SF-Push-LD/U2, Oerlikon Metco AG, Winterthur, Switzerland) was chosen as reference coating system. The commercially available cored wire SP112 (Corodur Fülldraht GmbH, Willich, Germany) was used as feedstock material. The chemical composition of the wire is given in Table 1.

**Table 1:** Chemical composition of the FeCrB/WC-Co feedstock material and the FeCrBSiMnC (SP112) wire, as given by the distributors [wt.-%]

| Alloy | Fe | Cr | B | C | W | Co | Si | Mn |
|---|---|---|---|---|---|---|---|---|
| FeCrBSiMnC (SP112) | Rest | 27.5 – 29 | 3.8 | 0.1 | - | - | 1.5 | 1.5 |
| FeCrB/WC-Co | Rest | 20 – 23 | 3.5 – 4 | 1 – 2 | 10 – 12 | 1 – 1.5 | | |

Industrially established process parameters were used to produce the reference coating, see Table 2. Subsequent to the coating process, the coatings were sealed in the "as-sprayed" state with an epoxy resin, to close open pores and to increase corrosion resistance. The industrially established sealer dichtol HM #2407 (DIAMANT Metallplastic GmbH, Mönchengladbach, Germany) was used for this purpose. The sealer was processed as specified by the manufacturer.

**Table 2:** WAS process parameters

| Parameter | Value |
|---|---|
| Current I [A] | 150 |
| Voltage U [V] | 30 |
| Pressure [bar] | 3.5 |
| Stand-off distance d [mm] | 150 |
| Surface speed [mm/s] | 600 |
| Wire feed rate w [g/min] | 115 |

For the application of novel HVAF coatings, the feedstock material FeCrB/WC-Co (Above Material Technology Co., Ltd., Beijing, China), which is composed of separate powder particles made of FeCrB and WC-Co, has been used. The chemical composition is given in Table 1. The powder includes a high content of Cr and B to achieve a good corrosion resistance and to promote the formation of hard phases, respectively. WC particles, embedded in a Co matrix, were added to the material composition to achieve a high wear resistance. Coating parameters for the HVAF coatings are given in Table 3. These parameters are based on the parameters developed in [15] and modified to achieve crack-free coatings with higher powder feed rates. The HVAF samples were produced using an AK-07 system (Kermetico Inc., Benicia, USA). An additional sealing of the HVAF coatings was omitted. In [15], the used FeCrB/WC-Co feedstock material had a grain size fraction of G = -45 +11 µm. In preliminary studies, a significant influence of the grain size fraction on crack formation was identified. For this reason, prior to the coating process, the feedstock material FeCrB/WC-Co was sieved to a grain size fraction of G = -32 +11 µm which is typical for the HVAF process. Due to the lower particle temperature compared to the HVOF process, it is assumed that the coarser grain fraction can cause non-uniform heating of the larger particles, which can lead to high residual stresses induced by the hard impact of non-uniform molten particles. The high compressive stress can result in cracks in the coating, especially for high powder feed rates.

**Table 3:** HVAF process parameters

a)

| Parameter | Value |
|---|---|
| Compressed air [bar] | 6.3 |
| Propane [bar] | 5.6 |
| Nitrogen [SLPM] | 18 |
| Stand-off distance d [mm] | 380 |
| Surface speed [mm/s] | 900, 4,500 |
| Powder feed rate p [g/min] | 40, 200 |
| Cooling air | 1.5 bar |

The experimental setup for the HVAF samples investigated in this study is illustrated in Figure 1 a). The substrates were fixed into the sample holder. The sample holder was mounted onto an electro motor, which allows a rotatory movement of the specimen holder. The HVAF gun only moves into z-direction to coat the samples. The samples were cooled from two sides. This experimental setup was used because preliminary studies showed that the layer thickness, which mainly depends on the powder feed rate and the surface speed of the particle-laden free jet, has a great influence on crack formation in FeCrB/WC-Co coatings applied by HVAF. The surface speed was increased depending on the powder feed rate. This results in the same layer thickness as in [15], with which crack-free coatings were already produced. Thicker layer thicknesses can lead to a higher local heat load, which in turn can lead to higher residual stress and thus to cracks in the coating. The heat propagation on the samples was additionally controlled with a cooling system to reduce the accumulating residual stress. Compared to study [15], the stand-off distance d was increased to d = 380 mm. At a higher spray distance, the particles dwell longer in the free jet and exhibit a more uniform melting, which in turn can reduce the induced stress due to non-uniform molten particles and related crack initiation. Subsequently, the wear and corrosion properties of the coatings are determined and compared to WAS sprayed reference coating.

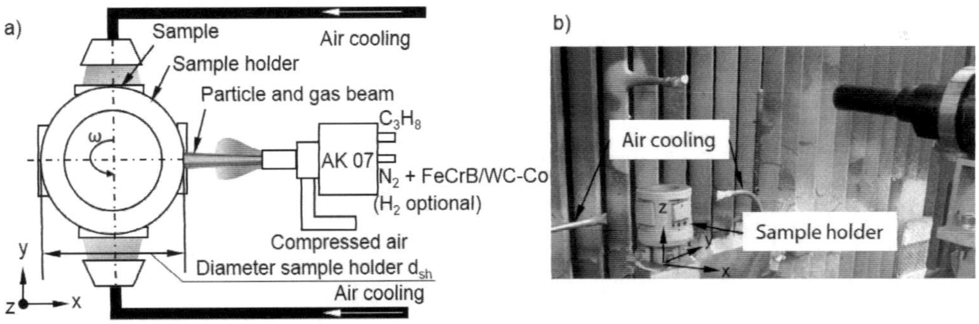


**Figure 1:** a) Sketch of the experimental setup
b) Experimental Setup

An overview of the sample names and the varied process parameters is given in Table 4. The different coating thicknesses of the HVAF samples were achieved by adjusting the number of transitions. Images of the cross sections were taken with a Zeiss Axiophot light microscope (Carl Zeiss AG, Oberkochen, Germany) to investigate and compare the microstructure of the HVAF and WAS samples. Surface roughness of the samples in the “as-sprayed” state was measured with the confocal laser scanning microscope (CLSM) VKX 200 (Keyence Corporation, Osaka, Japan). The phase composition of the FeCrB/WC-Co feedstock material and the HVAF coating systems was analysed with X-Ray Diffractometry (XRD) with a Seifert XRD 3000 from GE Sensing & Inspection Technologies GmbH (Hürth, Germany) using a Cu anode. The angle of incidence ω = 10°, step width

$\Delta 2\Theta = 0.05°$, holding time $t_h = 10$ s and measurement interval $2\Theta = 20°$ to $80°$ were kept constant for all measurements.

**Table 4:** Overview sample names and varied parameters

| Sample | Powder feed rate p [g/min] | Surface speed [mm/s] | Transitions | Note |
|---|---|---|---|---|
| WAS unsealed | 115 | 600 | 26 | |
| WAS sealed | 115 | 600 | 26 | Sealed with dichtol HM #2407 |
| HVAF 01 | 40 | 900 | 20 | |
| HVAF 02 | 40 | 900 | 10 | |
| HVAF 03 | 200 | 4,500 | 20 | |
| HVAF 04 | 200 | 4,500 | 10 | |

For the investigation of the wear properties, a POD tribometer (CSM Instruments SA, Freiburg, Germany) was used and the wear coefficient K was determined. The wear coefficient K describes the amount of energy that must be added to a system in order to obtain a certain amount of wear. To keep the initial conditions constant, the samples were ground and polished to a surface roughness of $Ra > 0.5$ µm. An $Al_2O_3$ ball with a diameter $ø_{cb} = 6$ mm was used as counter body to characterise the wear properties of the coatings. $Al_2O_3$ was chosen to conduct abrasive wear tests and to reduce adhesive wear. The used test parameters are shown in Table 5. All tests were performed without lubricant. Subsequent to the wear tests, the wear tracks were investigated using CLSM to determine the volume and diameter of track. With these values, the wear coefficient K was calculated according to equation (1) [2]. For every wear track, three measurement positions along the track were used to determine the mentioned parameters.

**Table 5:** POD test parameters

| Parameter | Value |
|---|---|
| Wear track diameter d | 5 mm |
| Wear track length s | 1,000 m |
| Normal load w | 10 N |
| Rotation speed | 100 mm/s |
| Test temperature | RT |

$$K \left[\frac{mm^3}{Nm}\right] = V \cdot \left[w(s \cdot \frac{d}{d_t})\right]^{-1} \qquad (1)$$

In order to analyse the corrosion resistance, polarisation tests and subsequent cross-section analyses of the corroded samples were performed to determine whether corrosion occurs only on the surface or also in the interface between substrate and coating. The polarization tests were performed in a 5 % NaCl solution with a pH value of $7.00 \pm 0.05$ at room temperature (Reference 600+, Gamry Instruments, Warminster, England). Prior to the polarization test, the samples were prepared in the same manner as for the POD test. A calomel electrode was used as reference electrode and the potential scan rate was set to 0.5 mV/s. The rest potentials and the corrosion current densities were investigated using the Tafel slope analysis.

## 3. Results and discussion

### Microstructure

In Figure 2, the cross sections of the WAS reference coating and the HVAF samples are shown. The WAS reference sample shows the typical lamellar structure of a WAS coating. Splat boundaries, pores, particles solidified in flight and few oxidized particles can be observed in the cross section. These splat boundaries and porosities can be a weak point regarding corrosion loading. The WAS sample has a coating thickness $d_{ct} \approx 720$ µm and is thus significantly higher than that of the HVAF samples. For the HVAF samples, the darker phases can be identified as WC-Co and the brighter phases as FeCrB matrix. No cracks were detected, even for a powder feed rate of p = 200 g/min and a coating thickness of $d_{ct} = 277$ µm. The microstructure of the samples HVAF 01 – HVAF 04 looks comparable, is very dense and exhibits a low oxide content. In the FeCrB phase, no grain boundaries between the individual particles of the feedstock material can be detected. Furthermore, only a very small number of unmolten particles were observed. The roughness of all HVAF samples in the as-sprayed state is in a similar range and significantly lower than the roughness of the WAS reference. The smoother surface of the HVAF samples is beneficial with respect to lower post-production costs.

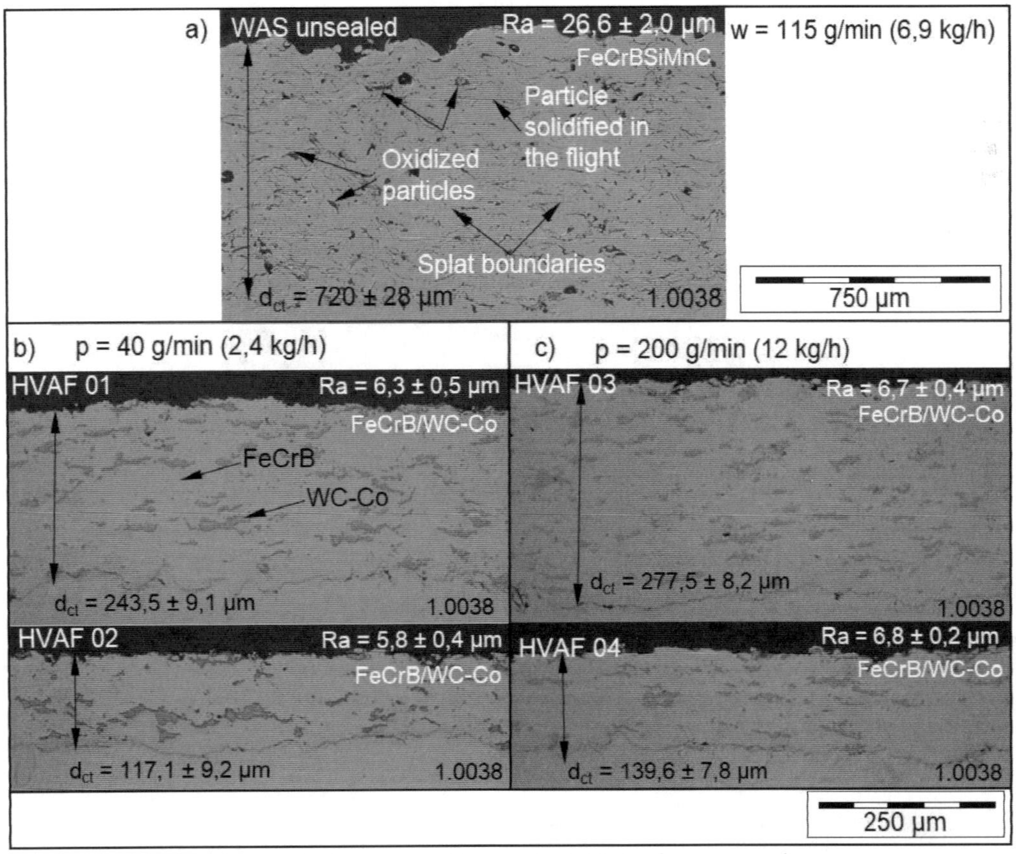


**Figure 2:** a) Cross section of the WAS reference coating
b) Cross section of the HVAF samples coated with 40 g/min
c) Cross section of the HVAF samples coated with 200 g/min

### Phase composition

The XRD measurements of the HVAF samples and the FeCrB/WC-Co feedstock material are shown in Figure 3. For the feedstock material FeCrB/WC-Co, peaks corresponding to FeCrB and WC were identified. The measured spectra of the HVAF sample and the feedstock material look comparable indicating that the spraying process did not significantly alter the phase composition.

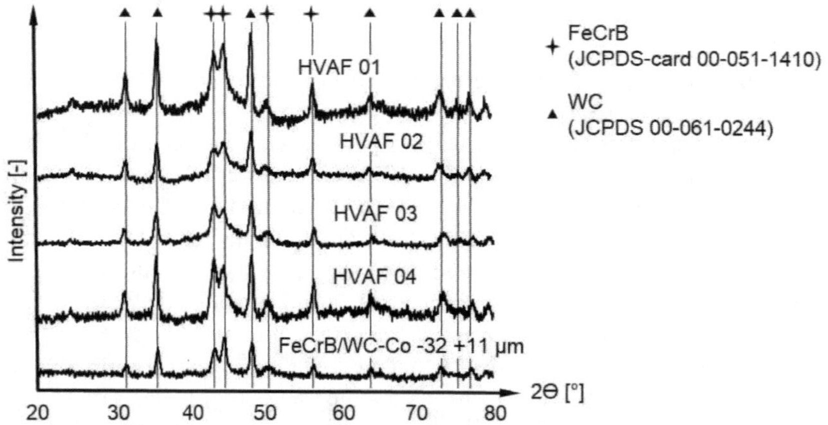


**Figure 3:** XRD analyses of the HVAF samples and the feedstock material FeCrB/WC-Co

**Wear test**

The analysis of the POD results is shown in Figure 4. In Figure 4 a), the wear tracks of the samples WAS sealed, HVAF 01 and HVAF 03 are shown. For all samples, non-homogenous wear tracks can be observed. It can be seen that the width of the wear track of the WAS sealed sample is wider than that of HVAF 01 and HVAF 03 while the width of HVAF 01 and HVAF 03 is comparable. The microscope images of the wear marks suggest that HVAF samples are less prone to wear then the investigated WAS samples. The wear tracks of the samples WAS unsealed, HVAF 02 and HVAF 04 did not exhibit any relevant differences to the corresponding wear track shown in Figure 4. The measured wear coefficient for all samples is in a similar range as the wear coefficient measured for the coating system developed in [15] by the authors, K = 5.0 $10^{-6}$ mm³/Nm.

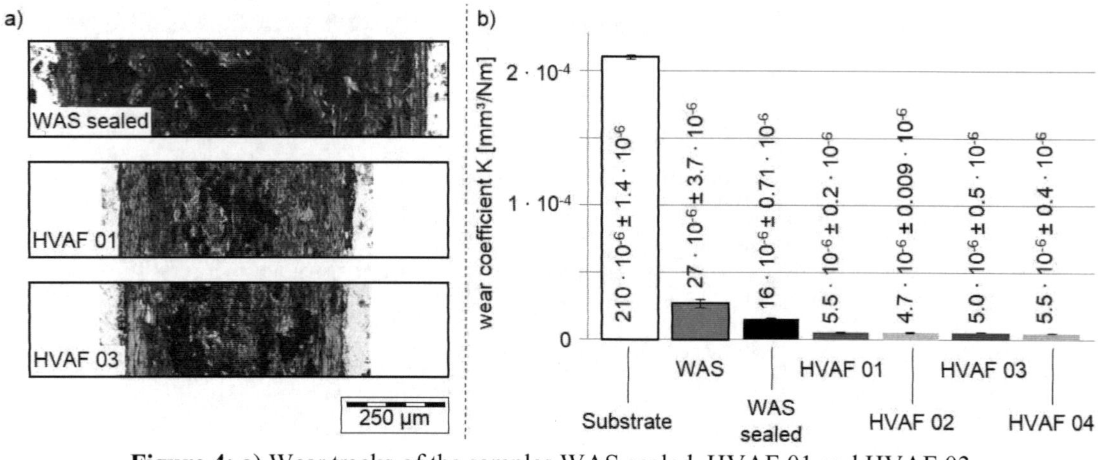


**Figure 4:** a) Wear tracks of the samples WAS sealed, HVAF 01 and HVAF 03
b) Wear coefficient K of all HVAF samples compared to WAS samples

In Figure 4 b), the wear coefficients K for all investigated samples are shown. WAS sealed has a lower wear coefficient K than the unsealed WAS sample. It is assumed that the lower wear coefficient can be traced back to the sealer, which prevents the coating from breaking out. The wear coefficient of all HVAF samples is significantly lower than the wear coefficient of the WAS samples. A slightly lower wear coefficient K among HVAF samples was measured in case of sample HVAF 02. This sample exhibits a wear coefficient of $K = 4.7 \cdot 10^{-6}$ mm³/Nm. The wear coefficients for the other HVAF

IOP Conf. Series: Materials Science and Engineering **373** (2018) 012012 doi:10.1088/1757-899X/373/1/012012

samples are slightly higher. It can be stated that the lower wear coefficients of HVAF coatings can be attributed to the high density of the coatings. In contrast, the WAS coatings exhibit a more porous microstructure, which can negatively influence the wear resistance. It can be also stated that the coating thickness and the powder feed rate have no significant influence on the wear behaviour of the HVAF samples in the considered tribological test.

**Corrosion test**

The corrosion tests showed significant differences between substrate and considered coating systems. The rest potential $U_R$ and corrosion current density $i_c$ is given in Table 6. The corresponding curves are designated in Figure 5. HVAF 01 exhibits the highest rest potential of $U_R = -251\ mV_{SCE}$ and lowest corrosion current density $i_c$. The thicker coating HVAF 02 has s comparable rest potential. The rest potentials of the samples produced with a powder feed rate of p = 200 g/min, HVAF 03 and HVAF 04, are significantly lower than the rest potentials of HVAF 01 and HVAF 02. The measured rest potential for the sealed WAS reference is in the same range as HVAF 03. An interesting aspect was observed in the samples HVAF 01 and HVAF 02 for higher overpotentials of $\eta \approx 530$ mV, the corrosion current density stays below 3.3 mA/cm². It is assumed that the powder feed rate influences the phase composition of HVAF samples and thus influences their corrosive properties. Although the exact reason is subject to further investigation, it can be stated that the powder feed rate has a significant influence on the corrosion resistance of the HVAF samples.

**Table 6:** Rest potential $U_R$ and corrosion current density $i_c$

| Sample | Rest potential $U_R$ [$mV_{SCE}$] | Corrosion current density $i_c$ [µA/cm²] |
|---|---|---|
| 1.0038 | -654 | 3.98 |
| WAS unsealed | -605 | 4.28 |
| WAS sealed | -419 | 5.44 |
| HVAF 01 | -251 | 0.50 |
| HVAF 02 | -293 | 0.39 |
| HVAF 03 | -435 | 4.52 |
| HVAF 04 | -478 | 4.98 |

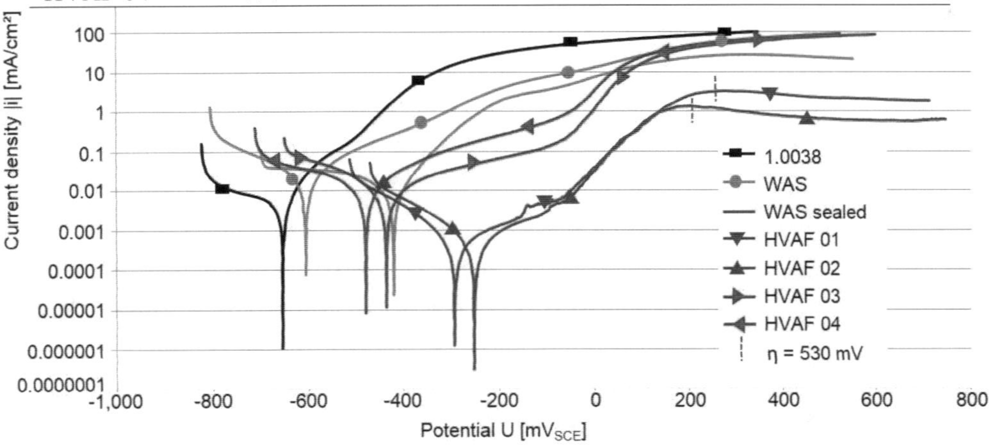


**Figure 5:** Corrosion current potential curves of the reference and the HVAF samples in 5 % NaCl-solution at room temperature

The corroded cross sections after the destructive polarization test are shown in Figure 6. The maximum corrosion depth was measured. For this purpose, the polished surface in the non-corroded area of the sample was selected as the reference plane. Figure 6 a) shows the cross section of the sealed WAS coating. No under-corrosion could be identified for this sample. Maximum corrosion depth measured in this test correspond to a value of $d_{cd}$ = 82 µm. Here it must be stated that in deeper layers, some spots with beginning of corrosion have been observed. Therefore, the given value for the corrosion depth shall be considered carefully.

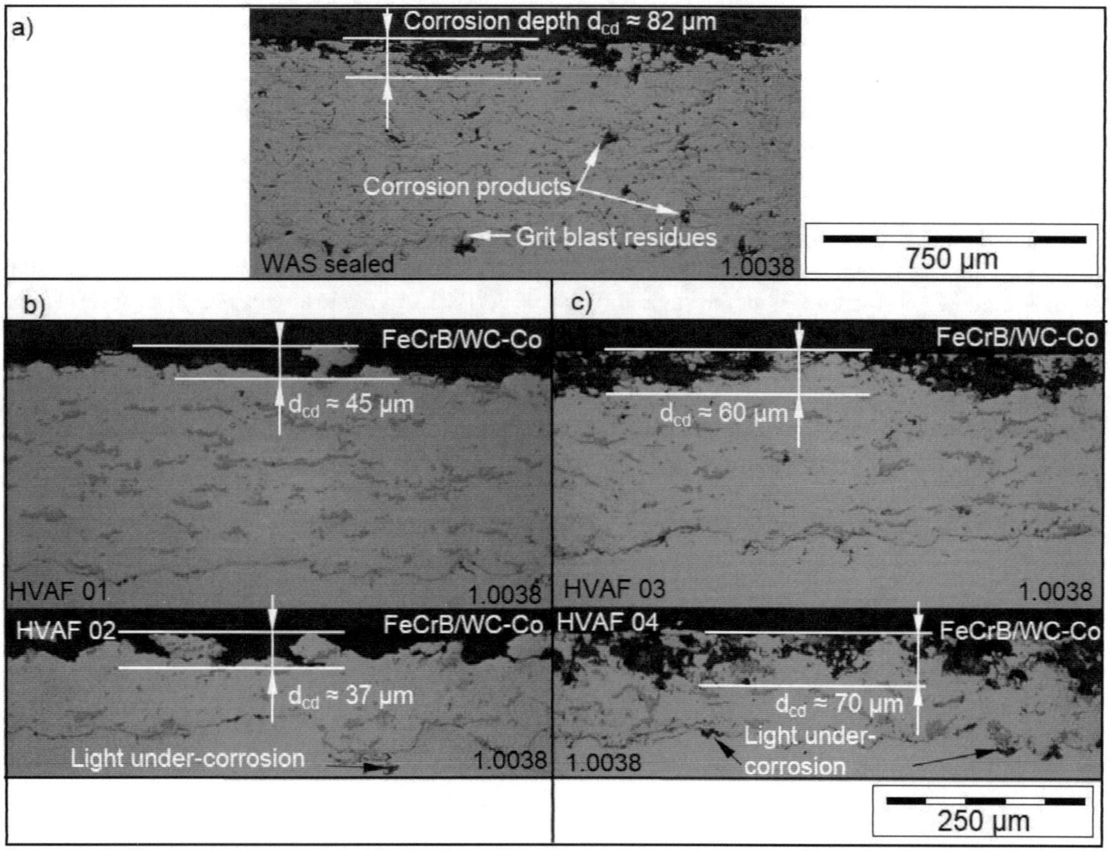


**Figure 6:** Cross sections after the corrosion test of the a) sealed WAS reference
b) HVAF 01 and HVAF 02
c) HVAF 03 and HVAF 04

The maximum corrosion depth $d_{cd}$ of the HVAF sprayed coatings are significantly lower. For the samples HVAF 01 and HVAF 02, a maximum corrosion depth $d_{cd}$ = 45 µm was measured. In case of the samples HVAF 03 and HVAF 04, the maximum corrosion depth increased to $d_{cd}$ = 60 – 70 µm, see Figure 6 b) and c). This is in accordance with the higher corrosion current densities of these samples measured using electrochemical polarisation test, see Figure 5. These results confirm once again that the powder feed rate influences the corrosion properties significantly. A reason for the difference in corrosion properties observed in case of low and high powder feed rates might be the reduced available heat energy per particle for high powder feed rates, which can decrease the melting degree of the particles, whereby nano-pores can occur at the splat boundaries. These pores cannot be detected with the used light microscope, therefore scanning electron microscope / energy dispersive X-ray spectroscopy (SEM/EDS) analyses are necessary. Another reason assumed for the changed corrosion properties at high powder feed rates is the altered phase composition of the coating due to the different degree of melting of the particles. The different powder feed rates might lead to local Cr depletion and, therefore, no protective passive layer can be formed, which protects the coating against corrosion. These local differences can be in nano dimension, whereby these differences cannot be detected in the XRD analysis used in this study. Further analyses such as particle diagnostic have to verify this hypothesis. In-depth analyses using the above-mentioned methods are planned in order to investigate, which mechanisms are mainly responsible for the altered corrosion properties at high powder feed rates. For the thick HVAF coatings, HVAF 01 and HVAF 03, no under-corrosion was detected. For the thin HVAF coatings, HVAF 02 and HVAF 04, light under-corrosion was detected in the interface. Normally, it is assumed at high overpotentials that the polarization curve of HVAF 02 has to approach the substrate's curve due to the observed under-corrosion, see Figure 5. One reason for not approaching the substrate's curve might be, that under-corrosion was only observed in very few places, which means the influence of the substrate cannot be observed in the polarization curve. Thus, the coating thickness and density of these samples is not sufficient to protect the substrate from corrosion at such high overpotentials, but it has to be stated that such high overpotentials are not realistic for most applications. Further investigation should check if under-corrosion would also occur if the overpotential is kept smaller. To sum up the results of the corrosion tests, it can be stated that the powder feed rate has a significant influence on the corrosion behaviour of the HVAF samples. For the samples WAS sealed and HVAF 03, almost the same rest potentials were measured. However, the corrosion depth $d_{cd}$ of all HVAF samples is lower than of the WAS sealed sample.

## 4. Conclusion and outlook

The goal of this study was to investigate the influence of high powder feed rates on coating properties of a cost-efficient wear and corrosion protection coating system, using the HVAF process and the novel FeCrB/WC-Co feedstock material. Therefore, coatings with a thickness of $d_{ct} \approx 130$ µm and $d_{ct} \approx 260$ µm and with a powder feed rate of $p = 40$ g/min and $p = 200$ g/min were produced. Subsequently, the microstructure, corrosion and wear properties of the new coating system were investigated and compared to a FeCrBMnSiC WAS sprayed reference protection coating system. The results can be summarized as follows:

- Dense, crack-free and low-oxide coatings can be produced with powder feed rates upto $p = 200$ g/min.
- The surface roughness Ra of the HVAF samples is significantly lower compared to the WAS sample. The powder feed rate and coating thickness of the HVAF samples have a negligible influence on the surface roughness.
- The XRD analyses show that the HVAF process did not alter the phase composition of the processed feedstock material significantly.
- With high powder feed rates of $p = 200$ g/min and a coating thickness of $d_{ct} \approx 270$ µm, the HVAF coatings achieve a rest potential comparable to that of the WAS sealed reference coatings.
- The powder feed rate has a significant influence on the corrosion behaviour of the HVAF samples. Corrosion resistance decreases with increased powder feed rate. Nonetheless, all HVAF samples exhibited a higher corrosion resistance than the sealed WAS reference.
- HVAF coatings show a superior wear behaviour under considered testing conditions in comparison to WAS coating independent of the powder feed rate.

Investigations under the considered testing conditions show that novel Fe-based coatings, exhibiting superior wear and corrosion resistance in comparison to conventional coating, can be realised by means of HVAF spraying. Here it must be noted that, depending on the real loading conditions, a sufficient coating thickness should be ensured to avoid under corrosion. In the near future the following investigations are planned to investigate the developed coating systems more detailed:

- Further analysis, e.g. SEM/EDS, are necessary to explain why the rest potential $U_R$ decreases with increasing the powder feed rate.
- Investigation of the corrosion behaviour of the HVAF samples for lower overpotentials, to compare the corrosion at the beginning of the anodic saturation.

- Processing of the same feedstock material with a grain size fraction G = -20 +3 µm, to investigate if an even smaller grain size fraction has a positive effect on the wear and corrosion properties.

## 5. References

[1] Schwenk A: HVOF-K2-Spritzen von Eisenbasislegierungen für Anwendungen in der Druckindustrie, Proceedings of 8. HVOF-Kolloquium 2009

[2] Osamanda A, Battenbough A and Staines A: Maßgeschneiderte HVOF-Beschichtungssyteme auf Fe-Cr und Fe-Cr-Ni Basis für den Verschleiß- und Korrosionsschutz 2012, *Thermal Spray Bulletin* **64**, 117-22

[3] Perdikaris C: Methods of Coating Yankee Dryer Drums, *European Patent Application*, 2001, **EP 1158066A1C**

[4] Wang Q, Zhang S, Cheng Y and Xiang J: Wear and corrosion performance of WC-10Co4Cr coatings deposited by different HVOF and HVAF spraying processes 2013, *Surface & Coating Technology* **218**, 127-36, DOI: https://doi.org/10.1016/j.surfcoat.2012.12.041

[5] Bolelli G, Berger L M, Boerner T and Koivuluoto H: Tribology of HVOF- and HVAF-sprayed WC-10Co4Cr hardmetal coatings: a comparative assessment 2015, *Surface & Coating Technology* **265**, 125-44, DOI: https://doi.org/10.1016/j.surfcoat.2015.01.048

[6] Hulka I, Serban V A, Niemi K and Vuoristo P: Comparison of structure and wear properties of fine-structured WC-CoCr coatings deposited by HVOF and HVAF spraying Processes 2012, *Solid State Phenomena* **188**, 422-27, DOI: 10.4028/www.scientific.net/SSP.188.422

[7] Hashmi S.: Comprehensive materials processing: thermal spray coating process 2014, *Newnes*, 230-266, ISBN: 978-0-08-096532-1

[8] Guo R Q, Zhang C, Chen Q and Yang Y: Study of structure and corrosion resistance of Fe-based amorphous coatings prepared by HVAF and HVOF 2011, *Corrosion Science* **53**, 2351-56, DOI: https://doi.org/10.1016/j.corsci.2010.12.022

[9] Bolelli G, Hulka I, Koivuluoto H and Lusvarghi L: Properties of WC-FeCrAl coatings manufactured by different high velocity thermal spray processes 2014, *Surface & Coating Technology* **247**, 74-89, DOI: https://doi.org/10.1016/j.surfcoat.2014.03.021

[10] Berger L M, Puschmann R, Spatzier J and Matthews S: Potential von HVAF-Spritzprozessen 2013, *Thermal Spray Bulletin*, 17-20

[11] Tenkula J, Hellman B, Majava J: Yankee Cylinder with Plasma-Sprayed Carbide Coating, 1992, *US Patent*, **US005123152A**

[12] Ghosh A K: Fundamentals of Paper Drying – Theory and Application from Industrial Perspective 2011, INTECH, 535-82, DOI: 10.5772/21594

[13] Vuoristo P, Nylén P: Industrial and Research Activities in Thermal Spray Technology in the Nordic Region of Europe 2007, *Journal of Thermal Spray Technology* **16**, 466-71, DOI: 10.1007/s11666-007-9136-2

[14] Vuoristo P: Thermal Spray Coating Processes 2014, *Comprehensive Material Processing* **1st Edition**, 229-76, DOI: http://dx.doi.org/10.1016/B978-0-08-096532-1.00407-6

[15] Bobzin K, Öte M, Knoch M A , Sommer J: Development of HVAF-sprayed novel Fe-based coatings for large area applications 2017, Proceedings of ITSC 2017 **336**, 339-45

## Acknowledgments

This work is based on the DFG project BO 1979/45-1. The authors gratefully acknowledge the financial support of the German Research Foundation (DFG).

# Investigation of the interfacial reactions between steel and aluminum coatings for hybrid casting

**K Bobzin, M Öte, S Wiesner and L Gerdt***

Surface Engineering Institute, RWTH Aachen University

*e-mail: gerdt@iot.rwth-aachen.de

**Abstract.** Coating of AA7075 was applied by means of cold gas spraying on steel substrates of 22MnB5 and DC04 as an interlayer for high pressure die casting of aluminum/steel hybrid components. The morphology and growth kinetics of intermetallic compounds formed at the interface between coating and steel has been investigated. Furthermore, the effect of alloying elements on the formation of the intermetallic phases was analyzed. The coated samples were heat treated by means of induction heating at the temperature T = 550 °C with different dwell times in the range of 10 s < t < 5 min. The reaction layer growth was examined by means of scanning electron microscope (SEM) with energy dispersive spectroscopy (EDS). Additionally, the intermetallic compounds were characterized by means of nanoindentation. Intermetallic compounds of AlFe phases occurred as the major constituent in the reaction zone for different combinations of coating and substrates.

## 1. Introduction

Trends in the modern automotive industry such as lightweight constructions give rise to the use of hybrid structures and therefore the demand for new joining techniques of dissimilar materials arises, in particular aluminum and steel. The hybrid structure of steel sheet and aluminum cast provides a favorable combination of good mechanical properties and reduced weight. For decades, aluminium and steel have remained the most significant materials in automotive industry. Joining of these materials, however, is still a challenging task due to dissimilar thermophysical and metallurgical properties and the formation of brittle intermetallic phases. Besides welding and brazing, the high pressure casting represents another technique for joining dissimilar metals. Casting of grey cast iron liners with aluminum [1] or Al inserts with Mg alloy [2] in the engine crankcase is already state of the art. Pressure casting of aluminum onto a steel insert has been also at the focus of research at RWTH Aachen University in recent years [3]. The cooperation between the foundry institute and industrial partners resulted in a "VarioStruct" prototype of the roof cross-beam of the vehicle [4]. Despite these successes, further improvement of the compound properties is required, as the problem of the brittle intermetallic phases and the gap formation at the aluminum/steel interface has not been completely solved. Due to enormous cooling rates after pressure casting and insufficient wetting of steel by aluminum, it is difficult to achieve a metallurgical bonding at the interface. In order to enable a gap-free material bonding between the steel and the cast aluminum alloy, different coatings for steel inserts are investigated in the current study. The main challenge is the application of coatings with a reliable adhesion on the steel substrate which facilitate metallurgical bonding with the aluminum melt during the pressure casting. The control of the reaction zone at the interface between aluminum and steel is of

IOP Conf. Series: Materials Science and Engineering **373** (2018) 012013 doi:10.1088/1757-899X/373/1/012013

special interest regarding the mechanical properties of the joint, in particular its strength. The majority of the studies analyse the growth behaviour of intermetallic phases during hot dip aluminizing of steel [5–8]. Some of them were focused on the effects of alloying elements on the morphology and growth kinetics of the intermetallic compounds [5,7]. During the high pressure die casting process the surface of the steel insert is heated up to a temperature of T = 400 – 520 °C, while the temperature of the molten aluminum is about T = 700 °C [3]. A coating of AA7075 with a low solidus temperature was selected to enable a metallurgical bonding with the cast alloy during high pressure die casting. First die pressure casting trials showed that a thin surficial layer of the coating could be partially molten, while residual coating remains solid nearby the interface of the substrate material. Thus, the diffusion processes and the formation of the intermetallic phases at the interface between coating and steel take place in the solid state. In the current study two different steel inserts of DC04 and 22MnB5 which are commonly used in the automotive industry were selected. In order to investigate the influence of the alloying elements of the steel on the morphology and growth kinetics of the intermetallic compounds, coated samples have been heat treated at temperatures of T = 550 °C with different dwell times in the range of 10 s < t < 5 min. In order to reproduce the high heating rates typical for high pressure die casting, induction heating has been chosen in the current study.

## 2. Experimental

The experimental procedure consisted of a cold spraying process, followed by heat treatment and the analysis of the coated samples. The principle scheme of the experimental setup is shown in Figure 1.

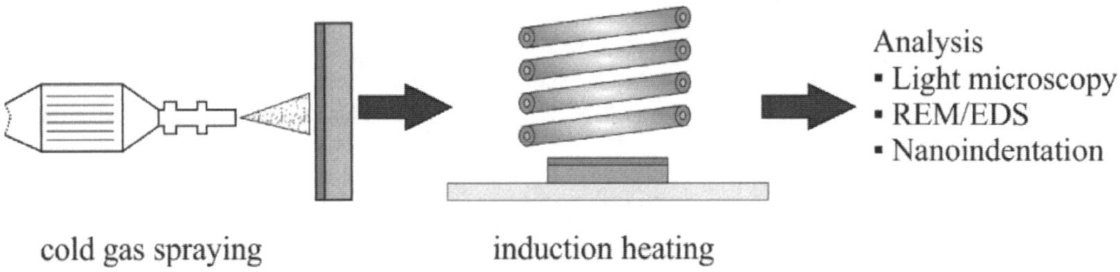


**Figure 1.** Principle scheme of the experimental procedure

### *2.1. Cold spraying*

In the present study, the conventional cold rolled deep-draw steel DC04 and steel 22MnB5 were selected as substrate materials. Alumina with the grit size of F150 was used as blasting abrasive for surface pre-treatment of the steel prior to cold spraying. The steel sheet was punched to rectangular plates with the dimensions of $l \times b \times d = 200 \times 30 \times 2\ mm^3$. Due to the low solidus temperature of the alloy AA7075, it was chosen as feedstock material for coating application to achieve a metallurgical bonding with the aluminum melt during the high pressure die casting by partial melting of the coating. The compositions of the materials are given in Table 1. AA7075 powder with a particle size distribution of -63 + 20 µm was delivered by TLS Technik GmbH (Bitterfeld, Germany). The Kinetics 8000 Cold spray system from Oerlikon Metco (Pfäffikon, Switzerland) was used for the deposition process with nitrogen as process and carrier gas.

**Table 1.** Chemical composition of the substrate and coating materials (wt.-%)

| | Si | Mn | Mg | C | S | P | Cu | Al | Fe | Zn | Cr |
|---|---|---|---|---|---|---|---|---|---|---|---|
| DC04 | | 0.40 | | 0.08 | 0.03 | 0.03 | | | bal. | | |
| 22MnB5 | 0.30 | 1.26 | | 0.23 | 0.01 | | | | bal. | | 0.12 |
| AA7075 | 0.40 | 0.30 | 2.50 | | | | 1.60 | bal. | | 5.50 | |

### *2.2. Heat treatment*
The coated samples were heat treated at the temperature of T = 550 ± 5 °C by using an induction heating system from iew Induktive Erwärmungsanlagen GmbH (Gumpoldskirchen, Austria) with five different dwell times varying between t = 10 s and t = 5 min. The temperature of the samples was controlled by a pyrometer.

### *2.3. Analytical methods*
In order to characterize the coating, the cold sprayed samples were metallographically prepared and examined in terms of their microstructure. The coating thickness was measured on the cross sections using a light microscope by Carl Zeiss (Oberkochen, Germany) with Axio Vision image analysis tool. Universal hardness HU were determined using a calibrated TI 950 TriboIndenter, Bruker Corporation (Billerica, Massachusetts, USA) by applying a Berkovich tip (indenter code TI-0083) with a nominal edge radius of approximately r = 20 nm using a loading force of F = 8 mN. The indentation was performed at five different positions with the same distance from the interface between steel and reaction zone to calculate the average value. In order to identify the reaction layers at the interface between steel and aluminium alloy after heat treatment, an energy dispersive spectroscopy analysis was performed by means of the scanning electron microscope Zeiss Leo 1530 with the EDS device Bruker Quantax 200.

## 3. Results and discussion
Figure 2 shows cross sections of the AA7075 coatings on two different substrate materials. Due to a good plasticity of the feedstock material the coatings of AA7075 exhibit a tight bonding and low porosity regardless of substrate material. The mean thickness of the coatings was approximately d = 140 μm.

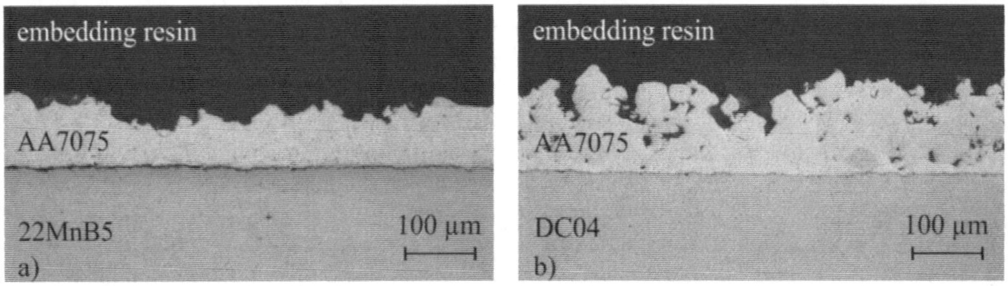


**Figure 2.** Cold sprayed coatings of AA7075 on a) steel 22MnB5 and b) steel DC04

### *3.1. Microstructural analysis of the coatings after heat treatment*

After the heat treatment at the temperature of T = 550 C for t = 10 s, a thin intermetallic layer could be observed at the interface between AA7075 coating and steel DC04 as well as at the interface between AA7075 coating and steel 22MnB5, Figure 3a, d. The morphology of the intermetallic compounds was similar for both substrate materials. When the heat treatment time reaches t = 60 s, some pores could be observed at the interface between intermetallic layer and the coating. This effect became more pronounced with increased dwell time of the heat treatment, Figure 3c, f. It can be attributed to the Kirkendall effect due to the higher mobility of Al atoms compared to Fe resulting in a predominated diffusion flux from the AA7075 coating towards the reaction zone. A similar behaviour was described by Springer et al. for friction stir welded compounds at the interface between DC01 and AlSi5 [9].

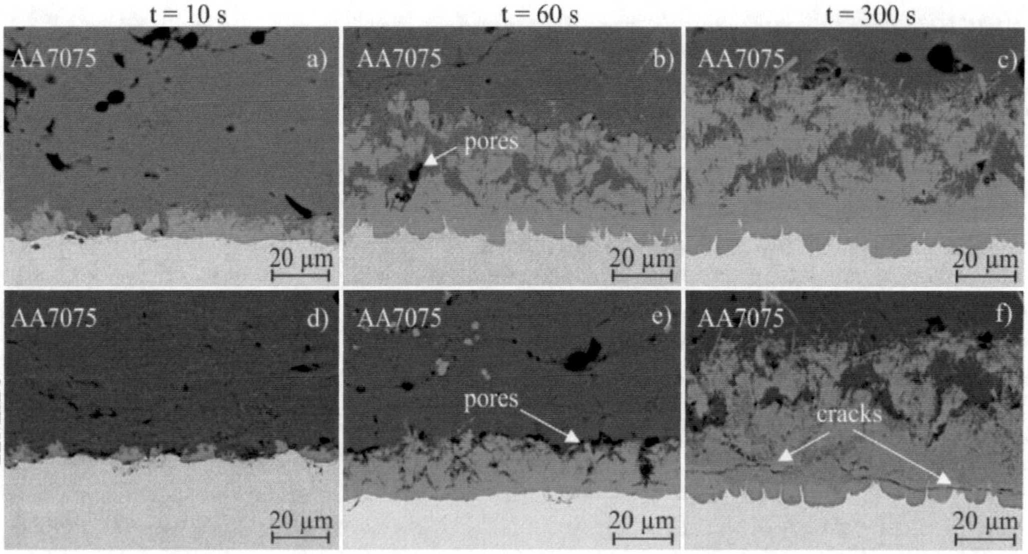


**Figure 3.** SEM images of the interface between AA7075 coating and steel DC04 after heat treatment at T = 550 °C for a) t = 10 c, b) t = 60 s, c) t = 300 s and between AA7075 coating and steel 22MnB5 after heat treatment at T = 550 °C for d) t = 10 s, e) t = 60 s, f) t = 300 s

Furthermore, some cracks occurred within the reaction zone between intermetallic layers. This phenomenon was considerably stronger for steel 22MnB5, Figure 3f. Similar crack propagation behaviour was described by Springer et al. during immersion of steel DC01 into aluminium melt AlSi5, accompanied by intermetallic phase formation [9]. Furthermore, Windmann et al. described crack formation in the reaction zone of Al-coated steel 22MnB5 during heat treatment at T = 920 - 1,000 °C, pointing out a preferred crack propagation along $Al_5Fe_2$ intermetallic phase [10]. The thermal stresses occurring during the cooling process could be a possible explanation for the crack formation. In Figure 4 thermal expansion coefficients of both steel grades and AA7075 coating is presented for a relevant temperature range. The occurrence of the cracks in the reaction zone can be explained by low fracture toughness of the AlFe intermetallic phases. Indeed, the difference between coefficients of thermal expansion between coating and steel 22MnB5 is slightly higher as in the case of steel DC04, however, it is too low to explain clearly such a different behaviour regarding the crack initiation.

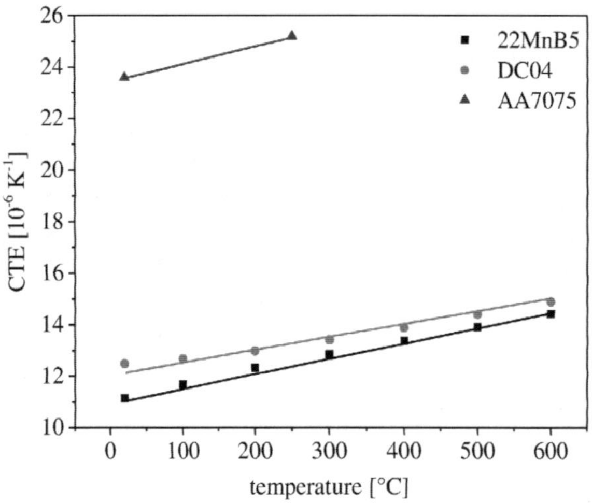


**Figure 4.** Coefficient of thermal expansion (CTE) for steel substrates and AA7075 alloy [13]

### *3.2. Analysis of the intermetallic compounds and their growth kinetics*

The morphology as well as chemical composition of the intermetallic compounds was investigated by means of SEM/EDS measurements. Figure 5 shows the line scan of the atomic fraction profile at the interface between the steel and the coating. A special characteristic of the reaction zone is that after longer heat treatment times, it could be subdivided into two regions with a low-iron region between them. Furthermore, there are two intermetallic layers with different compositions to be found in the reaction zone adjacent to the steel substrate.

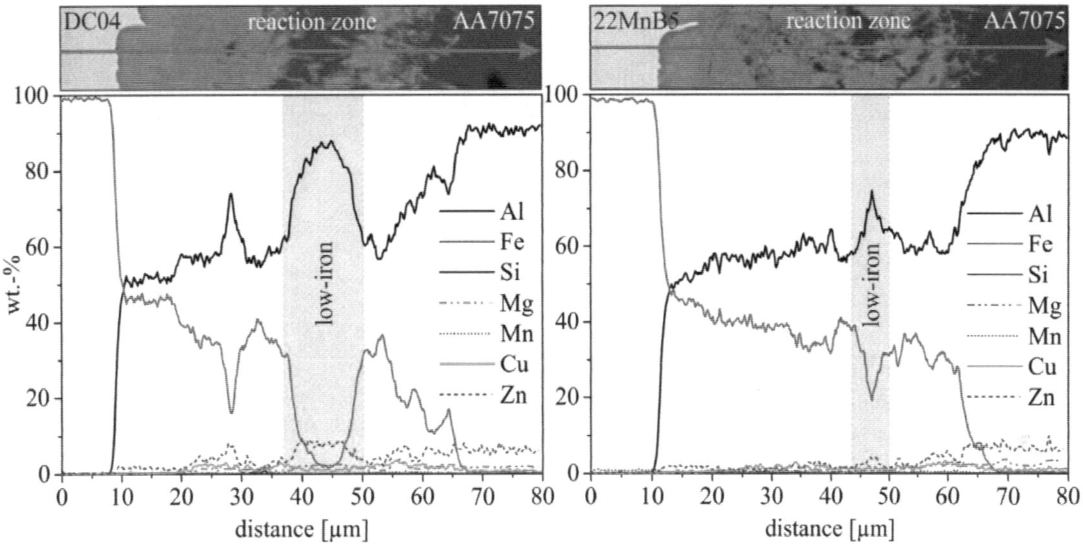


**Figure 5.** SEM/EDS analysis of the intermetallic compounds after heat treatment time of t = 5 min at the interface between AA7075 coating and a) DC04 steel and b) 22MnB5 steel

According to EDS analysis, a thin layer with a thickness of about $s_{IMP}$ = 8 µm at the steel interface has a composition corresponding to that of $FeAl_2$ or $Fe_2Al_5$, while the composition of the thicker one with a thickness of about $s_{IMP}$ = 20 µm indicates the formation of $Fe_4Al_{13}$ or $FeAl_3$. These results are in accordance with the data reported in the literature [6,10,11]. Some Zn content of about m(Zn) = 1,4 wt.-% could be observed in the reaction zone, however, the content of Zn and other key alloying elements such as Cu and Mg is significantly higher in the aluminum matrix. Thus, only aluminum and iron play an essential role in the formation of the intermetallic compounds. The main differences in chemical composition of both steel grades are related to the Mn and Si content. In Figure 5b it can be seen that the Mn content in the steel 22MnB5 remains at the same level as in delivery state and no significant diffusion into the reaction zone could be detected.

The growth kinetics of the intermetallic compounds was determined from the measurements of the reaction zone thickness. In Figure 6 the average thickness of the intermetallic phase after the heat treatment is plotted as a function of the square root of time. The linear regression has been used to describe the relationship between the thickness of the reaction zone and the square root of the heat treatment time. Commonly, this relation can be described by the parabolic law as for example in [5,7,12]

$$d = k \cdot t^{0.5} \quad (1)$$

In accordance with equation (1), the slope of the lines corresponds to the parabolic coefficient k. The growth kinetics of the intermetallic compounds for steel DC04 with k = 2.31 µm/s$^{0.5}$ is slightly faster than for the steel 22MnB5 with k = 1.73 µm/s$^{0.5}$. In the study of Springer et al., the reaction kinetics of the formation of AlFe and AlFeSi intermetallic compounds has been investigated for stir welding of steel DC01 and AlSi5 [12]. The measured coefficients k in this study were k = 0.201 µm/s$^{0.5}$ for the heat treatment temperature of T = 500 °C and k = 2.303 μm/s$^{0.5}$ for the heat treatment temperature of T = 600 °C.

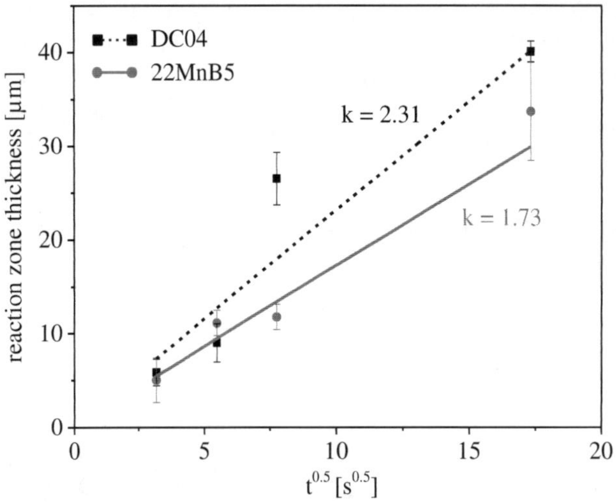


**Figure 6.** Mean thickness of the reaction zone between steel and AA7075 coating after the heat treatment at the temperature T = 550 °C

### *3.3. Hardness measurement*

Nanoindentation measurements were performed to further characterize the reaction zone at the interface between steel and coating. The deep drawing steel DC04 exhibits hardness values of HU = 2.09 ± 0.12 GPa. For both substrate materials the maximal hardness in the reaction zone was measured in the intermetallic layers adjacent to the steel: HU = 11.30 ± 1.13 GPa for steel 22MnB5 and HU = 12.73 ± 0.30 GPa for steel DC04. As expected, in the region with a low iron content between the two intermetallic layers the hardness was considerably lower. In the region of the reaction zone bordering on the AA7075 coating, the hardness of the intermetallic layers was HU = 7.71 ± 0.89 GPa in case of steel DC04 and HU = 5.22 ± 1.30 GPa in case of steel 22MnB5. The results of the hardness measurement reflect the structure and morphology of the reaction zone observed by means of SEM/EDS analysis. The measured hardness of the intermetallic layer assumed as $Fe_2Al_5$ with a hardness of around HU = 12 GPa is a little bit higher as commonly reported in the literature, for example by Ogura et al. with HU = 7.8 GPa [14] or by Dybkov et al. with HU = 8.9 ± 0.9 GPa [11]. This can be possibly explained by the presence of alloying elements and in particular Zn as a substitute for Al in the $Fe_2Al(M)_5$ intermetallic phase with M = Zn, Cu or Mg.

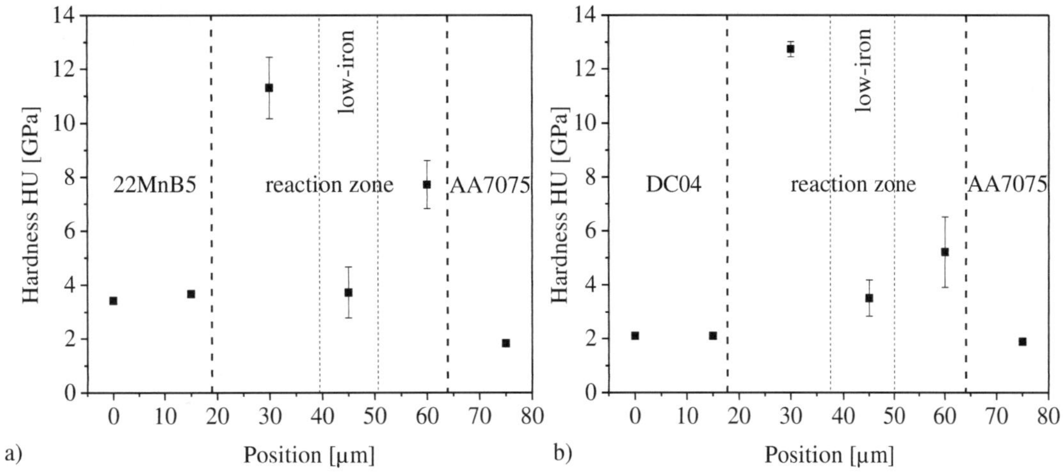


**Figure 7.** Hardness of the reaction zone after heat treatment time of t = 5 min at the interface between a) steel 22MnB5 and AA7075 coating, b) steel DC04 and AA7075 coating

## 4. Conclusions

In the current study, the formation of the reaction zone between cold gas sprayed AA7075 coating and steel DC04 or steel 22MnB5 during the heat treatment at T = 550 °C was experimentally examined. It could be shown that despite different chemical composition of both steel grades, the structure and morphology of the intermetallic layers at the interface between steel and coating were nearly identical. The following findings can be reported based on the results of the current study:

- For both substrate materials, the intermetallic phases of $FeAl_2$, $Fe_2Al_5$, $Fe_4Al_{13}$ or $FeAl_3$ detected by means of SEM/EDS analysis were found to be the main components of the reaction zone between steel and AA7075 coating.
- The reaction zone consists of two AlFe intermetallic layers separated by a thin zone with a low-iron content. Furthermore, pores occurred at the interface between the reaction zone and coating, possibly due to the Kirkendall effect.
- Growth kinetics of the intermetallic compounds for steel DC04 with parabolic coefficient of growth $k = 2.31\ \mu m/s^{0.5}$ was slightly faster than for the steel 22MnB5 with $k = 1.73\ \mu m/s^{0.5}$.
- For steel 22MnB5, an occurrence of the crack formation in the reaction zone after cooling from heat treatment temperature was distinct compared to the steel DC04. It can be a crucial factor for the reliability of high pressure die cast Al/steel hybrids.

## 5. References

[1] Nisslé, S., Dassler, S., Noster, U., Mundl, A. (2008). Verbundguss, Neue Auslegungsmöglichkeiten für Gussteile durch Funktionsintegration oder lokale Verstärkung. Tagungsband der 5. Ranshofener Leichtmetalltage. LKR-Verlag, Ranshofen, 35–44.
[2] Fent, A. (2006). Magnesium-Aluminium-Verbundkurbelgehäuse- Einsatz von Neutronen in der Bauteilentwicklung. Garching.
[3] Oberschelp, C. (2012). Hybride Leichtbaustrukturen für den Karosseriebau - gusswerkstofforientierte Anwendungsuntersuchungen für das Druckgießen. Diss., Gießerei-Inst. der RWTH, Aachen.
[4] Variostruct: multi-material-components. http://www.variostruct.de/?L=1.

[5] Fragner, W., Papis, K., Wosik, J., Uggowitzer, P. J. (2008). Herausforderungen und Lösungsmöglichkeiten bei der Herstellung von Verbundgussteilen. Vortragstext, 5. Ranshofener Leichtmetalltage 2008, LKR-Verlag, herausgegeben von U. Noster, F. Riemelmoser u. P. J. Uggowitzer, Ranshofen, 75–85.
[6] Kobayashi, S., Yakou, T. (2002). Control of intermetallic compound layers at interface between steel and aluminum by diffusion-treatment. Materials Science and Engineering: A **338**/1-2, 44–53.
[7] Ei-Mahallawy, N. A., Taha, M. A., Shady, M. A., Ei-Sissi, A. R., Attia, A. N., Reif, W. (2013). Analysis of coating layer formedon steel strips during aluminising by hot dipping in Al-Si baths. Materials Science and Technology **13**/10, 832–840.
[8] Nazari, K. A., Shabestari, S. G. (2009). Effect of micro alloying elements on the interfacial reactions between molten aluminum alloy and tool steel. Journal of Alloys and Compounds **478**/1-2, 523–530.
[9] Springer, H., Kostka, A., dos Santos, J. F., Raabe, D. (2011). Influence of intermetallic phases and Kirkendall-porosity on the mechanical properties of joints between steel and aluminium alloys. Materials Science and Engineering: A **528**/13-14, 4630–4642.
[10] Windmann, M., Röttger, A., Hahn, I., Theisen, W. (2017). Mechanical properties of $Al_XFe_Y$ intermetallics in Al-base coatings on steel 22MnB5 and resulting wear mechanisms at press-hardening tool steel surfaces. Surface and Coatings Technology **321**, 321–327.
[11] Dybkov, V. I. (1990). Interaction of 18Cr-10Ni stainless steel with liquid aluminium. J Mater Sci **25**/8, 3615–3633.
[12] Springer, H., Kostka, A., Payton, E. J., Raabe, D., Kaysser-Pyzalla, A., Eggeler, G. (2011). On the formation and growth of intermetallic phases during interdiffusion between low-carbon steel and aluminum alloys. Acta Materialia **59**/4, 1586–1600.
[13] Springer Materials. http://materials.springer.com/ (letzter Zugriff am 23.12.2017).
[14] Ogura, T., Ueda, K., Saito, Y., Hirose, A. (2011). Nanoindentation Measurement of Interfacial Reaction Layers in 6000 Series Aluminum Alloys and Steel Dissimilar Metal Joints with Alloying Elements. Mater. Trans. **52**/5, 979–984.

**Acknowledgments**
This research is supported by the German Research Foundation (DFG) under contract No. BO 1979/39-1. Special thanks are addressed to our project partners from the institute of metal forming (IBF) and foundry institute (GI) from RWTH Aachen University.

IOP Conf. Series: Materials Science and Engineering **373** (2018) 012014 doi:10.1088/1757-899X/373/1/012014

# Spray pattern analysis in TWAS using photogrammetry and digital image correlation

**W Tillmann[1], H G Rademacher[2], L Hagen[1], M Abdulgader[1] and M El Barad'ei[1]**

[1] Lehrstuhl für Werkstofftechnologie, TU-Dortmund, Germany
[2] Institut für Forschung und Transfer, RIF e.V., Germany

**Abstract.** In terms of arc spraying processes, the spray plume characteristic is mainly affected by the flow characteristic of the atomization gas at the nozzle inlet and intersection point of the wire tips, which in turn affect the particle distribution at the moment of impact when molten spray particles splash onto the substrate. With respect to the route of manufacturing of near net-shaped coatings on complex geometries, the acquisition of the spray patterns is pressingly necessary to determine the produced coating thickness. Within the scope of this study, computer fluid dynamics (CFD) simulations were carried out to determine the distribution of spray particles for different spray parameter settings. The results were evaluated by three-dimensional spray spot analyses using an optical measurement based on photogrammetry and digital image correlation. The optical measurement represents a promising and much faster candidate to measure spray patterns compared to the tactile measurement system but with an equal accuracy. For given nozzle configurations and spray parameter settings, numerous spray patterns were examined to their shape factors, demonstrating the potential of an online analysis, which encompasses a "fast sample loop" and a data processing system to generate a three-dimensional surface of the spray spot profile.

## 1. Introduction

In terms of twin wire arc spraying (TWAS), the in-flight particle conditions and gas flow characteristics across the expanded spray plume get mainly determined by the flow characteristics of the atomization gas at the nozzle inlet and intersection point of the electrode tips, as well as the electrode phenomena during atomization [1-6]. Thus, in addition to electrodynamic aspects, the presence of the electrodes (wires) and contact-tubs in the gas flow has an enormous impact on the resulting flow regimes. Moreover, the configuration of the nozzle design (i.e. nozzle inlet, or air cap) affect both, the resulting flow regimes at the electrode tips, and the gas flow characteristics across the expanded spray plume. In order to optimize the nozzle design in a TWAS process, numerous studies have focused on the flow regimes by means of computational fluid dynamic (CFD) simulations using different nozzle configurations [6-13]. According to these studies, the major objective was focused on the gas flow characteristics inside and outside the spray gun, as well as gas-particles interactions. Although many numerical studies have been conducted to investigate the flow characteristics, there have been only few examinations on the spray plume divergence and spray pattern. Studying different nozzle concepts, Hussary et al [11] evaluated different designs with respect to the trajectories of spray particles, velocity distributions and the deposit distribution. It was found that the spray plume divergence decreases significantly with an increasing atomization gas pressure. The authors also reported that the incident angle of the secondary gas, since it used for the setup, affect the spray plume divergence significantly. Same results were found in [12, 13]. According to Gedzevicius et al. [12], the supply of a secondary gas, depending on the local injection, has not only a focusing effect, but also

IOP Conf. Series: Materials Science and Engineering **373** (2018) 012014 doi:10.1088/1757-899X/373/1/012014

leads to a further disintegration (secondary atomization) of the molten spray particles. However, the authors [13] previously demonstrated that the spray plume divergence get distinctly influenced by the nozzle design used for spraying.

With respect to complex geometries, there is a great need to apply near net-shape coatings in order to minimize costs for post-processing. Nowadays, complex geometries obtain component features such as various radii. In this respect, the spray pattern plays an important role as a focused spray plume can lead to a controlled coating deposition. Previous studies [14, 15] already emphasized the relevance of the acquisition of three-dimensional spray patterns (footprints) in order to assess the particle distribution at the moment of impact onto the substrate. The footprints can be described by a mono-modal spherical-shaped or irregular-shaped distribution. Moreover, the footprints can be characterized by the spray pattern area, diameter, eccentricity or circularity. It is therefore necessary to determine these characteristics in order to evaluate suitable spray process parameters, and nozzle designs, respectively. Horner et al. [15] studied the spatial characteristics of deposited spray patterns such as spray area, eccentricity, flatness, and deposition rate for various spray parameters. It was experimentally shown that the spray area get mainly influenced by the atomization gas pressure. Thus, a large-scaled spray area results from low atomization gas pressure and low secondary gas pressure. In contrast, the spray pattern eccentricity was most dependent on secondary gas pressure, whereas the spray pattern flatness was most dependent on atomization gas pressure and arc current used during spraying. To conclude, the study demonstrated that an amendment on the operating conditions in TWAS leads to a great change of the resulting footprint.

With respect to the aforementioned studies [14, 15], the footprints were produced on large-scaled metal plates. A high resolution surface profiling system was used to measure the three-dimensional footprints in order to characterize the particle concentration at the moment of impact onto the substrate. As a result, a great experimental expenditure is necessary to scrutinize the footprints experimentally.

Within the scope of this study, CFD simulations were conducted to determine the distribution of spray particles for different spray parameter settings. The results were evaluated by three-dimensional footprint analyses using an optical measurement based on photogrammetry and digital image correlation. The footprints were analysed for various spray operating conditions examining the potential of an online analysis, which encompasses a "fast sample loop" and a data processing system to generate a three-dimensional surface of the spray spot profile.

## 2. Computer-Fluid-Dynamics (CFD) Simulation

The 3D-CFD analysis is performed inside the spray gun and for 110 mm standoff distance along the spraying axis. An inlet port is assigned for the entrance of the atomization gas into the spray gun as shown in figure 1. The function of the two connecting tubs, which are connected to the electrode ports is to charge the wires electrically to different poles (anode and cathode). The wires are intersecting at the back of the spray gun, as the connecting tubs are inserted in two intersecting channels. The atomization gas is directed towards the arc zone.

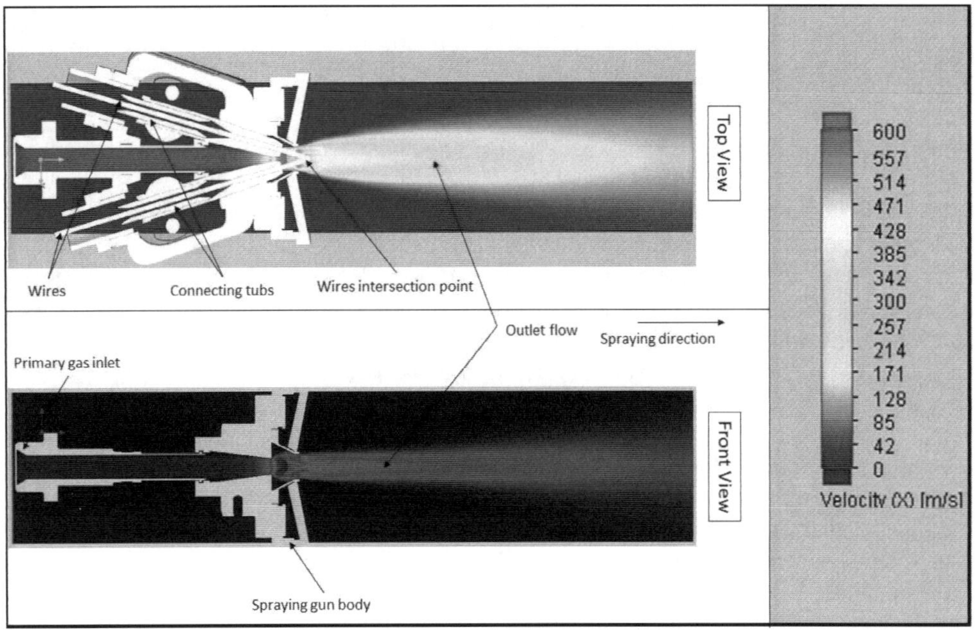


**Figure 1.** Section views of a 3D-CFD model of the twin-wire arc spraying (TWAS) process showing the output gas velocity

The internal gas flow is shaped according to the outer solid surfaces of the assemblies inside the spray gun, and the external flow analysis is only bounded by the defined computational domain boundaries. For shorter computation time, the brick cells were used for gridding the computational domain. But for achieving higher simulation accuracy [6], finer grid size is used for the narrow channels inside the spray gun and along the centre core of the spray flow.

A TWAS was completely modelled by using the Flow Simulation Analysis tool in SolidWorks® software according to the following computational domain parameters, the input fluid pressures are 4, 6, 8, 10 and 12 bar air pressures and the solid surfaces of the models are assumed to be ideally adiabatic. The ambient initial conditions were assumed, and the standard k-ε turbulence model is used in the analysis. The used mesh settings are based on a prior mesh sensitivity convergence analysis to obtain the proper mesh setting, which ensures a compromise between the simulation accuracy and the required computational time.

The numerical simulation study is required for time and cost saving measures to estimate the particle concentration in the moment of impact on the substrate.

## 3. Experimental Setup

Rectangular (150 mm x 150 mm x 16 mm) C45 steel specimens were employed as the substrate material in order to produce spray pattern profiles. For the preparation of the coating deposition a 15 mm wide frame of the substrate was shielded from grit blasting and spraying during the manufacturing of the samples. This polished area has been used within the scanned profile as a reference height and to compensate the small rigid body rotations due to an imperfect placement of the specimen. After that, the surfaces were grit-blasted and cleaned in an ultrasonic ethanol bath.

The spray torch of the Durspray 450 arc spraying system (Durum Verschleißschutz, Germany; OEM: T-Spray) was carried by a robot type IRB 4600 (ABB, Sweden) to achieve a perpendicular alignment at the centre of the surface at the desired distance of 110mm. With regard to the feedstock material, an iron-based solid wire 316L Si (T-Spray, Germany) was used. The PGP was varied in a range as seen above. No SG was used during the study as interactions (flow characteristics, secondary atomization behaviour) would become more complex. For the experimental setup, the voltage and current were kept on a constant level (voltage = 34 V, current = 110 A, spray-time 3s).

After applying the coating for 3 seconds the contour of the specimens was measured with a coordinate measuring machine type Prismo Vast (Zeiss, Germany). The measuring accuracy of this machine is about 2µm and a maximum of 200 measuring points per second can be recorded. In the present case, approximately 1.7s were required per measurement point due to larger probing distances. One hour was needed to measure one footprint with 2120 points. Also, the transport and mounting of the specimens and configuration of the machine were time consuming.
The shielded field was used to define a plane as the zero level for reference. Through the grit-blasting, the start level for the coating is about 0.075mm to 0.15mm below the zero level. The spatial resolution is essentially not only limited by the equidistant measuring grid, but also by the 0.5 mm sapphire ball-tip. The measurements used two different measuring grids. In a range of 18x18mm around the centre of the specimen, on which the TCP of the coating robot was aligned, a grid of 0.5mm x 0.5mm was used. For the remaining area (100mm x 100mm) a grid of 4mm x 4mm was used.
As an alternative, an optical measuring system was used. The optical measuring method is based on the principle of photogrammetry. The footprint is synchronously photographed by two Nikon D800E SLR cameras, which were mounted on a rail. For calibration, a special checkerboard pattern was used. The purpose is to determine the properties of the camera lenses, as well as the alignment of the cameras to each other and thus enable a true-to-scale measurement. The digital image correlation (DIC) is used to assign an area of the footprint in the two camera images. For this method, the surface must have a high-contrast random pattern. This is given by the coating itself in good approximation.
The software used for this measurement is Istra4D (Dantec Dynamics, Denmark). The calculated surface contour can be exported as measuring coordinates comparable to a coordinate measuring machine. But these values refer to the camera coordinate system. To make them comparable a best plane fit transformation must be performed. There is a source of error in this case, because this step is not only applied on the protected edge, but also on the entire considered sample.

## 4. Results

To analyse the measuring data the software OriginLab 2017 (OriginLab Corp., USA) and for comparison of the two different measurement systems the software GOMinspect (GOM, Germany) was used.
The simulation does not give direct information about the footprint but a distribution of the spray plume velocity. This is equated to a footprint shape. Figure 2 shows a comparison of the simulated data and the tactile measurement of a coated footprint with the same parameters. The top view looks similar, but the side view shows a significant dip. Accordingly, the differences between simulation and measurement become clear. The simulation can give a hint to choose appropriate parameters but a measurement is always necessary if the actual footprint shape is necessary. The circularity is a property which can be extracted from this data. Figure 3 shows the comparison of the circularity values of the simulation data and the tactile measurement. Varying the PGP between 4 and 12 bar directly influences the circularity of the footprint. The tendency is equal with both datasets.
The main disadvantage of the tactile measurement systems is the long measurement time, and that the result accuracy depends on the stylus geometry and the measurement procedure. The values between the measuring points are interpolated. Therefore especial high gradients as they occur with the crater-like structure of the footprints produced by the TWAS process lead to big deviations. In order to reduce this influence, the measuring grid has to be reduced, which would increase the measuring time considerably. To reduce this disadvantage a new approach is used. A camera based measurement system can capture the whole footprint at once. Also the measurement point density is clearly larger, as shown in figure 4. Both footprints are comparable, but there is a little difference in the absolute value. This deviation is a result of the previously mentioned error of the calculated reference plane and by the deviation due to the interpolation between the coarse densities of the measuring points of the tactile measuring system. Figure 5 shows an error map calculated with the software GOMinspect. The influence of gradients and an offset-error due the reference plane is recognizable.

IOP Conf. Series: Materials Science and Engineering **373** (2018) 012014 doi:10.1088/1757-899X/373/1/012014

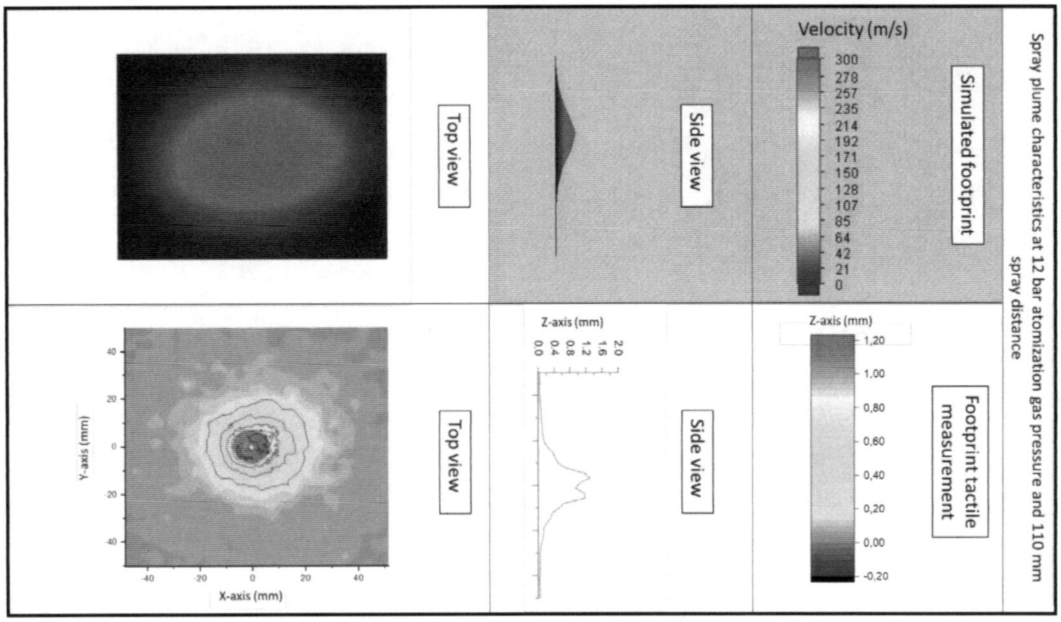


**Figure 2.** Comparison between a simulation footprint and a digitalized footprint profile by using a 3 D tactile measurement system

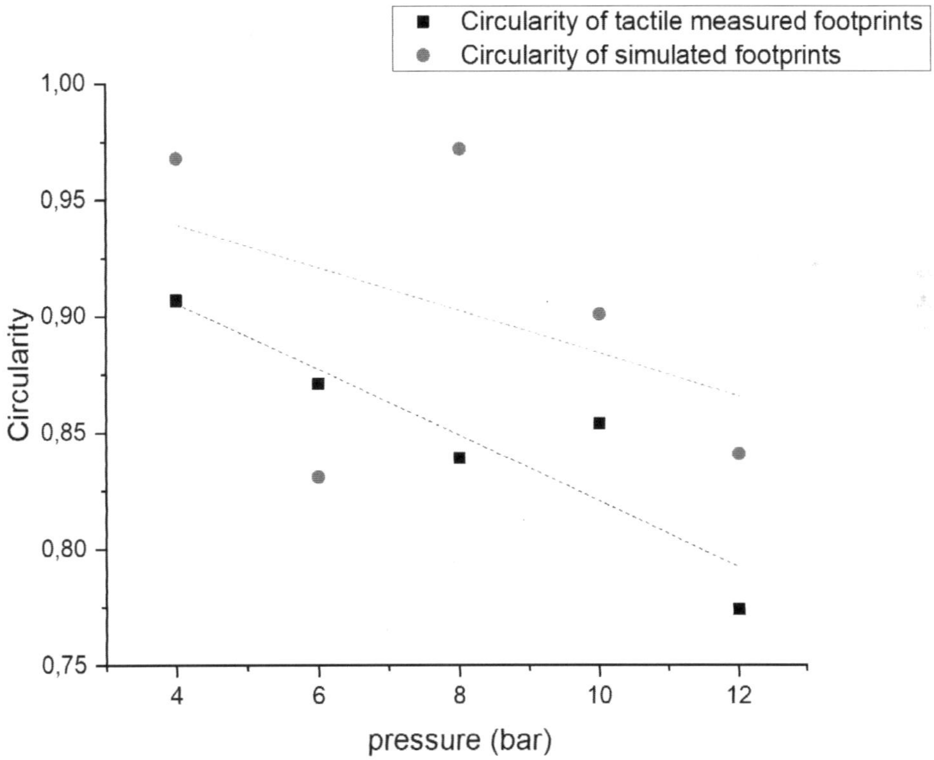


**Figure 3.** Comparison of the circularity of the simulated and the tactile measured footprints

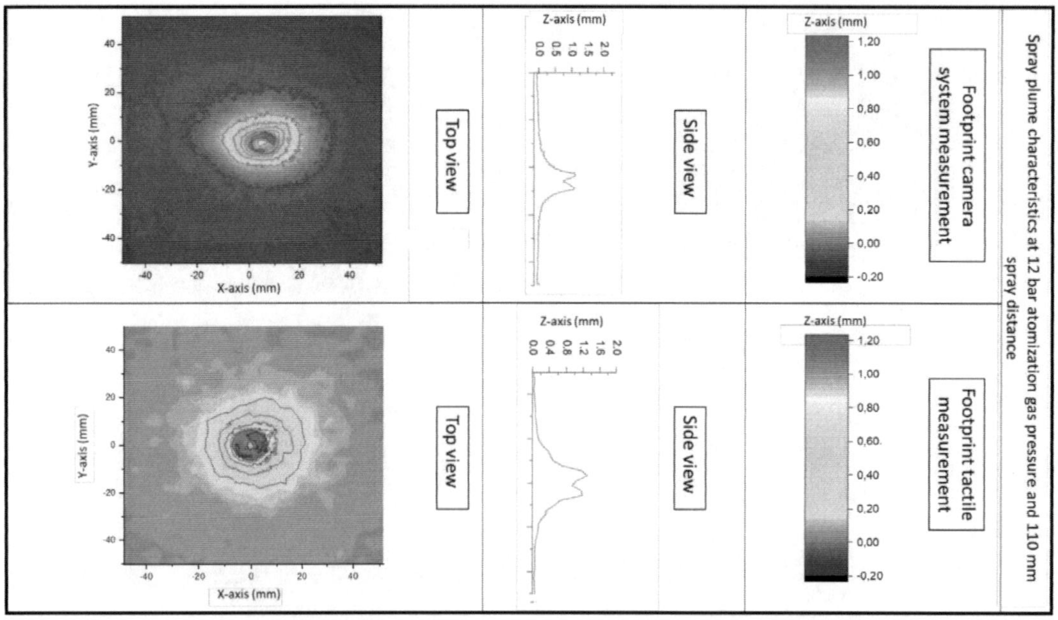


**Figure 4.** Comparison between a digitalized footprint profile by the camera system and a digitalized Footprint profile by a 3 D tactile measurement system

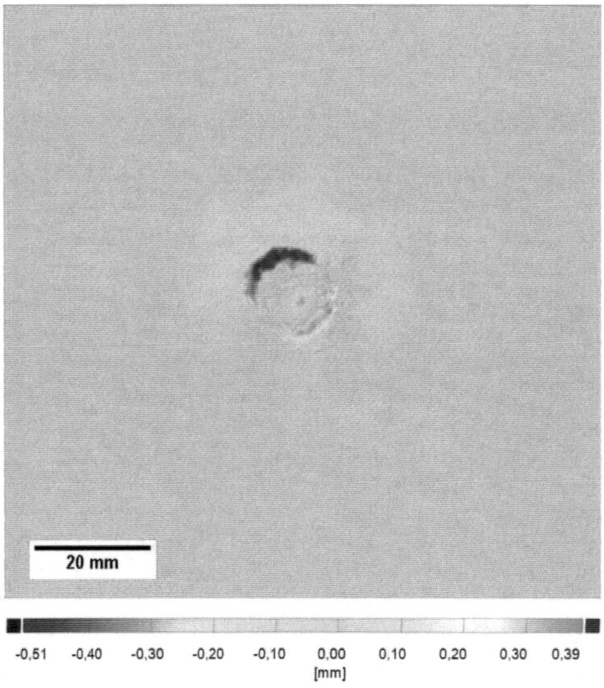


**Figure 5.** Measurement deviation between the tactile and the optical system

Footprints with smoother contour produced, for example, by the HVOF process, are having a much lower error in the same measuring setup. But for the circularity criteria, the absolute values are not important. A direct comparison of the simulation, tactile and optical measurement shows this in figure 6.

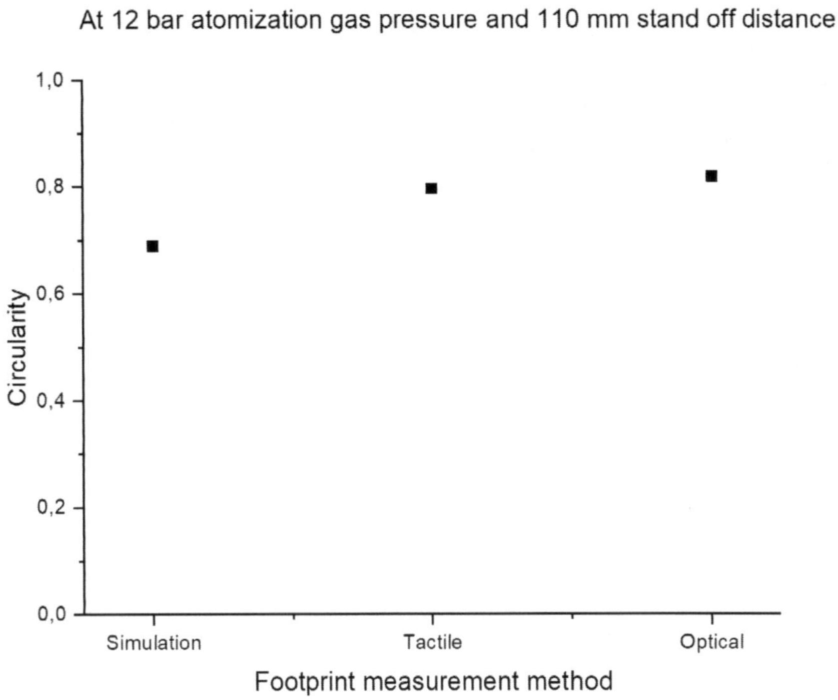


**Figure 6.** Comparison of the calculated circularity of the footprint

## 5. Conclusion and Outlook

In this paper a CFD-simulation is shown as a possibility to evaluate the influence of a pressure variation to the spray plume with a given setup. The resulting footprints were verified using tactile measurements by comparing the circularity value. A big disadvantage of the tactile measurement is the time-factor. For a better detection of small edge areas, a finer grid would be desirable. But this would require a significantly longer measuring time. A new and innovative measurement system is introduced in order to overcome these disadvantages. The new camera-based system is significantly faster than the tactile measurement system, and the system can obtain very close results to the tactile measurement. Further works shall deal with the improvement of the optical measurement method and an optimization of the simulation.

## 6. References

[1] Wilden J and Wank A and Schreiber F 2000 Wires for Arc- and High Velocity Flame Spraying - Wire Design, Materials and Coatings Properties (*Thermal Spray: Surface Engineering via Applied Research*) pp. 609-617

[2] Pourmousa A and Abedini A and Mostaghimi J and Chandra S 2004 Particle Diagnostics in Wire-Arc Spraying System (*Thermal Spray: Advances in Technology and Application*) pp. 962-968

[3] Planche M P and Liao H and Coddet C 2004 Relationships between In-flight Particle Characteristics and Coating Microstructure with a Twin Wire Arc Spray Process and Different Working Conditions (*Surf. Coat. Technol.* 182) pp. 215-226

[4] Jandin H and Liao H and Feng Z Q and Coddet C 2003 Correlations between Operating Conditions, Microstructure and Mechanical Properties of Twin Wire Arc Sprayed Steel Coatings *Mater. Sci. Eng.* 349 pp. 298-305

[5] Wang X and Zhuang D and Pfender E and Heberlein J and Gerberich W 1994 Effect of Atomizing Gas Pressure on Coating Properties in Wire Arc Spray *Thermal Spray Industrial Applications* pp. 587-592
[6] Tillmann W and Abdulgader M and Anjami N and Hagen L 2015 Studying the Effect of the Air-Cap Configuration in Twin-Wire Arc-Spraying Process on the Obtained Flow Characteristics Using Design of Experiment Oriented Fluid Simulation *J. Therm. Spray Technol.* (vol 24) pp. 1-2, 46-54, DOI: 10.1007/s11666-014-0183-1
[7] Chen Y and Liang X and Liu Y and Xu B 2009 Numerical Analysis of the Effect of Arc Spray Gun Configuration Parameters on the External Gas Flow *J. Mater. Process. Technol.* 209(18-19), pp. 5924-5931
[8] Chen Y and Liang X and Wei S and Chen X and Xu B 2011 Numerical Simulation of the Twin-Wire Arc Spraying Process: Modeling the High Velocity Gas Flow Field Distribution and Droplets Transport *J. Therm. Spray Technol.* (vol 21(2)) pp. 263-274
[9] Toma S L and Bejinariu C and Baciu R and Radu S 2013 The Effect of Frontal Nozzle Geometry and Gas Pressure on the Steel Coating Properties Obtained by Wire Arc Spraying *Surf. Coat. Technol.* (vol 220) pp. 266-270
[10] Bolot R and Planche M P and Liao H and Coddet C 2008 A three-dimensional model of the wire-arc spray process and its experimental validation *Journal of materials processing technology* (vol 200(1)) pp. 94-105
[11] Hussary N and Schein J and Heberlein J 1999 Control of jet convergence in wire arc spray systems *Tagungsband Conference Proceedings* ed E Lugscheider and RA Kammer pp. 17-19
[12] Gedzevicius I and Valiulis A V 2006 Analysis of Wire Arc Spraying Process variables on coatings properties *Journal of Materials Processing Technology* (vol 175) pp. 206-211
[13] Gedezevicius I and Bolot R and Liao H 2003 Application of CFD Wire-Arc Nozzle Geometry Improvement *Proceedings of the International Thermal Spraying Conference (ITSC,* Orlando/USA) pp. 977-980
[14] Tillmann W and Hagen L and Abdulgader M and Rademacher H G and Schmidt A and Müller H and Wiederkehr T 2014 Investigations of Technological Developments in Nozzle Design for Twin Wire Arc Spraying by Means of Spray Plume Characteristics *Proceedings of the International Thermal Spraying Conference (ITSC,* Barcelona / Spain) pp. 455–46
[15] Horner A L and Hall A C and McCloskey J F 2015 The Effect of Process Parameters on Twin Wire Arc Spray Pattern Shape *Coatings* (vol 5) pp. 115-123.

**Acknowledgments**
Parts of the investigation were conducted at the Institute of Machining Technology of the TU Dortmund. In this regard, we thank Tobias Siebrecht for measuring the samples with the tactile coordinate measuring machine Zeiss Prismo Vast.
The authors gratefully acknowledge the financial support of the program "Zentrales Innovationsprogramm Mittelstand" (ZIM) by the Federal Ministry of Economic Affairs and Energy (BMWi) on the basis of a decision by the German Bundestag within the research project "Funkt. Oberflächen – RyHoDo / Düsenkonfiguration und Validierung Sekundärgasdüsenkappen".

# Design of high strength polymer metal interfaces by laser microstructured surfaces

**P Steinert[1], A Dittes[2], R Schimmelpfennig[1], I Scharf[2], T Lampke[2] and A Schubert[1]**

[1] Professorship Micromanufacturing Technology, Chemnitz University of Technology, Chemnitz, Germany
[2] Institute of Materials Science and Engineering, Chemnitz University of Technology, Chemnitz, Germany

**Abstract.** In the areas of automotive, aeronautics and civil structures, lightweight construction is a current and a future need. Thus, multi material design has rapidly grown in importance, especially hybrid materials based on fiber reinforced plastics and aluminum offer great potential. Therefore, mechanical interlocking is a convenient way of designing the interface. Laser structuring is already used to generate a variety of surface topographies leading to high bond strengths. This paper investigates different laser structures aiming on highest joint strengths for aluminum and glass fiber reinforced polyamide 6 interfaces. Self-organizing pin structures comprised by additional micro/nano features as well as drilled hole structures, both ranging on the micrometer range, are compared to corundum blasting as a standard method for surface conditioning. For the presented surface structures, thermal joining and ultrasonic assisted joining are regarded towards their potential for an optimum joint design.

## 1 Introduction

Lightweight construction is a supreme demand for a wide range of applications in order to save energy by weight reduction. High lightweight potential can be tapped through combining miscellaneous materials into hybrid composites. By an appropriate material selection, advantages such as low density, high bending resistance, high load-capacity and especially low weight can be achieved, for example through the use of sandwich structures [1]. In the automotive industry, especially polymer-metal hybrids (PMH) are used due to their beneficial strength-to-weight ratio and their easy processability.

Modern strategies such as adhesive bonding or micromechanical interlocking are promising due to their integrability into cost efficient series production. Adhesive bonding is based on generating mechanical and chemical bonds [2]. Mechanical interlocking is achieved by generating macroscopic, microscopic as well as nanoscopic surface structures at the metal adherend using mechanical, chemical, electrochemical or physical processes [3, 4]. In addition, the form fit can be combined with improved chemical interaction through adapted surface chemistry. For this, an increased formation of primary and secondary bonds with the polymer adherend is useful [5]. Thus, laser structuring may lead to adhesion promotion by multiple mechanisms, such as providing geometrical undercuts, increasing the surface area resulting in an increased number of chemical bonding sites and a suitable change in the surface chemistry.

Even there is no accepted scientific knowledge on the contribution of the elementary adhesion mechanisms to the increase in bond strength in aluminum-plastic joints, laser structuring is used in order to achieve interfaces for highest mechanical loading. Hence, it is necessary to identify appropriate structures and the corresponding manufacturing strategies to fulfill practical demands.

## 2 Laser surface modification

Laser structuring allows a high design flexibility in terms of structuring the surface. Different scientific working groups could provide valuable findings in the area of surface modification by laser ablation up to now. It was shown for example that microscopic pin structures can be realized on the surfaces of different metallic materials as for example steel, titanium, cooper as well as aluminum [6]. In this context a self-organized process that is characterized by local temperature differences and cooling speeds leads to the formation of micro pins with dimensions in the nanometer or micrometer range [6, 7].

Potential for the application of these microscopic structures was identified primarily for the modification of surface wettability and for the interface design of technical bonds [8]. By the help of periodic micro pins on titanium surfaces manufactured by a nanosecond laser system, a significant increase in adhesive bond strength compared to planar samples could be confirmed by Baburaj et al. [9]. The authors found that the excellent performance of the microstructured samples was caused by increased micromechanical interlocking, a significant surface area increases by an order of magnitude as well as the structure deformation under mechanical load and thermal expansion [9]. A modified surface chemistry with submicron-sized oxide particles was also identified as favorable for the increase of the bond strength [9, 10]. However, the comparison to conventional methods for surface preparation was not considered so far.

A further efficient strategy for surface preparation of bonds, the suitability of bore arrays by single pulse and percussion drilling was studied. By the help of this approach, only a slight increase of the strength of metal-ceramic bonds could be verified in comparison to abrasive blasting as surface preparation method [11, 12]. In the case of metal-metal adhesive joints [13] and metal-plastic joints [14] an increase of bond strength was documented for an increasing hole number per area. This can be traced to the increasing surface area that supports adhesion at the interface. By using laser drilled holes in the surfaces the sufficient filling of the structures was one of the main challenges. Compressed air in the microstructures can thereby weaken the connection between substrate and adhesive or counter body, respectively [14, 15].

## 3 Experimental design: generation of microstructures and joining processes

### *3.1 Materials*

PMH joints were prepared from the aluminum alloy EN-AW-6082 T6 artificially aged (Gemmel Metalle) and the fiber reinforced plastic (FRP) Durethan BKV30 (Lanxess). The FRP is reinforced by glass short fibers using a fiber mass content of 30 %. The materials are presented in Table 1. Prior joining, the surfaces of the aluminum adherends were structured through different treatment methods (see 3.2) and cleaned ultrasonically assisted in ethanol.

Table 1: Material data

| Material | Elastic modulus [GPa] | Yield Strength [MPa] | Ultimate Strength [MPa] | Elongation to failure [%] |
|---|---|---|---|---|
| Durethan BKV30 | 6.1 | - | 105 | 6 |
| EN-AW-6082 T6 | 70 | $R_{P0.2} = 250$ | $R_M = 300$ | 10 |

### *3.2 Surface microstructuring*

*Laser ablation*

For the experimental work the aluminum alloy EN-AW-6082 was used. The ablation of the material was carried out by a nanosecond laser system from Spectra Physics® with the following specifications:

- Laser medium: Nd:$YVO_4$
- Wavelength: 532 nm
- Pulse duration: 10 ns
- Mean power: max. 13 W
- Focus diameter: ≈ 15 µm

During laser processing the material behavior depends significantly on the energy input into the surface. In order to characterize the ablation behavior depending on the intensity *I*, parameter experiments were carried out by the variation of the pulse frequency, the focus position, the pulse distance and the number of passes. In the preliminary parameter experiments the laser intensity *I* was varied in the range from about 1.0 J/cm² to 100 J/cm², calculated on the basis of manufacturers data sheet. As scanning strategy, line scanning with a line distance equal to the pulse distance was performed.

Figure 1: Selected examples showing ablated surfaces of EN-AW-6082 material, ablation with formation of micro pins, $I \approx 5.5$ J/cm² (A), ablation of micro holes, d ≈ 10 µm, h ≈ 22 µm (B)

According to the wide range of tested energy densities, a diverse material behavior can be observed. Beside the tendency that higher intensity leads to higher surface roughness and stronger oxidation of the ablated surface, the formation of micro pins with high aspect ratio could be verified for different parameter combinations. It was found that the conditions for the generation of the structures prevail in the range of a defined laser intensity of $I \approx 3 \ldots 6$ J/cm².

For the comparison of the wetting behavior and filling conditions during the joining processes, hole drilling was used as a second approach for the surface modification of the metallic parts. In this context, the microstructures were generated on two different geometrical scales. Small scaled holes were produced by percussion drilling with 10 pulses each. This resulted in structure dimensions shown in Figure 1B. The large scale holes were magnified by factor 10 and manufactured by spiral drilling, resulting in d ≈ 100 µm and h ≈ 120 µm. magnified by factor 10 and manufactured by spiral drilling, resulting in d ≈ 100 µm and h ≈ 120 µm. In both cases, the structure density, that denotes the portion of the structured area in the total area, was set to ≈ 11 %. A visual comparison between the manufactured surfaces with formation of micro pin structures and surfaces with defined micro holes is presented in Figure 1.

*Corundum blasting*

To reference the joint strengths achieved by the different laser structures, aluminum specimens were corundum blasted, since it is a widely used pretreatment for PMH joints. The aluminum specimens were structured by the blasting abrasive EK 024 (Wiwox) for 10 s. A pressure of 0.2 MPa was used and the abrasive was shot onto the aluminum surface under an angle of 70° from a distance of 0.3 m.

### *3.3 Joining procedures*

*Hot pressing*

The FRP and the aluminum were joined by a hot pressing process using a constant pressure of 0.4 MPa during the whole joining process, which was found to ensure satisfying penetration of the polymer melt into the laser generated microstructures of the aluminum adherend. Conductive heat transfer from a copper heating into the aluminum adherend and therefrom to the FRP was utilized. The temperature progress inside the aluminum bulk material next to the joining zone was measured by a thermocouple. A temperature ramp was applied until a maximum temperature of 205 °C was reached. Then, the heating was turned off, the temperature increased to a peak value of 230 °C and the specimens cooled down to 100 °C via air cooling. The specimens were subsequently conditioned to a normal climate according to DIN EN ISO 1110 [16] in order to adjust the moisture absorbed by the polyamide 6 based matrix prior mechanical testing.

*Ultrasonic joining*

An alternative method for the production of a joint between FRP and metallic materials such as aluminum is the ultrasonic assisted (US) joining with a vibration direction acting parallel to the surface of the workpiece. The normal force acts perpendicular superimposed on the vibration direction. An anvil fixes the lower joining part while the upper joining part is set in motion by the oscillating tool, the sonotrode (Figure 2). Due to the interaction of the oscillations of the tool system in the ultrasonic range and the acting normal force, the boundary friction between the joining partners leads to a localized heat input into the joining area and a melting of the thermoplastic polymer or the matrix occurs. The process times for generating a joint produced by means of ultrasonic assisted joining are highly dependent on the material combination used and the state of the surface structure of the metallic joining partner [17]. For the sample geometry used, the joining time is about 1.6 seconds. The following settings were used as machine parameters, the energy was 300 Ws, the amplitude of the oscillating system was 30 μm and the air pressure for the pneumatic press was 0.25 MPa, which corresponds to an acting joining pressure of 2,6 MPa at the joining area.

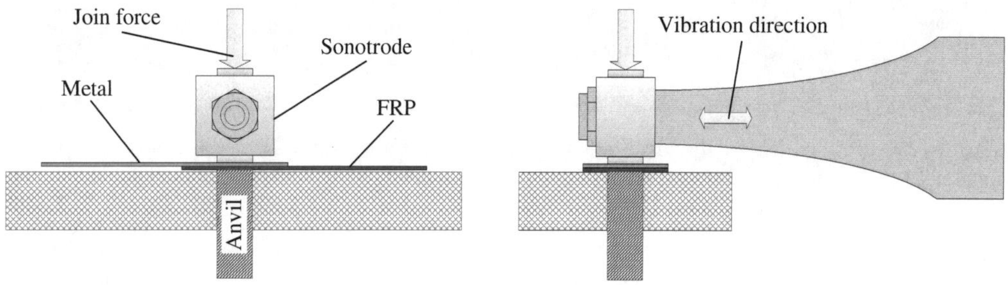


Figure 2: Principle of ultrasonic joining with vibration direction parallel to the workpiece surface and perpendicular acting joining force with arrangement of specimen [18]

*3.4 Static single leg tensile shear test*

To evaluate the PMH joint strengths achieved through the different surface structures and joining procedures, single leg lap shear tests according to DIN EN 1465 [19] were performed at a test speed of 1 mm/min. In contrast to the standardized geometry, the overlap length was reduced to 5 mm in order to enable testing of higher shear strengths, while still achieving interfacial failure due to an adapted stress distribution among the specimen [20]. The specimen geometry is depicted in Figure 3.

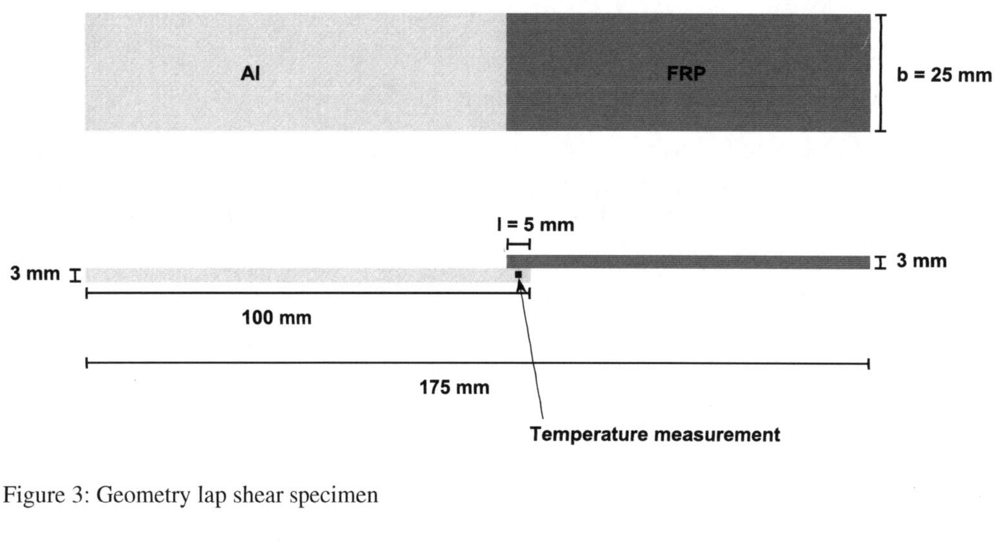


Figure 3: Geometry lap shear specimen

The maximum shear strength value is calculated following equation (1).

$$\sigma_{max} = \frac{F_{max}}{A} = \frac{F_{max}}{b \bullet l} \tag{1}$$

## 4 Results and discussion

### *4.1 Results with thermal joining*

Cross-sectional images were taken at the joining zones of the PMH specimens in order to evaluate the potential of each structure and the thermal joining process for generating a mechanically performant joint (Figure 4). The micro pin structures and corundum blasted surfaces were completely penetrated and surrounded by polymer. Above, no air cells are located at the interfacial zone. Thus, the quality of the joints for both structures is concluded to be high, even though an improvement is possible by positioning the reinforcing fibers closer to the geometrical structure to support interfacial load transfer. In contrast, the joint quality is lower for both hole structures. For the major (100 µm) holes, all of the holes were penetrated by polymer melt, but not completely filled. Thus, a gap exists between aluminum and the polymer, which is likely a consequence of locked in air and unequal thermal shrinkage of the adherends. No fibers penetrating the major holes were observed at the cross-sectional images. The joint quality of the minor (10 µm) hole structure is the least, since only a low percentage of the holes is penetrated by plastic melt. Overall, none of the three ablated structures was penetrated by glass fibers or an agglomeration of fibers next to geometrical features of the structures took place.

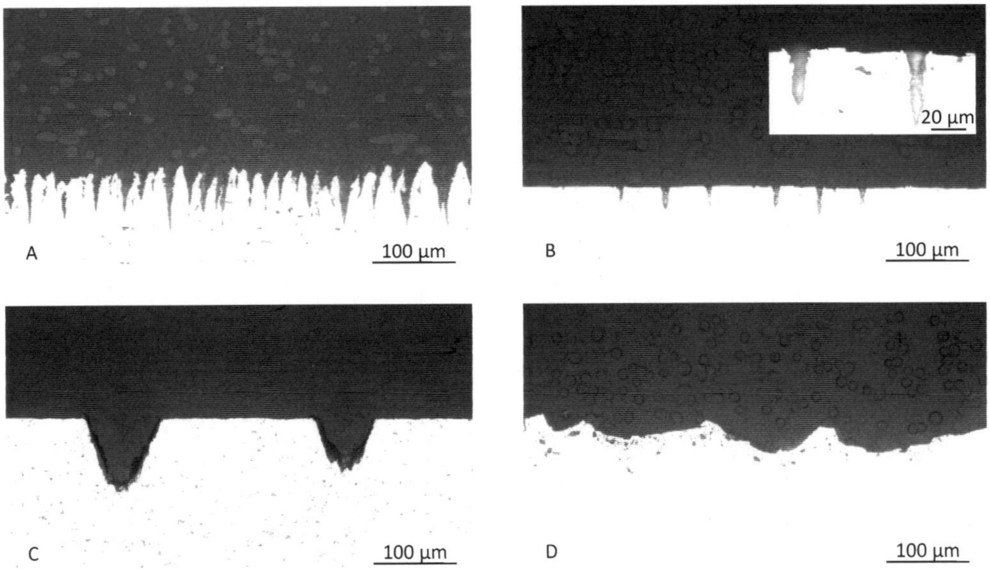


Figure 4: Cross-sectional images of thermally joined aluminum-FRP hybrids: micro pin (A), hole 10 µm (B), hole 100 µm (C) and corundum blasted structures (D).

The presented structures were designed in order to achieve high strength PMH interfaces. The micro pin structure fulfills this demand, since shear stresses above 30 MPa were sustained (Figure 5). Thus, a strong improvement compared to widely used corundum blasting (12 MPa) is achieved. An investigation of the fractured surfaces showed, that in general the micro pins remain undamaged, while cohesive failure inside the FRP took place at the joining zone (Figure 6A). Two microscopic fracture modes are stochastically distributed at the fractured surface, forming small microscopic zones each. No clear arrangement of the zones was observed. On one side, there are nearly flat areas of polymer residing between the micro pins. The flat areas end around the top height of the micro pins and no indication of remaining reinforcing fibers was observed. Thus, these areas likely developed through shear away of small volumes of locally unreinforced polymer. On the other side, crater-shaped polymer residues protrude out of the micro pin structure. It is assumed, that the local presence of glass fibers next to the micro pins led to increased local load transfer causing cohesive failure as well as polymer-fiber delamination.

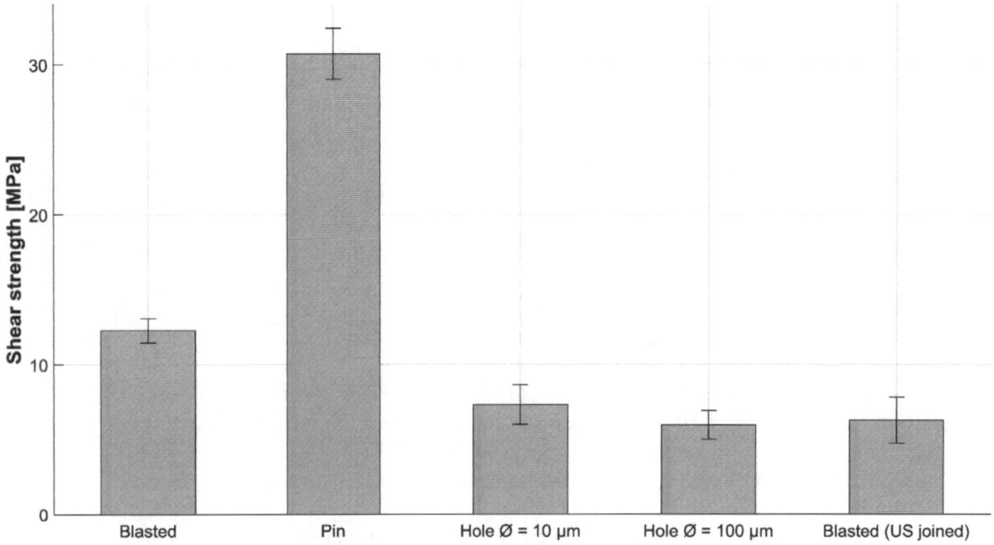


Figure 5: Maximum shear strengths for diverse surface geometries of aluminum-FRP hybrids.

By means of corundum blasting, rough geometrical features suitably orientated towards shear loads were generated. No polymer residues were observed at the fractured surface, since the polymer was drawn-off due to local plastic deformation. The joint strength achieved by the hole structures were comparatively low, which is a consequence of incomplete polymer penetration into holes and the low structure density. Thereby, the load transfer from aluminum into the FRP causes stress concentrations at the few undercut sides. Thus, cohesive failure takes place and led to sheared away polymer at the topside of the holes at relatively low stress levels (Figure 6B). In order to achieve higher joint strengths, the structural density has to be increased and fibers have to penetrate the holes to support the load transfer.

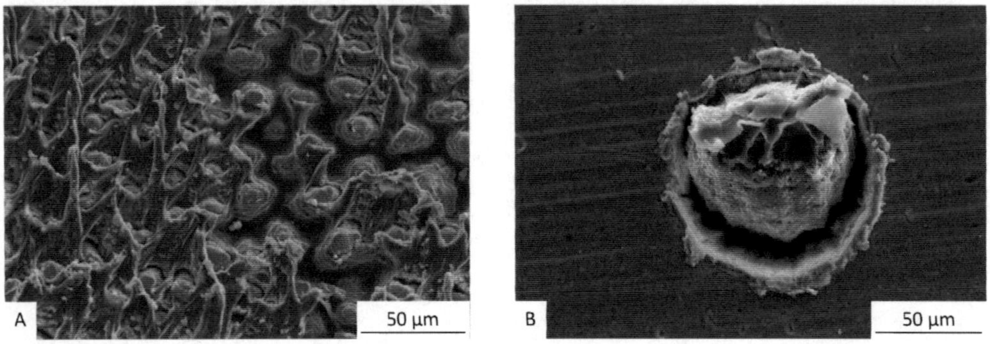


Figure 6: Scanning electron microscope images taken at the fractured surface of aluminum indicating plastic residues: micro pin (A) and major hole (100 µm) structure (B)

*4.2 Results with ultrasonic joining*

Using the ultrasonic assisted joining method leads to a local melting of the polymer matrix and an adhesive bond between polymer melt and metal surface. In combination with the pressure acting at the same time on the joining area, the polymer melt penetrates into the cavities of the microstructures.

Depending on the microstructure of the surface, the adhesive part of the bond strength can be increased, resulting in an increase in the tensile shear strength.

To achieve an increase in the load transfer through the joining zone, an involvement of the reinforcing fibers into the structures is required. Thus, a weakened interlayer only consisting of unreinforced polymer can be minimized or avoided and an increased joint strength can be achieved [17]. Figure 7 shows cross-sectional images of ultrasonically joined aluminum-FRP hybrids. It can be seen that the microstructures are filled with polymer matrix as well as for the thermal joining process. However, the adhesive bonding of the polymer to the metallic joining partner is not uniform. Furthermore, in the case of the selected joining parameters, a shearing of the micro pins occurred (Figure 7A). The generated fragments are placed in the cavities of the microstructure and partially close them. A similar effect can also be observed with the corundum blasted samples (Figure 7B). In contrast to the micro-pin structure, a smoothing of the surface roughness was observed. This also negatively affected the achievable bond strength and resulted in a significant reduction in tensile shear strength compared to thermally bonded samples with the same microstructure. In addition, an involvement of the fiber component with the microstructures can be excluded. Further investigations are necessary to counteract the mentioned effects and to enable an incorporation of the reinforcing fibers into the joint.

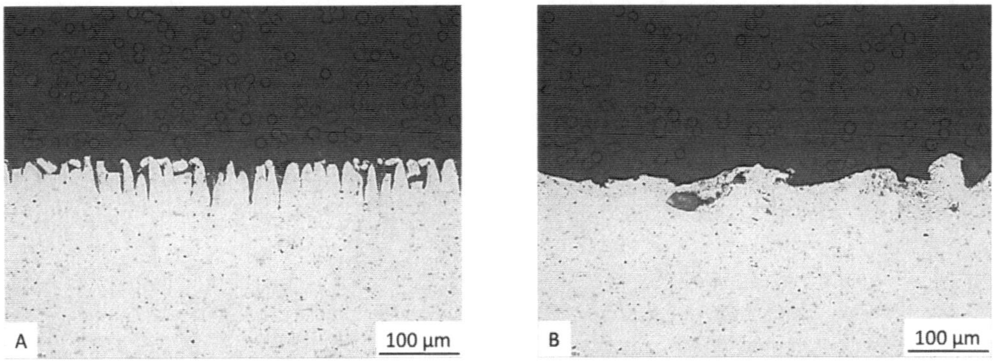


Figure 7: Cross-sectional image of ultrasonically joined aluminum-FRP hybrids: micro pins (A), corundum blasted (B).

## 5 Conclusions

Based on the experimental results the following conclusions were drawn:

- Formation of pin microstructures can be realized by laser processing of aluminum using a well-defined energy input
- Shear stresses above 30 MPa were achieved with outwardly oriented micro pin structures and thermal joining what seems reasoned by appropriate polymer penetration and higher thermal shrinkage of the polymer compared with the metal adherend
- Joint strength achieved by hole structures were comparatively low, which is a consequence of incomplete polymer penetration and interlinking into the holes and the comparatively low structure density
- None of the tested microstructures can be penetrated by load-supporting glass fibers using the regarded joining technologies
- Ultrasonic joining with the used parameters was found to be unappropriated for the joining of polyamide 6 based FRP materials with aluminum alloy EN-AW-6082
- Thermal joining was found to be more appropriate for the considered material combination, however full penetration of the hole structures by polymer melt was not yet achieved

More study will be necessary to understand and to influence the complex mechanisms of pin structure formation during laser processing. Adapted geometrical properties and arrangements of these structures might improve the strength of metal-plastic joints further. In addition to that, future work must be

IOP Conf. Series: Materials Science and Engineering **373** (2018) 012015 doi:10.1088/1757-899X/373/1/012015

focused on the contribution of the diverse adhesion mechanisms to the measured strength and the fracture behavior.

## 6 References

[1] L. Librescu and T. Hause, 2000 "Recent developments in the modeling and behavior of advanced sandwich constructions: a survey," *Composite Structures 48,* pp. 1-17

[2] M. Göring, K. Schreiter, A. Schuberth, T. Windberg, H. Jung, S. Anders, P. Müller, D. Nickel, D. Nestler, L. Kroll, B. Wielage, T. Lampke and S. Spange, 2017 "Amino Group Bearing Organic–Inorganic Hybrid Materials for Joining Aluminum Alloys and Thermoplastic Fiber-Reinforced Parts," *Advanced Materials Interfaces 4,* p. 1601115

[3] B. Wielage, G. Alisch, T. Lampke and D. Nickel, 2008 "Anodizing - A key for surface treatment of aluminium," *Key Engineering Materials 384,* pp. 263-281

[4] A. Heckert and M. F. Zaeh, 2014 "Laser Surface Pre-treatment of Aluminium for Hybrid Joints with Glass Fibre Reinforced Thermoplastics," *Physics Procedia 56,* pp. 1171-1181

[5] G. Ramarathnam, M. Libertucci, M. M. Sadowski and T. H. North, 1992 "Joining of polymers to metal," *Welding Journal 71,* pp. 483-490

[6] B. K. Nayak and M. C. Gupta, "Self-organized micro/nano structures in metal surfaces by ultrafast laser irradiation," *Optics and Lasers in Engineering 48,* p. 940–949.

[7] G. Tang and A. Abdolvand, 2013 "Structuring of titanium using a nanosecond-pulsed Nd:YVO4 laser at 1064 nm," *The International Journal of Advanced Manufacturing Technology 66,* p. 1769–1775

[8] D. H. Kam, S. Bhattacharya and J. Mazumder, 2012 "Control of the wetting properties of an AISI 316L stainless steel surface by femtosecond laser-induced surface modification," *Journal of Micromechanics and Microengineering 22,* p. 105019

[9] E. Baburaj, D. Starikov, J. Evans, G. A. Shafeev and A. Bensaoula, 2007 "Enhancement of adhesive joint strength by laser surface modification," *International Journal of Adhesion and Adhesives 27,* p. 268–276

[10] A. J. Harris, B. Vaughan, J. A. Yeomans, P. A. Smith and S. T. Burnage, 2013 "Surface preparation of silicon carbide for improved adhesive bond strength in armour applications," *Journal of the European Ceramic Society 33,* p. 2925–2934

[11] M.-K. Son and H.-C. Choe, 2011 "Evaluation of Interfacial Bonding Strength between Laser Textured Metal Coping and Porcelain," *Procedia Engineering 10,* p. 2286–2291

[12] I. Südmeyer, T. Hettesheimer and M. Rohde, 2010 "On the shear strength of laser brazed SiC–steel joints: Effects of braze metal fillers and surface patterning," *Ceramics International 36,* p. 1083–1090

[13] H. Man, K. Chiu and X. Guo, 2010 "Laser surface micro-drilling and texturing of metals for improvement of adhesion joint strength," *Applied Surface Science 256,* p. 3166–3169

[14] J. Byskov-Nielsen, J. V. Boll, A. H. Holm, R. Højsholt and P. Balling, 2010 "Ultra-high-strength micro-mechanical interlocking by injection molding into laser-structured surfaces," *International Journal of Adhesion and Adhesives 30,* p. 485–488

[15] L. Romoli, F. Moroni and M. M. A. Khan, 2017 "A study on the influence of surface laser texturing on the adhesive strength of bonded joints in aluminium alloys," *CIRP Annals - Manufacturing Technology 66,* p. 237–240

[16] DIN EN ISO 1110: Plastics - Polyamides - Accelerated conditioning of test specimens. German version EN ISO 1110:1997.

[17] R. Schulze, S. Jahn, H. Zeidler, T. Lindner and A. Schubert, 2016 "Multi material ultrasonic joining using microstructured joining partners," *16th Euspen Conference Proceedings,* pp. 511-512

[18] J. Wodara and H. Herold, 2004 "Ultraschallfügen und -trennen," in *Vol. 151 of Fachbuchreihe Schweisstechnik*, DVS-Verlag, Düsseldorf

[19] *DIN EN 1465: Adhesives–determination of tensile lap-shear strength of bonded assemblies. German version EN 1465:2009-07.*

[20] E. Saborowski, M. Scholze, T. Lindner and T. Lampke, 2017 "A numerical and experimental comparison of test methods for the shear strength in hybrid metal/thermoplastic-compounds," *IOP Conf. Series: Materials Science and Engineering 181,* p. 012031

**Acknowledgement**

This study was supported by the Federal Cluster of Excellence EXC 1075 "MERGE Technologies for Multifunctional Lightweight Structures". Funding by the German Research Foundation (Deutsche Forschungsgemeinschaft, DFG), Germany is gratefully acknowledged.

IOP Conf. Series: Materials Science and Engineering **373** (2018) 012016 doi:10.1088/1757-899X/373/1/012016

# Influence of the welding temperature and the welding speed on the mechanical properties of friction stir welds in EN AW-2219-T87

**A Bachmann, M Krutzlinger and M F Zaeh**

Institute for Machine Tools and Industrial Management, Technical University of Munich, Boltzmannstrasse 15, 85748 Garching, Germany

**Abstract.** Friction Stir Welding (FSW) is an innovative joining technique, which has proven to produce high quality joints in high strength aluminum alloys. Consequently, it is commonly used to manufacture lightweight aerospace structures with stringent requirements. For these structures, it is necessary to ensure a high ultimate tensile strength (UTS). Various studies have reported that the UTS is significantly influenced by the welding parameters. Samples welded with different parameter sets showed a considerably different UTS, despite being free from detectable welding defects (e.g. tunnel defect, voids, or lack of penetration). Based on the observations in the literature, a hypothesis was posed. The welding temperature along with the welding speed determine the UTS of the weld. This study aims to prove this hypothesis experimentally by using temperature-controlled FSW to join plates of EN AW-2219-T87 in butt joint configuration. The welded samples were examined using visual inspection, metallography, X-ray imaging, and uniaxial tensile tests. Finally, a statistical analysis was conducted. Hereby, the hypothesis was confirmed.

## 1. Introduction

The use of aluminum alloys has significantly grown in the second half of the 20th century. This was mainly driven by the density of aluminum, which is relatively low compared to other engineering materials, such as steel [1]. The aerospace industry employs high strength aluminum alloys in order to reduce the structural mass of airplanes and rockets. However, one major drawback of these lightweight materials is their limited fusion weldability. When high-strength aluminum alloys are welded, they tend to the formation of pores and hot cracks, both of which are typical for fusion-welding processes. Friction Stir Welding (FSW) is an innovative solid-state joining process that was invented in 1991. Since the solidus temperature of the material is usually not exceeded during FSW, the formation of hot cracks and pores can be avoided. For this reason, FSW has quickly evolved as a key joining process in the aerospace industry [2]. An important example of a high-strength aluminum alloy that is employed for rocket structures is EN AW-2219-T87 [3]. It contains a high amount of copper (approximately 6 %), which yields advantageous mechanical properties by precipitation hardening. However, the weld seam remains the weakest element of structures made of EN AW-2219-T87. In order to enable the next steps of mass reduction, it is crucial to understand and control the mechanisms that deteriorate the mechanical properties of friction stir welds in EN AW-2219-T87.

The micro structure of friction-stir-welded and arc welded samples of EN AW-2219-T851 was compared by CAO & KOU. They observed the formation of an eutectic phase in the arc welded samples, which was probably formed due to melting and subsequent solidification. In the friction stir welds, θ-phase particles ($Al_2Cu$) were found in the welding nugget (WN), but no evidence of bulk melting could be determined. However, large clusters of θ-particles appeared towards the end of the weld and in the vicinity of the probe tip. This led to the conclusion that the formation of the particle clusters is associated to the material flow around the probe [4].

ELANGOVAN & BALASUBRAMANIEN investigated the influence of the geometrical features of the probe and the rotational speed on the mechanical properties. The best results were obtained with a square probe and a rotational speed of 1600 $min^{-1}$. The superiority of the square probe was confirmed in a second study, in which the shape of the probe and the welding speed were varied [5; 6].

The corrosion behavior of friction stir welded EN AW-2219-T87 samples was investigated by SUREKHA ET AL. and SRINIVASAN ET AL. A dependency of the corrosion resistance from the rotational speed and the welding speed was observed. It was assumed that the effect was caused by the dissolution of copper precipitates [7]. Furthermore, a higher corrosion resistance of the welding zone compared to the base material was reported [8].

The microstructure of a friction-stir-welded EN AW-2219-T87 sample was analyzed by ARORA ET AL. In the welding nugget, a grain refinement was observed along with a 40 % dissolution of the θ-particles. It was assumed that both are caused by dynamic recrystallization. The dissolution of θ-particles was also observed in the thermomechanically affected zone (TMAZ), although it was not as significant as in the WN. Furthermore, it was reported that grain coarsening occurred in the TMAZ [9].

A superior ultimate tensile strength (UTS) and less variance in the UTS of friction stir welds compared to arc welds in EN AW-2219-T87 were reported by KAHNERT ET AL. The tensile tests were conducted under cryogenic conditions (20 K – 77 K), which mirrors the environment in rocket fuel tanks [3].

The influence of the rotational speed on the mechanical properties of friction-stir-welded EN AW-2219-T87 samples was investigated by SCHNEIDER ET AL. It was observed that an increased rotational speed leads to an increased UTS and hardness of the welds. The authors concluded that the increased rotational speed supports the dissolution of θ-particles and the coarsening of θ'-particles [10].

DOUDE ET AL. investigated the influence of the rotational speed and the welding speed on the mechanical properties of friction-stir-welded EN AW-2219-T87 samples. A parameter window for defect-free friction stir welds was identified. Within the parameter window, an increased rotational speed increased the UTS as well as the hardness of the weld. As the rotational speed mainly influences the welding temperature, it was assumed that the welding temperature governs the metallurgical effects. It was concluded that during FSW of heat-treatable aluminum-copper alloys, the welding temperature has to be sufficient to enable the dissolution of the θ-particles and to prevent their coarsening [11].

A similar conclusion was made by RAO ET AL., who investigated the influence of the tool geometry on the UTS of friction-stir-welded EN AW-2219-T87 samples. The highest UTS was achieved using a hexagonal probe, which also led to the highest welding temperature [12].

A thorough investigation on the formation and the behavior of copper precipitates during FSW of EN AW-2219-T87 was conducted by KANG ET AL. It was observed that the θ'- and θ''-particles are fully dissolved in the WN, whereas θ-particles are only partially dissolved. In the heat affected zone (HAZ), the θ''-particles were also dissolved completely, whereas θ'-particles were only dissolved partially. This effect was explained by the decreasing temperature perpendicular to the welding direction. It was assumed that this effect is also the reason for the varying hardness values measured in this direction. As reported by CAO & KOU, large clusters of θ-particles were detected in the WN. From the analysis of the microstructure it was concluded that diffusion processes are not responsible for the formation of the particle clusters. It was assumed that their formation is associated to the material flow around the probe. However, tensile tests of the welded samples showed no impact of the particle clusters on the mechanical properties [13; 14].

To summarize, the welding temperature and the welding speed are believed to be the determining factors for the mechanical properties of the welds. Both parameters influence the diffusion processes during and immediately after welding. This assumption has been supported by the results obtained from the analysis of the microstructure. However, it has not been proven that constant mechanical properties can be ensured by welding with different tools at the same welding temperatures. Furthermore, the correlation between welding temperature, welding speed, and resulting UTS is unknown.

## 2. Hypotheses and methodology

### *2.1. Hypotheses*

Related works have suggested that the dissolution and coarsening of copper precipitants determines the mechanical properties of friction stir welds in EN AW-2219-T87. The dissolution as well as the coarsening of precipitants are mainly driven by time- and temperature-dependent processes. Hence, the thermal cycle during FSW has to be controlled in order to regulate the mechanical properties of the weld. The scope of this paper is to prove that this can be achieved by temperature-controlled FSW. The subsequent hypotheses, which have yet to be proven, are:

- The welding temperature and the welding speed have a significant influence on the mechanical properties of friction stir welds. Therefore, a parameter set composed by a welding temperature and a welding speed can be transferred to a different welding task resulting in a similar UTS.
- The tool rotation serves to induce the required heat energy and to ensure the necessary material flow during FSW. It is supposed that a rotational speed ensuring a suitable welding temperature provides also an adequate material flow. Hence, it is admissible to regulate the welding temperature by adjusting the rotational speed.

### *2.2. Approach*

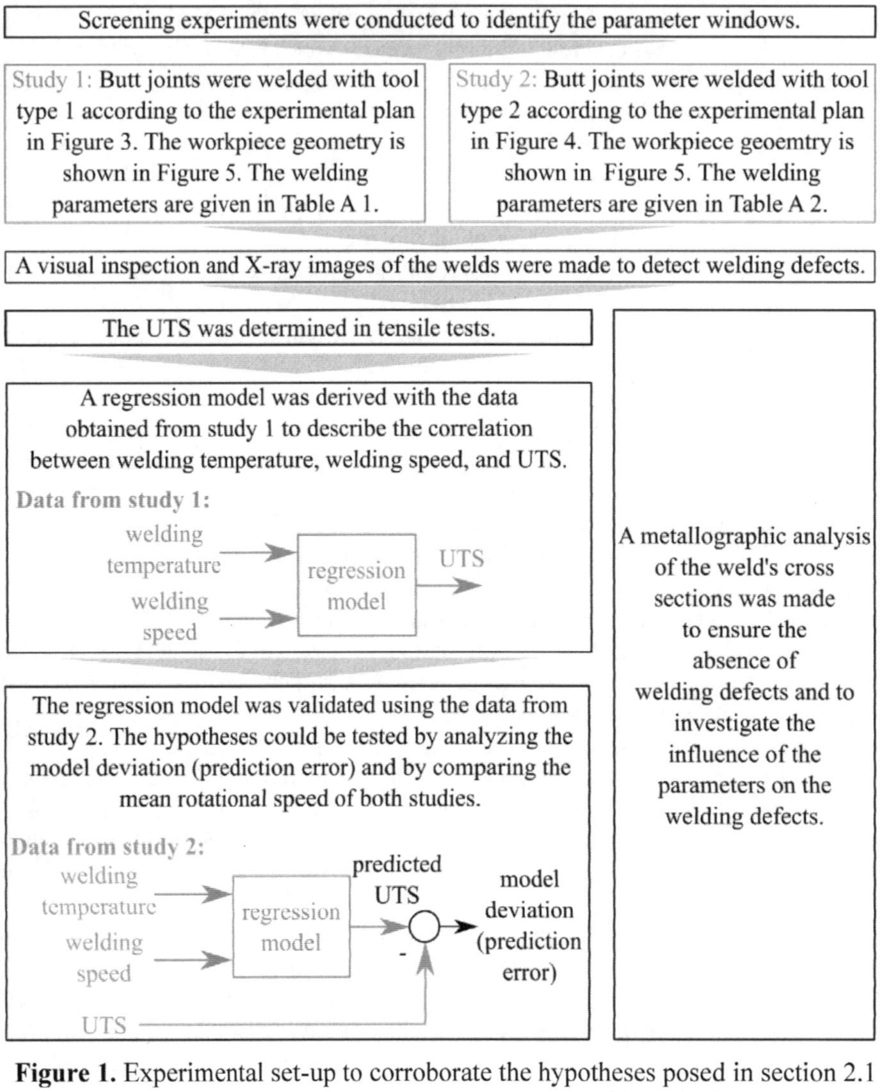


**Figure 1.** Experimental set-up to corroborate the hypotheses posed in section 2.1

Two sequential parameter studies (study 1 and study 2) were conducted to corroborate the hypotheses posed in section 2.1. The welding speed and the welding temperature were varied in both studies. The temperature control system presented in BACHMANN ET AL. [15] was used to control the welding temperature. Two tools of the same geometry, but with different dimensions were used during each study. Thereby, it was ensured that similar welding temperatures were caused by significantly different rotational speeds. The results of the parameter study 1 were used to identify a regression model, which describes the correlation between the welding temperature, the welding speed and the UTS. The data from parameter study 2 were employed to both the validation of the model and the verification of the hypotheses. The complete experimental procedure is depicted in Figure 1.
First, the parameter windows for the tools were identified in preliminary screening trials. Subsequently, the actual welding experiments were conducted according to an experimental plan. The welded plates were inspected visually and X-rayed to reveal volumetric defects in the welds. Thereby, an influence of detectable welding defects (e.g. voids, tunnel defects, or lack of penetration) on the resulting UTS could be avoided. The plates that passed the non-destructive testing (NDT) were selected to extract specimens for the uniaxial tensile tests and for metallography. Finally, the UTS and the input parameters were used to derive a regression model, which was employed to predict the UTS for the study 2. The model was deemed to be valid if the residuals and the prediction error were within the same order of magnitude. This corroborates the hypothesis, because it is expected that the different tool sizes result in significantly different rotational speeds to provide the same temperatures. The different methods and the equipment used for each step of the experimental procedure are described in the following.

*2.2.1. Experimental set-up and screening.* The welding machine used during the studies is based on an industrial robot. It is described and depicted in BACHMANN ET AL. [15]. Furthermore, the temperature control system described by BACHMANN ET AL. [15] was used throughout the studies for this paper. The temperature measuring system was described by COSTANZI ET AL. [16]. Two tools of the same geometry but of different sizes were employed in this study. The tool geometries and their dimensions are given in Figure 2 and Table 1.

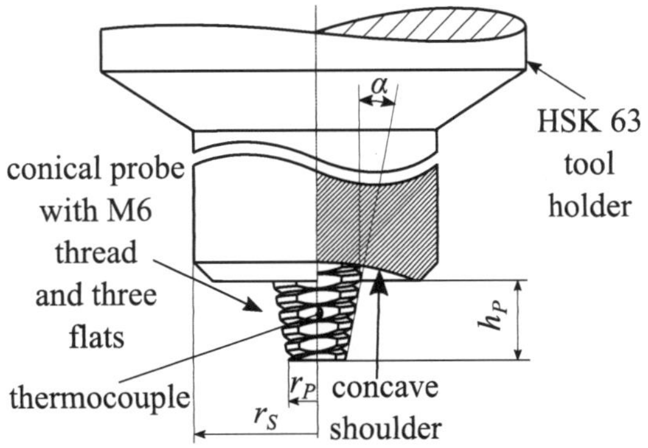


**Figure 2.** Tool geometry with the most important dimensions of the FSW tools used in the experiments [15]

Preliminary screening experiments were conducted to find a suitable downward force $F_z$ as well as to determine the parameter window, in which macroscopic welding defects do not occur. Hereby, bead-on-plate welds were made on 4 mm thick plates of EN AW-2219-T87. The welds were inspected visually and a bending test was conducted to ensure full penetration of the plates. Cross-sections of the welds were extracted to verify the absence of tunnel defects. The resulting parameter windows are depicted in Figure 3 for tool type 1 and Figure 4 for tool type 2, respectively. The parameter window for tool type 1 extends to a welding speed of $v_S$ = 400 mm/min. The maximum welding temperature was set to $T_{S,max}$ = 540 °C, which is beneath the solidus temperature $T_{solidus}$ = 548 °C of EN AW-2219-T87 [1]. The minimum welding temperature was determined in the screening experiments. It marks the limit to the onset of lack of consolidation or tunnel defects. The maximum rotational speed of the spindle was 3000 min$^{-1}$. This led to the limitations in parameter window 2 (see Figure 4). Due to the smaller tool diameter, higher rotational speeds were required by the temperature controller than could be provided by the spindle motor. The suitable downward force for tool type 1 was to $F_z = 9000$ N. Thereby, welds with little flash and without surface lack of fill could be produced. The screening experiments with tool type 2 showed similar results for equal welding temperatures and welding speeds using a downward force of $F_z = 7000$ N.

**Table 1.** Numerical values of the tool dimensions (see Figure 2)

| | Probe radius $r_P$ in mm | Shoulder radius $r_P$ in mm | Conical probe angle $\alpha$ in ° | Probe length $h_P$ in mm |
|---|---|---|---|---|
| **Tool type 1** | 2.350 | 10 | 11 | 3.87 |
| **Tool type 2** | 1.835 | 8 | 10 | 4.15 |

*2.2.2. Welding experiments.* Two parameter studies labeled study 1 and study 2 were conducted. In both studies the welding temperature and the welding speed were varied systematically, the tilt angle of the tool was kept constant at 2.5° throughout the experiments. The geometry of the samples, which were welded in butt joint configuration, is shown in Figure 5. The faying surfaces of the plates were cleaned thoroughly with isopropanol before welding. A rigid clamping device was used to avoid gaps between the joining partners. The weld length on each plate was 450 mm. In order to reduce the experimental effort, the welding temperature was changed in the middle of the weld (see Figure 5). The temperature was changed (transition zone) over a distance of 55 mm in the center of the weld. This part was not used for further investigation.

Tool type 1 was employed for study 1. More heat could be delivered to the process zone due to the larger diameters of the shoulder and of the probe. Consequently, higher welding temperatures could be achieved even at increased welding speeds. The parameter window is depicted in Figure 3. A modified D-optimal experimental plan was used due to the restrictions posed by the arising welding defects at high welding speeds and low welding temperatures. The parameter sets, which were tested in study 1, are depicted as circles in Figure 3. The trials using the center point as well as two random points were repeated to determine the reproducibility of the results of the welding experiments.

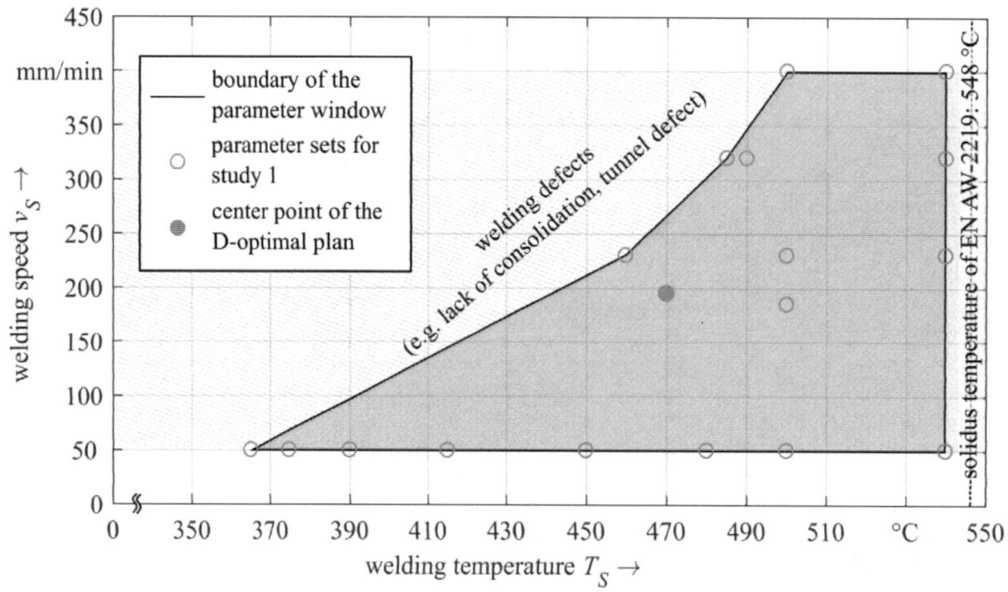


**Figure 3.** Parameter window 1 and parameter sets for the study 1

Tool type number 2 was employed during study 2. Owing to the smaller shoulder and probe diameter, the amount of heat produced with tool type 2 was significantly lower. This led to further restrictions of the parameter window, because the rotational speed required from the controller was higher than the maximum rotational speed provided by the spindle. This limitation arising from the welding machine is marked in Figure 4. As a consequence, the parameter window 2 was split and three experimental plans

IOP Conf. Series: Materials Science and Engineering **373** (2018) 012016 doi:10.1088/1757-899X/373/1/012016

were derived: Two D-optimal plans covered the parameter window at high and low temperature levels and a full factorial plan was used for the intermediate temperature levels. The parameter window for the study 2 and its division into the three experimental plans is shown in Figure 4. The parameter sets from study 2 are marked as green asterisks. The center point of the full factorial sub plan (marked as dashed asterisk in Figure 4) was repeated twice to assess the reproducibility of the welding process.

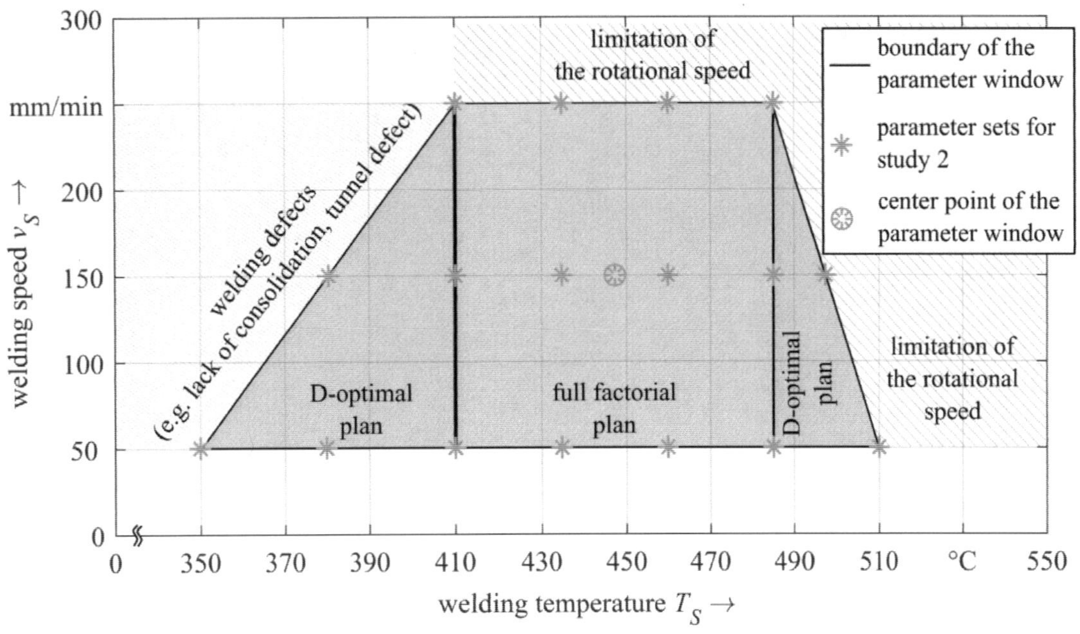


**Figure 4.** Parameter window 2 and parameter sets for the study 2

*2.2.3. Visual inspection.* The welds were inspected visually by employing a criteria set. The primary focal points were: surface appearance (regular surface, surface galling), regularity of the width, flash formation, surface cracks, appearance of the weld root (suck back or excessive convexity), and appearance of the key hole. Tunnel defects can often be identified by inspecting the keyhole, as the tunnel often evolves up to the lateral surface of the key hole.

*2.2.4. X-ray imaging.* The welded plates were examined using X-ray attenuation imaging. Zones containing abnormalities in the X-ray image were marked on the plates. These areas were not used for the subsequent analysis to avoid that the macroscopic welding defects influence the results.

*2.2.5. Tensile tests.* The tensile tests were conducted approximately six weeks after welding. It was assumed that potential post-weld aging was terminated after this period. The specimens required for the tensile tests were extracted from the welded plates as shown in Figure 5. The positions of the specimens were selected according to the restrictions given by DIN EN ISO 25239-3. One plate was used to investigate two parameter sets, as the parameters were changed in the center of the weld. The width of the sheets differed to facilitate the accessibility of the robot and to enable an adequate clamping.

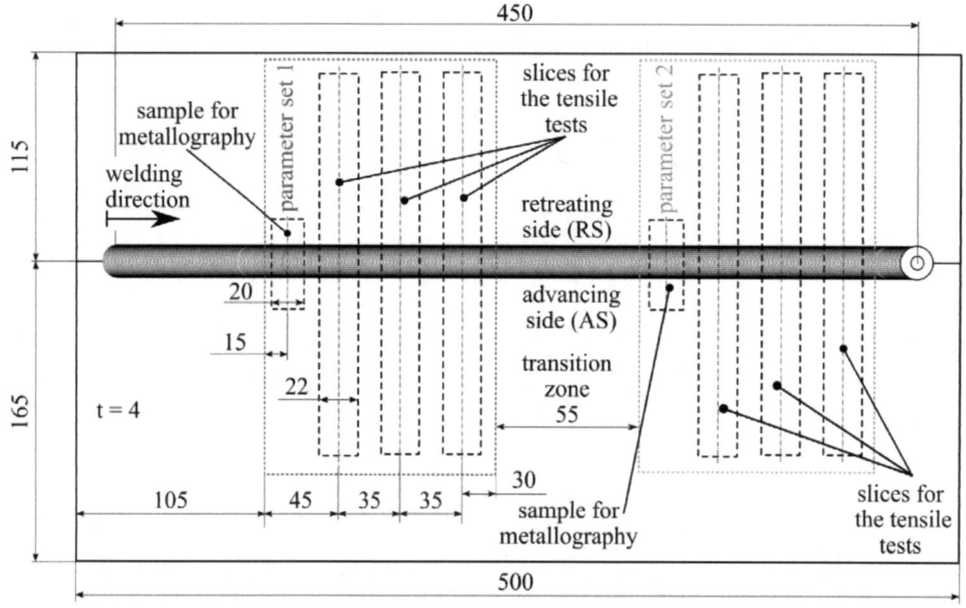


**Figure 5.** Dimensions of the welded plates with the position of the specimens for the tensile test and the samples for metallography in mm

Three samples were extracted for each parameter set to enable the estimation of the process variance. The slices obtained from the plate were milled to the final specimen geometry shown in Figure 6. The specimen geometry was adapted from the shape E in DIN 50125. Hereby, the length and the width of the specimen were increased to avoid slipping during the tensile test. The uniaxial tensile tests were performed according to the guidelines in DIN EN ISO 6892-1.

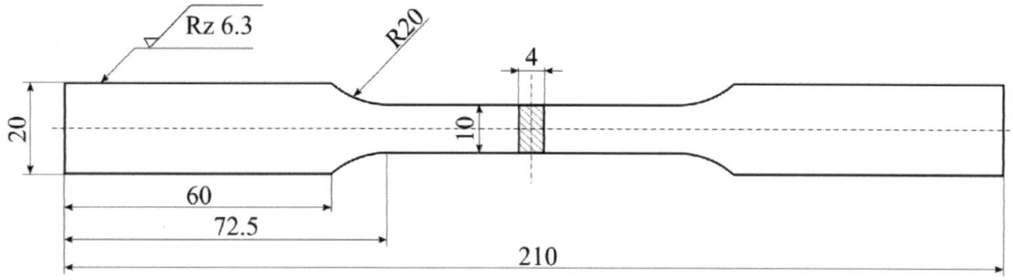


**Figure 6.** Specimen for the tensile tests

*2.2.6. Statistical analysis and regression.* The software Matlab was used to derive a multiple linear regression model, which describes the correlation between the welding temperature $T_S$ (input), the welding speed $v_S$ (input), and the UTS $R_m$ (output). The software tool provides sample functions to fit the model and determines the fit quality. Usually, polynomial models are used due to their simplicity. In this paper, the differences between the model and the experimental data from study 1 were defined as residuals. The differences between model and experimental data from study 2 were defined as the prediction error, because the data from study 2 were not employed for modelling. During the fitting step, the coefficients of the polynomial were adjusted to minimize the mean squared error *MSE* of the residuals [17]. The model quality was assessed by employing the coefficient of determination $R^2$ [18]. The root mean squared error *RMSE* and the mean absolute error *MAE* are statistical measures that are used in this paper to quantify the differences between experiment and the model [17]. The hypothesis posed in section 2.1 was tested by comparing the *RMSE* and the *MAE* of the residuals and the prediction error, respectively. The equations used to calculate the statistical measures are described in the appendix.

## 3. Results and discussion

*3.1. UTS and regression.*

The resulting UTS data of both studies are given in Tables A 1 and A 2. The difference between minimum and maximum was approximately 150 MPa. Various trials from study 1 were replicated: the parameters from experiments no. 1, 12, and 9 were repeated in trials no. 21, 18, and 14 and yielded relative standard deviations of 0.85 %, 1.85 %, and 0.27 %, respectively. The center point (experiment no. 30) of the full factorial sub-plan in study 2 was repeated in trial no. 38, and no. 39. The resulting standard deviation was 0.8 %. This demonstrates the high reproducibility of FSW, because the standard deviation of the UTS remains within the same magnitude, regardless of whether the specimens were obtained from two different plates welded with the same parameters or from the same plate. The lowest UTS values, which resulted from samples welded with parameters from the lower boundary of the parameter window 2, showed a high standard deviation. The same characteristic was observed for the sample welded at the highest temperature and at the highest welding speed. This indicates that the process limits were reached.

A comparison of the resulting UTS for both studies shows good agreement (see Figure 7). Figure 7 illustrates the UTS for different welding temperatures at a welding speed of 50 mm/min. The values obtained from study 1 are marked as blue circles, the values obtained from study 2 are marked as green asterisks. Furthermore, the mean rotational speeds required to achieve the welding temperatures are depicted. Figure 7 shows that the differences between the rotational speeds from study 1 and from study 2 increase with increasing welding temperature. In contrast, the differences in UTS for similar welding temperatures are small. The general trend of the UTS from study 1 agrees with the results from study 2. The only exception is the sample welded with $T_S = 485$ °C (experiment no. 40). The standard deviation of the rotational speed is not depicted because the rotational speed was adjusted by the controller to maintain the set welding temperature. Accordingly, the standard deviation represents the activity of the controller rather than variations in the results by identical process parameters.

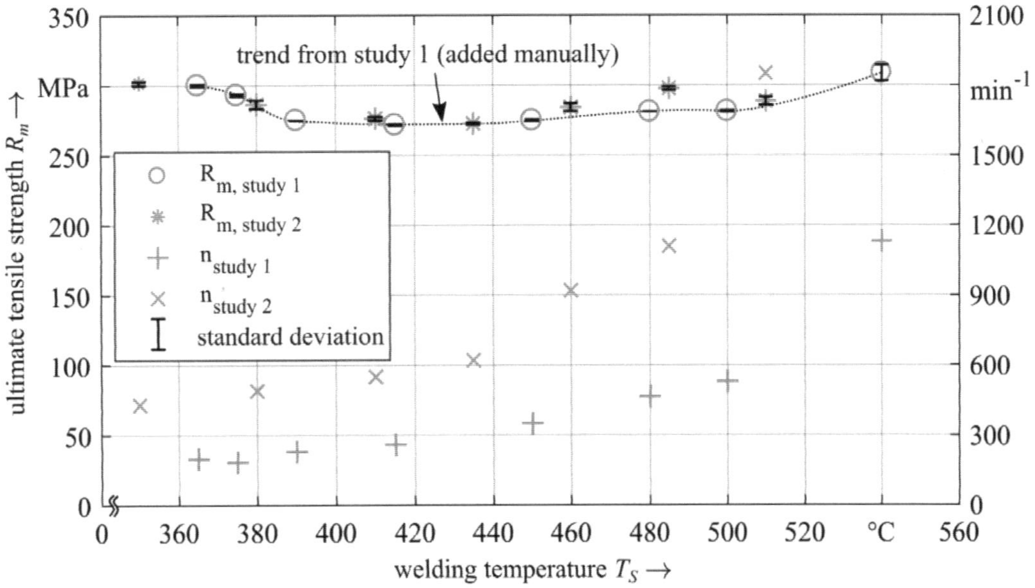


**Figure 7.** Mean ultimate tensile strength and mean rotational speeds for different welding temperatures at welding speed $v_s = 50$ mm/min

The data from study 1 were used to derive a regression model. The welding speed and the welding temperature were employed as input parameters, while the UTS served as the output parameter of the regression model. When multiple plates were welded with the same parameter sets the mean value was employed. In total 18 data sets were used to derive the model. A polynomial model of second order with second order interactions provided the best fit quality. The model quality was quantified by the coefficient of determination $R^2$. A lower model order led to an insufficient model quality, while a higher order was assumed to result in overfitting. The polynomial of the identified model is:

$$R_m(T_S, v_S) = a_0 + a_1 \cdot v_S + a_2 \cdot T_S + a_3 \cdot v_S \cdot T_S + a_4 \cdot v_S \cdot T_S^2 + a_5 \cdot T_S^2 + a_6 \cdot v_S^2, \quad (1)$$

with the linear coefficients $a_0$, $a_1$, $a_2$ , $a_3$, $a_4$, $a_5$, and $a_6$. The coefficients of Eq. (1) as well as the coefficient of determination $R^2$ and the *RMSE* of the model residuals are given in Table 2. The p-values of the factors were smaller than 2 %. Consequently, all coefficients of Eq. (1) were determined to be significant [18].

**Table 2.** Model parameters

| $a_0$ in MPa | $a_1$ in MPa·min$^{-1}$/mm | $a_2$ in MPa/°C | $a_3$ in MPa min$^{-1}$ (mm·°C) | $a_4$ in MPa/(min$^{-1}$·°C$^2$) | $a_5$ in MPa/°C$^2$ | $a_6$ in MPa·(min$^{-2}$/mm$^2$) | $R^2$ | *RMSE* in MPa |
|---|---|---|---|---|---|---|---|---|
| 1413.6 | -7.1646 | -5.0445 | 0.030386 | -3.0349·10$^{-5}$ | 6.7401·10$^{-4}$ | 5.4899·10$^{-3}$ | 0.98 | 3.71 |

The model was used to predict the UTS for the parameter sets of study 2. The absolute residuals and the absolute prediction errors are depicted in Figure 8. It can be observed that the prediction errors of the parameter sets of study 2, which were outside the parameter window of study 1 (i.e. no. 27, no. 28, no. 34, and no. 35), were significantly higher than those which were inside.

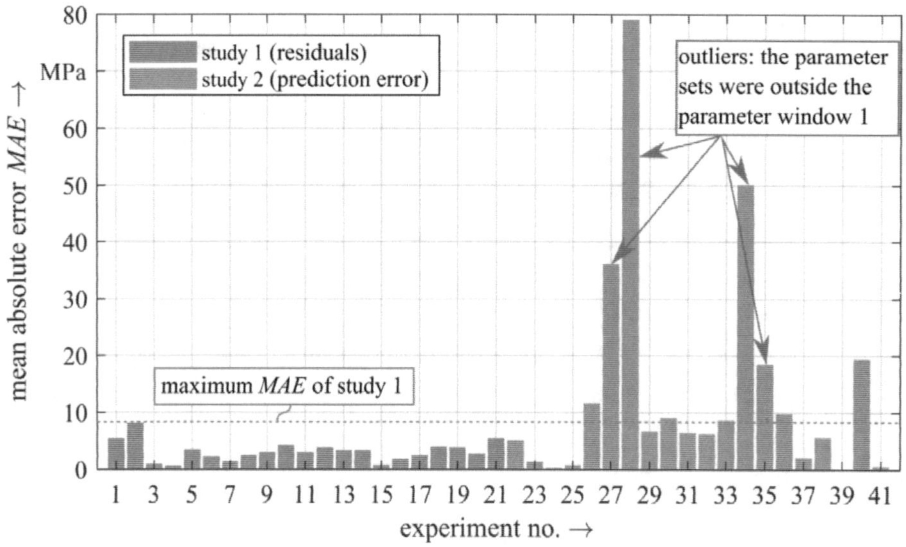


**Figure 8.** Mean absolute error (*MAE)* of both studies

Initially, the limited extrapolation capability of regression models was considered to be the reason for the high absolute prediction error. According to Table A 2, experiment no. 27 and no. 34 showed a relatively high standard deviation. This was interpreted as evidence of an unstable welding process. Hence, the cross-sections and the X-ray images were more intensely investigated. This revealed a tunnel defect on the advancing side in three parameter sets (i.e. no. 27, no. 34, and no. 35). The X-ray images of experiment no. 28 could not be interpreted clearly: it appeared as if the tunnel defect in the weld, which occurred at the parameter set no. 27, ceased in the transition zone of the plate. However, a reliable statement could not be made. Hence, the fractured surface of the tensile test specimens was analyzed. Thereby, a lack of consolidation was found on the advancing side of the weld.

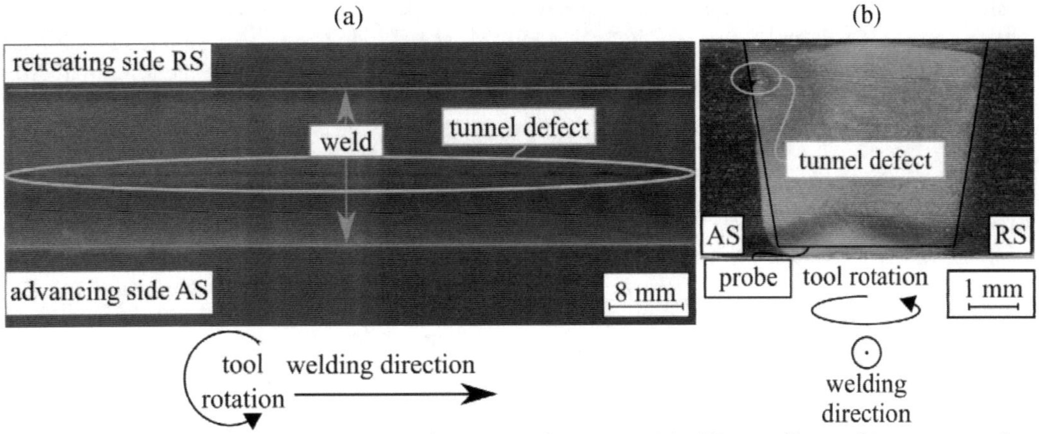


**Figure 9.** (a) X-ray image of the weld from experiment no. 34; (b) metallographic cross-section of the weld from experiment no. 34

Consequently, the results of experiments no. 27, no. 28, no. 34, and no. 35 were considered as outliers. The only remaining abnormality was experiment no. 40. It showed a relatively high not explainable prediction error. Therefore, the data from experiment no. 40 was included in the subsequent statistical analysis, whereas the outliers were considered separately.

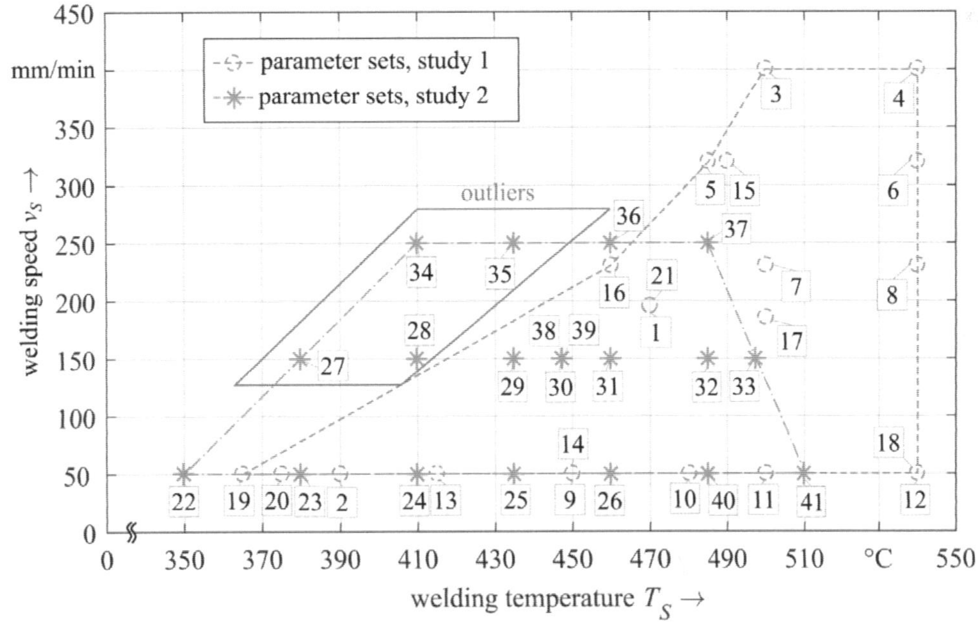


**Figure 10.** Parameter sets of study 1 and study 2 with identified outliers

The *RMSE* and the *MAE* for the prediction error as well as for the residuals are given in Table 3. Both measures of the prediction error without outliers are in the same range as the residuals. This means that the model describes the correlation between welding temperature, welding speed, and UTS of the weld accurately at a mean absolute prediction error of 6.03 MPa within the boundaries of the parameter window 1. The model is valid also for study 2, although a different tool size, a different downward force, and a different batch of raw material was used. The applicability of the model to a different experimental set-up leads to the conclusion that the relevant process parameters were identified. This means that the welding temperature und the welding speed are the determining factors for the UTS of friction stir welds in EN AW-2219-T87. This corroborates the hypothesis posed in section 2.1 and implies that a required

UTS can be ensured by maintaining an adequate welding temperature and welding speed. Furthermore, the results depicted in Figure 8 suggest that the rotational speed *n* does not influence the UTS directly. The results indicate that the rotational speed primarily serves as a heat source. The metallurgical effects, which occur during the welding process, are most likely caused by the thermal cycle and must therefore depend on temperature and time.

**Table 3.** *RMSE* and *MAE* calculated with the residuals and the prediction error

| *RMSE* of the residuals in MPa | *MAE* of the residuals in MPa | *RMSE* of the prediction error with outliers in MPa | *MAE* of the prediction error with outliers in MPa | *RMSE* of the prediction error without outliers in MPa | *MAE* of the prediction error without outliers in MPa |
|---|---|---|---|---|---|
| 3.710 | 3.430 | 23.81 | 13.99 | 7.720 | 6.030 |

## 4. Conclusions

Two parameter studies were conducted in order to investigate the influence of the welding temperature and the welding speed on the UTS of friction stir welds in EN AW-2219-T87. The results were used to derive a regression model, which describes the correlation between the two welding parameters and the UTS. This led to the following conclusions:

- The UTS of friction stir welds in EN AW-2219-T87 is determined predominantly by the welding temperature and the welding speed.
- The rotational speed primarily influences the welding temperature. The results suggest that the thermal cycle plays an important role for the UTS of the welds. It could be shown that samples joined using similar welding speeds and welding temperatures show similar UTS values, even though a significantly different rotational speed was employed.

The underlying effects, which cause the variance in the UTS, will be the subject of future investigations.

## 5. Appendices

The mean square error was determined according to

$$MSE = \frac{1}{n}\sum_{i=1}^{n}\left(R_{m,model,i} - R_{m,experiment,i}\right)^2, \tag{2}$$

with the UTS calculated by the model $R_{m,model}$, the UTS obtained from the experiment $R_{m,experiment}$, the total number of experimental points *n*, and the index variable *i* [17]. The coefficient of determination $R^2$ was calculated by

$$R^2 = 1 - \frac{\sum_{i=1}^{n}\left(R_{m,model,i} - R_{m,experiment,i}\right)^2}{\sum_{i=1}^{n}\left(R_{m,experiment,i} - \bar{R}_{m,experiment}\right)^2}, \tag{3}$$

with the mean value of the UTS $\bar{R}_{m,experiment}$ from the experiments. The mean value of the UTS was obtained by

$$\bar{R}_{m,experiment} = \frac{1}{n}\sum_{i=1}^{n} R_{m,experiment,i}. \tag{4}$$

The RMSE can be determined by the root of the MSE as follows:

$$RMSE = \sqrt{MSE} = \sqrt{\frac{1}{n}\sum_{i=1}^{n}\left(R_{m,model,i} - R_{m,experiment,i}\right)^2}. \tag{5}$$

IOP Conf. Series: Materials Science and Engineering **373** (2018) 012016 doi:10.1088/1757-899X/373/1/012016

*The MAE* was calculated by

$$MAE = \frac{\sum_{i=1}^{n} \Delta R_{m,abs,i}}{n}, \tag{6}$$

*where*

$$\Delta R_{m,abs,i} = \left| R_{m,model,i} - R_{m,experiment,i} \right|. \tag{7}$$

**Table A 1.** Experimental plan of parameter study 1

| Exp. no. | Welding temperature $T_S$ in °C | Welding speed $v_S$ in mm/min | Mean rotational speed $n$ in min$^{-1}$ | Downward force $F_z$ in N | Tilt angle β in ° | Tool type - | Mean UTS $R_m$ in MPa | Standard deviation of the UTS $S_{Rm}$ in MPa |
|---|---|---|---|---|---|---|---|---|
| 1 | 470 | 195 | 530 | 9000 | 2.5 | 1 | 315.67 | 4.16 |
| 2 | 390 | 50 | 230 | 9000 | 2.5 | 1 | 275.00 | 0.00 |
| 3 | 500 | 400 | 965 | 9000 | 2.5 | 1 | 333.33 | 5.51 |
| 4 | 540 | 400 | 1608 | 9000 | 2.5 | 1 | 339.67 | 14.43 |
| 5 | 490 | 320 | 835 | 9000 | 2.5 | 1 | 331.67 | 0.58 |
| 6 | 540 | 320 | 1510 | 9000 | 2.5 | 1 | 349.67 | 0.58 |
| 7 | 500 | 230 | 795 | 9000 | 2.5 | 1 | 331.00 | 1.00 |
| 8 | 540 | 230 | 1385 | 9000 | 2.5 | 1 | 343.00 | 1.00 |
| 9 | 450 | 50 | 350 | 9000 | 2.5 | 1 | 274.67 | 0.58 |
| 10 | 480 | 50 | 465 | 9000 | 2.5 | 1 | 281.00 | 0.00 |
| 11 | 500 | 50 | 530 | 9000 | 2.5 | 1 | 281.33 | 0.58 |
| 12 | 540 | 50 | 1100 | 9000 | 2.5 | 1 | 304.67 | 1.15 |
| 13 | 415 | 50 | 260 | 9000 | 2.5 | 1 | 271.67 | 0.58 |
| 14 | 450 | 50 | 356 | 9000 | 2.5 | 1 | 275.00 | 1.00 |
| 15 | 485 | 320 | 750 | 9000 | 2.5 | 1 | 324.75 | 1.00 |
| 16 | 460 | 230 | 600 | 9000 | 2.5 | 1 | 307.33 | 4.62 |
| 17 | 500 | 185 | 700 | 9000 | 2.5 | 1 | 320.00 | 1.00 |
| 18 | 540 | 50 | 1160 | 9000 | 2.5 | 1 | 312.33 | 6.03 |
| 19 | 365 | 50 | 200 | 9000 | 2.5 | 1 | 300.00 | 1.00 |
| 20 | 375 | 50 | 185 | 9000 | 2.5 | 1 | 293.00 | 1.00 |
| 21 | 470 | 195 | 530 | 9000 | 2.5 | 1 | 315.67 | 0.58 |

**Table A 2.** Experimental plan of parameter study 2

| Exp. no. | Welding temperature $T_S$ in °C | Welding speed $v_S$ in mm/min | Mean rotational speed $n$ in min$^{-1}$ | Downward force $F_z$ in N | Tilt angle β in ° | Tool type - | Mean UTS $R_m$ in MPa | Standard deviation of the UTS $S_{Rm}$ in MPa |
|---|---|---|---|---|---|---|---|---|
| 22 | 350 | 50 | 430 | 7000 | 2.5 | 2 | 301.46 | 1.15 |
| 23 | 380 | 50 | 490 | 7000 | 2.5 | 2 | 286.43 | 3.00 |
| 24 | 410 | 50 | 550 | 7000 | 2.5 | 2 | 276.31 | 1.39 |
| 25 | 435 | 50 | 620 | 7000 | 2.5 | 2 | 272.61 | 0.65 |
| 26 | 460 | 50 | 920 | 7000 | 2.5 | 2 | 284.20 | 2.87 |
| 27 | 380 | 150 | 650 | 7000 | 2.5 | 2 | 238.15 | 12.12 |
| 28 | 410 | 150 | 760 | 7000 | 2.5 | 2 | 202.86 | 5.48 |
| 29 | 435 | 150 | 960 | 7000 | 2.5 | 2 | 296.08 | 1.28 |
| 30 | 447,5 | 150 | 1020 | 7000 | 2.5 | 2 | 302.70 | 2.01 |
| 31 | 460 | 150 | 1080 | 7000 | 2.5 | 2 | 304.69 | 1.63 |
| 32 | 485 | 150 | 1470 | 7000 | 2.5 | 2 | 314.46 | 3.28 |
| 33 | 497,5 | 150 | 1670 | 7000 | 2.5 | 2 | 322.24 | 0.29 |
| 34 | 410 | 250 | 1000 | 7000 | 2.5 | 2 | 224.04 | 25.60 |
| 35 | 435 | 250 | 1180 | 7000 | 2.5 | 2 | 275.21 | 3.06 |
| 36 | 460 | 250 | 1680 | 7000 | 2.5 | 2 | 320.20 | 5.54 |
| 37 | 485 | 250 | 1810 | 7000 | 2.5 | 2 | 322.71 | 5.65 |
| 38 | 447,5 | 150 | 1030 | 7000 | 2.5 | 2 | 299.27 | 0.43 |
| 39 | 447,5 | 150 | 1030 | 7000 | 2.5 | 2 | 297.72 | 1.21 |
| 40 | 485 | 50 | 1110 | 7000 | 2.5 | 2 | 297.76 | 1.2 |
| 41 | 510 | 50 | 1850 | 7000 | 2.5 | 2 | 288.64 | 2.87 |

**Table A 3.** Tensile tests to characterize the base material

| Specimen | UTS $R_m$ in MPa |
|---|---|
| Base material in rolling direction | 476 |
| Base material perpendicular to the rolling direction | 471 |
| Base material perpendicular to the rolling direction | 469 |

## 6. References

[1] Ostermann F 2014 *Anwendungstechnologie Aluminium (‚Application technology of Aluminum‘),* (Berlin, Heidelberg: Springer)

[2] Colligan K J 2010 The Friction Stir Welding Process *Friction stir welding,* ed D Lohwasser and Z Chen (Cambridge, UK: Woodhead Publishing Limited) pp 15 – 41

[3] Kahnert M, Mestek M, Windisch M, Tessier I and Okualla M 2012 Preparation of Friction Stir Welding of the Aluminum Alloy AA 2219 for Launcher Applications *9th International Friction Stir Welding Symposium* (Huntsville, Alabama) May 15th–17th 2012

[4] Cao G and Kou S 2005 Friction Stir Welding of 2219 Aluminum: Behavior of (AI2Cu) particles *Welding Journal* **84** 1–7

[5] Elangovan K and Balasubramanian V 2007 Influences of pin profile and rotational speed of the tool on the formation of friction stir processing zone in AA2219 aluminium alloy *Materials Science and Engineering* A **459** 7–18

[6] Elangovan K and Balasubramanian V 2008 Influences of tool pin profile and welding speed on the formation of friction stir processing zone in AA2219 aluminium alloy *Journal of Materials Processing Technology* **200** 163–75

[7] Surekha K, Murty B S and Prasad Rao K 2009 Effect of processing parameters on the corrosion behaviour of friction stir processed AA 2219 aluminum alloy *Solid State Sciences* **11** 907–17

[8] Srinivasan P B, Arora K S, Dietzel W, Pandey S and Schaper M K 2010 Characterisation of microstructure, mechanical properties and corrosion behaviour of an AA2219 friction stir weldment *Journal of Alloys and Compounds* **492** 631–37

[9] Arora K S, Pandey S, Schaper M and Kumar R 2010 Microstructure Evolution during Friction Stir Welding of Aluminum Alloy AA2219 *Journal of Materials Science and Technology* **26** 747–53

[10] Schneider J, Stromberg R, Schilling P, Cao B, Zhou W, Morfa J and Myers O 2013 Processing effects on the friction stir weld stir zone *Welding Journal* **92** 11s–9s

[11] Doude H, Schneider J, Patton B, Stafford S, Waters T and Varner C 2015 Optimizing weld quality of a friction stir welded aluminum alloy *Journal of Materials Processing Technology* **222** 188–96

[12] Rao C V, Reddy G M and Rao K S 2015 Influence of tool pin profile on microstructure and corrosion behaviour of AA2219 Al–Cu alloy friction stir weld nuggets *Defence Technology* **11** 197–208

[13] Kang J, Feng Z-C, Frankel G.S, Huang I W, Wang G-Q and Wu Ai-Ping 2016 Friction Stir Welding of Al Alloy 2219-T8 *Metallurgical and Materials Transactions* A **47** 4553–65

[14] Kang J, Feng Z-C, Li J-C, Frankel G S, Wang G-Q amd Wu Ai-Ping 2016 Friction Stir Welding of Al Alloy 2219-T8 *Metallurgical and Materials Transactions* A **47** 4566–77

[15] Bachmann A, Gamper J, Krutzlinger M, Zens A and Zaeh M F 2017 Adaptive model-based temperature control in friction stir welding *The International Journal of Advanced Manufacturing Technology* **93** 1157–71

[16] Costanzi G, Bachmann A and Zaeh M. F 2017 Entwicklung eines FSW-Spezialwerkzeugs zur Messung der Schweißtemperatur (‚Development of a FSW tool to measure the Welding Temperature‘) *DVS Berichte vol 337* (Düsseldorf: DVS Media GmbH) 2017 119–125

[17] Kosfeld R, Eckey H F and Tuerck M 2016 *Deskriptive Statistik* (‚*Descriptive statistics*‘), (Wiesbaden: Springer)

[18] Siebertz K, van Bebber D and Hochkirchen T 2010 *Statistische Versuchsplanung* (‚*Design of Experiments*‘), (Heidelberg: Springer)

**Acknowledgments**

The IGF-research project no. 19.516 N of the “Research Association on Welding and Allied Processes of the DVS” has been funded by the AiF within the framework for the promotion of industrial community research (IGF) of the Federal Ministry for Economic Affairs and Energy because of a decision of the German Bundestag.
The authors would like to thank the German Research Foundation (DFG) for funding the project “Temperature Control in FSW”. Additional gratitude is given to Amanda Zens, Manuel Kessler, and Philippe Schua for their assistance.

IOP Conf. Series: Materials Science and Engineering **373** (2018) 012017 doi:10.1088/1757-899X/373/1/012017

# Intermetallic layers in temperature controlled Friction Stir Welding of dissimilar Al-Cu-joints

**R Marstatt[1,*], M Krutzlinger[2], J Luderschmid[1], G Constanzi[2], J F J Mueller[1], F Haider[1] and M F Zaeh[2]**

[1] Chair for Experimental Physics 1, University of Augsburg, Universitaetsstr. 1, 86159 Augsburg, Germany
[2] Institute for Machine Tools and Industrial Management (*iwb*), Technical University of Munich, Boltzmannstr. 15, 85748 Garching, Germany

[*] e-mail: roland.marstatt@physik.uni-augsburg.de

**Abstract.** Friction Stir Welding (FSW) can be performed to join dissimilar metal combinations like aluminium and copper, which is of high interest in modern production of electrical applications. The amount of intermetallic phases in the weld seam is significantly reduced compared to traditional fusion welding technologies. Because the solidus temperature is typically not reached during FSW, the growth of intermetallic phases is impeded and the intermetallic layer thicknesses typically remains on the scale of a few hundred nanometres. These layers provide a substance-to-substance bond, which is the main joining mechanism. Latest research confirms that the layer formation is most likely driven by the heat input during processing. Hence, the welding temperature is the key to achieve high quality joints. In this study, aluminium and copper sheets were welded in lap joint configuration using temperature-controlled FSW. An advanced in-tool measurement set-up was used to determine precise temperature data. Scanning electron microscopy (SEM) was used to analyse metallurgical aspects (e.g. structure and composition of the intermetallic phases) of the joints. The results show a correlation between the welding temperature and the thickness of the intermetallic layer and its structure. The temperature control significantly improved the correlation compared to previous studies. This leads to an enhanced understanding of the dominating joining mechanisms.

## 1. Introduction and state of the art

The recent changes in automotive design and construction require welding technologies, which are capable of joining dissimilar metals. One material combination of high importance in this regard is aluminium and copper. Both materials are widely used to produce efficient electrical powertrains. Friction Stir Welding (FSW) became an established process to join dissimilar materials [1]. The high electrical currents in automotive applications demand for high electrical conductivities within the connections of aluminium-copper joints.

The correlation between the welding parameters and the welding conditions on the one hand as well as the mechanical properties of the joints and their metallurgical structure on the other hand have been analysed in numerous studies. Detailed reviews concerning these cause-effect relationships were provided by [2–4].

The formation of layers of intermetallic compounds (IMC) has a significant influence on the joint properties. The layer thicknesses vary mostly due to the restrictions of the experimental set-up resulting in thermal effects such as heat accumulation for instance. Amongst others, the joint configuration, a tool offset and the parameter settings are the main influences on the process. The amount of IMCs decreases for an increasing feed rate and a decreasing rotational speed as reported by [5] for lap joints of aluminium 1060 and commercially pure copper. The authors suggested that neither low nor high heat input conditions result in sufficient joint strengths. The IMCs were observed near the workpiece interface within a region that was termed by the authors as black area. A reduced tensile strength for lap joints of aluminium Al5083 and commercially pure copper was concluded by [6]. This was ascribed to increased amounts of IMCs and micro cracks due to a high heat input. Several IMCs at different positions within cross sections of the seams for butt joints of aluminium AA2024-T3 and pure copper Cu10100 were detected [7]. Both studies did not specifically report a detailed analysis of the interface area. According to [8–14], an IMC layer is formed at the interface of the materials due to interdiffusion. Continuous IMC layers of about 1 $\mu m$ thickness were observed for butt joints of aluminium 1060 and commercially pure copper [8, 9]. Similar results were reported for aluminium-copper butt joints (5A02 with T2 and AA1100-H14 with cp, respectively) [10, 11]. It was concluded that an excellent metallurgical bonding is achieved by very thin IMC layers leading to enhanced joint strengths. Butt joints of aluminium 1050 and commercially pure copper were welded by [12]. Here, the tool was only stirring in the aluminium. The authors detected a thin IMC layer of about 200 $nm$, which was supposed to form after the tool passed. Detailed analyses on the formation kinetics of the IMCs were performed by [13]. They reported layer thicknesses up to 4 $\mu m$ for butt joints of aluminium 6082-T6 and pure copper. Since the layer was thicker than 2.5 $\mu m$, which was described as the upper limit to achieve a sufficient joint strength by [9], 4 $\mu m$ could result in a reduced joint strength. However, the joint strength was not measured by [13] and the used materials from both studies differ significantly. The correlation between the welding temperature and the growth of the IMC layers for lap joints of aluminium EN AW-1050 and copper CW008A was analysed in [14]. Increasing the rotational speed led to increased and saturated welding temperatures. A similar trend was observed for the dependency between the rotational speed and the joint strength. The authors proved the thicknesses of the IMC layers (all below 1 $\mu m$) to correlate with the welding temperature via an Arrhenius law.

The formed IMCs were specified by selected area electron diffraction (SAED) in transmission electron microscopy (TEM). It was found by [9; 12; 13] that the IMCs contain the θ-phase ($Al_2Cu$) and the $\gamma_2$-phase ($Al_4Cu_9$). Although [12] and [13] only observed these phases, the $\eta_2$-phase (AlCu) as a third component was detected in [9]. The stoichiometry of the IMC double layer was analysed by [14] for a set-up that is comparable to the current study. Investigations based on energy-dispersive X-ray spectroscopy (EDS) suggested the formation of the phases AlCu and $AlCu_3$.

The electrical resistance of friction-stir-welded aluminium-copper joints was investigated by [15] and [16]. Increasing resistances were measured for increasing heat inputs whereas the rotational speed was kept constant for welding butt joints of AA5754 and CW11000 [15]. It was suggested that this could result from the formation of IMCs. However, experiments on lap joints of aluminium ASTM 6060 T5 and copper ASTM B110 were performed in [16]. The authors could not observe a significant influence of the rotational speed on the electrical resistance and concluded that there is no indication of a deleterious effect of formed IMCs on the electrical conductivity of the joints.

It is evident that the thickness of IMC layers significantly affects the mechanical and physical properties of friction-stir-welded joints. Hence, strategies to influence or even control the thickness are crucial to achieve tailored joint properties. One approach is to offset the probe from the materials' interface as suggested by [12] to maintain thin layers of about 200 $nm$. The formation of IMCs in butt joints of aluminium AA6082-T6 and copper Cu-DHP was inhibited by offsetting the tool in [17]. However, the mechanical properties of the joints were not improved. Another approach is to control the welding temperature as demonstrated by [18] for lap joints of aluminium EN AW-6082-T651 and copper CW008A on a FSW robot in force-controlled mode. A significantly reduced flash formation due to the controlled heat input was observed.

Summarizing the studies discussed above, the following conclusions can be drawn:

- FSW is suitable to join aluminium and copper.
- The formation of IMCs influences the joint properties.
- Thin IMC layers provide a strong metallurgical bonding.
- Tool offsetting prohibits mechanical intermixing and IMCs in the nugget.
- Temperature-controlled FSW facilitates a defined heat input.

## 2. Experimental and analytical set-up

The effects of temperature-controlled FSW on the thickness of IMC layers are discussed in this study. For this purpose, lap joints of aluminium and copper were produced with defined welding temperatures. The temperature control was achieved by a PI controller and combined with a position control during the process to ensure a defined distance between the probe tip and the interface of the materials.

### *2.1. Experimental set-up*

The sample dimensions and the lap configuration of the aluminium-copper joints were defined according to [14] (see Figure 1). The aluminium sheets of the commercially pure alloy EN AW-1050 have dimensions of $245\ x\ 100\ x\ 4\ mm$ and the copper sheets of the commercially pure alloy CW008A $245\ x\ 100\ x\ 2\ mm$. The overlap length was set to $40\ mm$. The copper sheet was positioned on the retreating side (RS) and the aluminium sheet on the advancing side (AS) of the tool. The tool dimensions were $14\ mm$ for the shoulder diameter and $5\ mm$ for the conical probe ($3.6\ mm$ probe tip diameter). The probe length was applied to ensure only stirring in the aluminium sheet with a probe-tip-to-interface distance of $d_{PT-I} = 0.1\ mm$ for a tilt angle of 2° and a shoulder plunge depth of $0.1\ mm$. The tool probe was machined with three equally distributed flats and threaded. The experiments were performed on a CNC milling machine Heller MCH250.

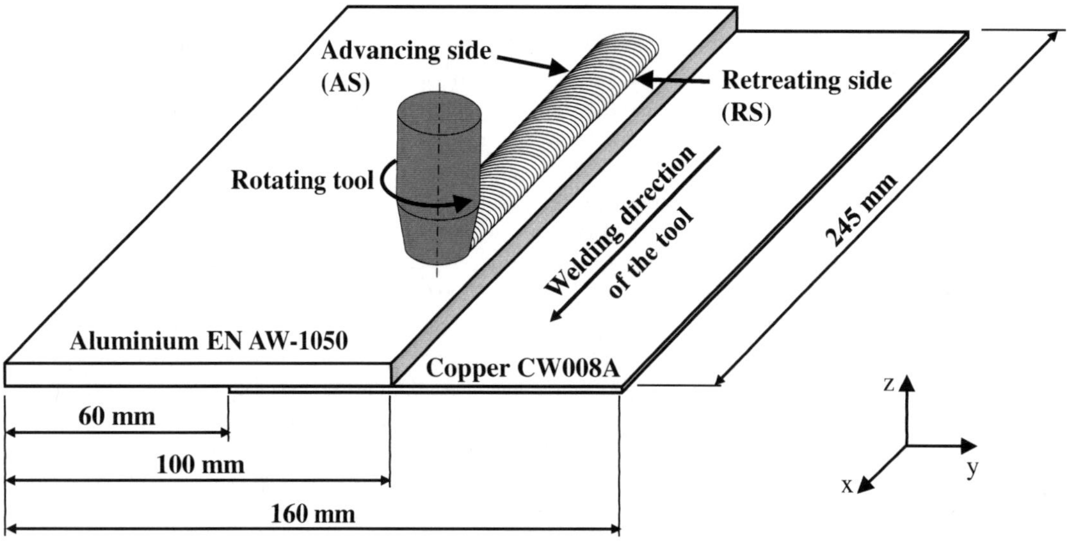


**Figure 1.** Schematic diagram of the sample dimensions and of the lap-welded aluminium-copper joints according to [14].

A closed-loop approach was used to implement a temperature-controlled FSW process. The controller system, which consists of the temperature measurement system according to [19], the PI controller and the milling machine, allowed to combine temperature as well as position control. The probe was equipped with a thermocouple of type K (diameter $0.5\ mm$) to measure the actual welding temperature during FSW (see Figure 2). Since the thermocouple was positioned at the middle of one flat and regarding the high thermal conductivity of the base materials, the interface temperature is assumed to almost equal the welding temperature. The PI controller calculated the required rotational speed $n$ to adjust the welding temperature to the set temperature.

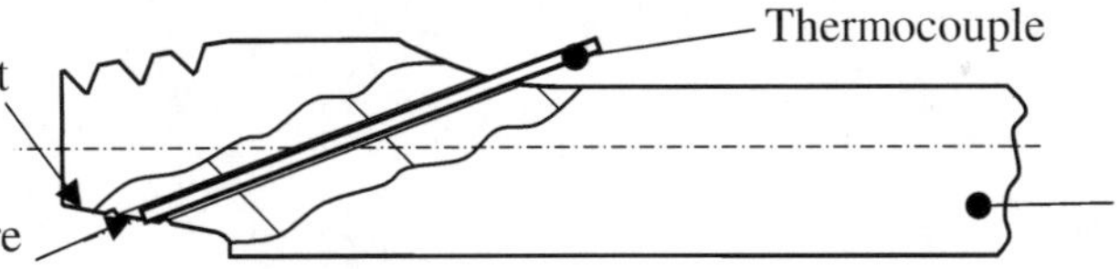


**Figure 2.** Position of the thermocouple to measure the welding temperature for temperature control.

The experiments were conducted for six different levels of the welding temperature; the lowest and highest temperature level were repeated (see Table 1). The boundaries of the welding temperature were defined based on experiments at constant rotational speeds of $n = 800\ min^{-1}$ and $n = 2800\ min^{-1}$ (see [14]). The feed rate was set to $v = 300\ mm/min$.

**Table 1.** Settings of the welding temperature for the experiments.

| Sample | Welding temperature in $°C$ |
|---|---|
| 1 | 410 |
| 2 | 410 |
| 3 | 430 |
| 4 | 465 |
| 5 | 500 |
| 6 | 535 |
| 7 | 570 |
| 8 | 570 |

*2.2. Analytical set-up*

The specimens for analyses with scanning electron microscopy (SEM) were cut out perpendicular to the welding seam at $85\ mm$ and $135\ mm$ relative to the starting edges of the sheets. Grinding down to FEPA-P4000 and polishing with diamonds of $3\ \mu m$ and $1\ \mu m$ were conducted to prepare the specimens for the SEM. The final polishing was applied with a solution of $SiO_2$ nanoparticles and $H_2O_2$ in water. For the analyses, a ZEISS Merlin equipped with an Oxford EDS system was used.

## 3. Interface analysis

The investigation of the joining interface revealed the existence of interlayers consisting of two IMC phases. Figure 3 shows the SEM images of three interface areas welded at temperatures of $410\ °C$, $465\ °C$ and $540\ °C$. The layer thicknesses ranged within some hundred nanometres depending on the parameter settings. For the phase adjacent to copper, EDS analyses based on SEM using $4.5\ keV$ acceleration voltage resulted in a composition of $37.4\ at\%$ Al and $62.6\ at\%$ Cu. The phase layer next to the aluminium consisted of $64.0\ at\%$ Al and $36.0\ at\%$ Cu. The stoichiometries imply that the IMCs $Al_4Cu_9$ and $Al_2Cu$ were formed, which was described in [9; 12; 13]. Since the analysis of nanolayers in SEM via EDS can be error-prone to a large extent, the IMCs could also be AlCu and/or $AlCu_3$. The formation of those IMCs was mentioned in other studies (e.g. [14]). TEM sample preparation of the interface by ion polishing is a challenging task. As all attempts lead to insufficient sample quality, distinct identification of the IMCs via TEM diffraction as conducted in [12] or [13] was not successful. Cutting electron transparent lamella by focused ion beam (FIB) is planned for the future.

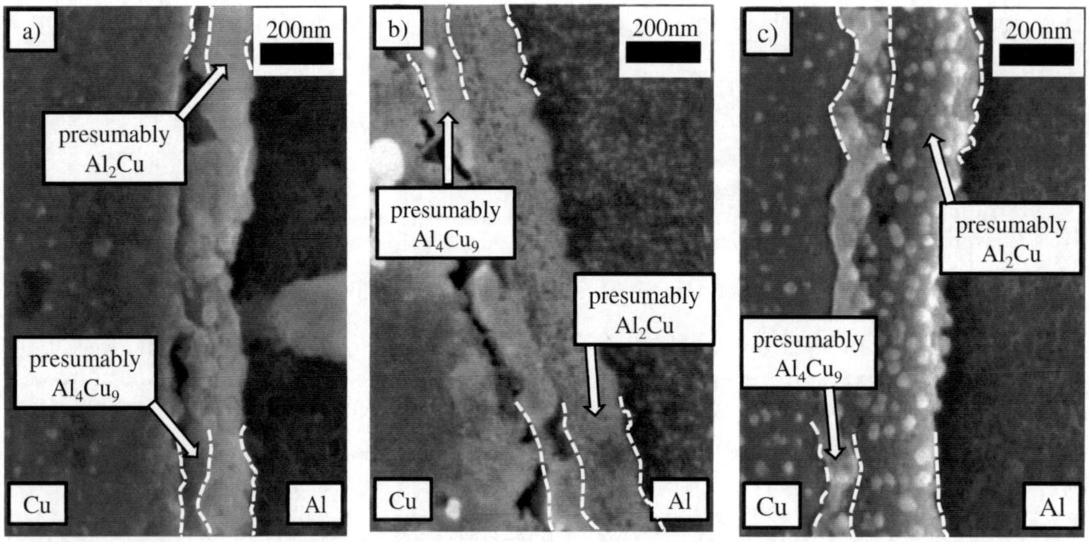


**Figure 3.** SEM images of the interface at a) 410 °$C$, b) 465 °$C$ and c) 540 °$C$ nominal temperature.

To gather valid information on the thickness of the IMC layers, the SEM specimens were measured at different positions of the cross section. Thus, a sound mean value and the standard deviation of the layer thickness for each parameter setting could be derived. Figure 4 shows the measured layer thicknesses drawn logarithmically against the inverse temperature in units of $K$. The margins surrounding the mean values are the standard deviations of the respective measures. A linear behaviour is clearly visible in this Arrhenius plot.

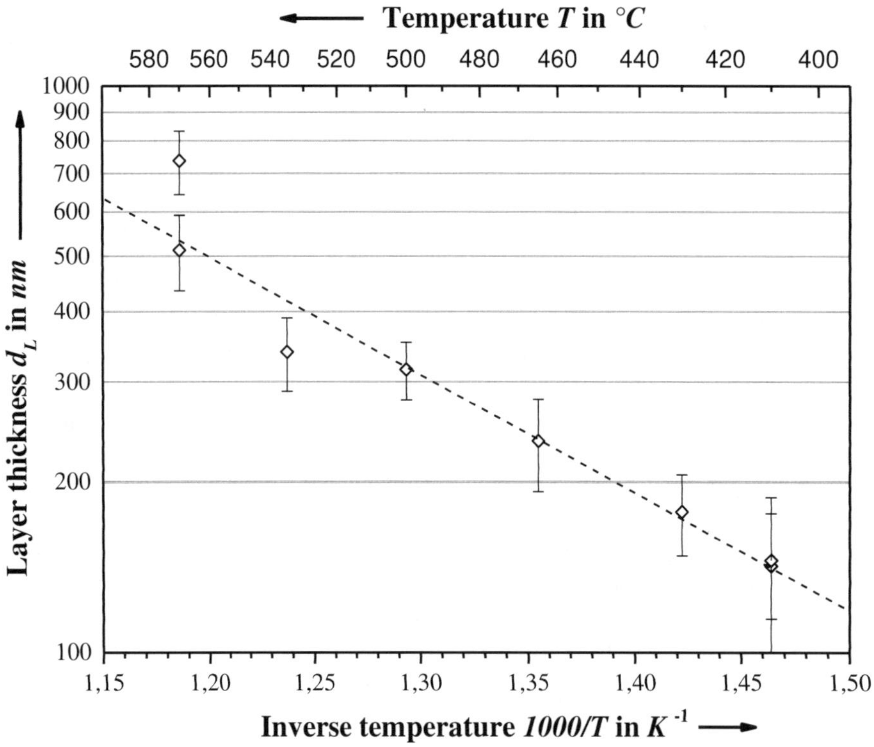


**Figure 4.** Arrhenius plot of welding temperature dependent doublelayer thickness.

The thickness $d(t)$ of the intermetallic layers grows by interdiffusion according to

$$d(t) = \sqrt{2Dt}, \tag{1}$$

where $t$ is the time of diffusion and $D$ is the rate constant of diffusive layer growth. The growth process is assumed to be thermally activated as

$$D(T) = D_0\, e^{-\frac{E_{act}}{k_B T}}. \tag{2}$$

$T$ is the temperature, $E_{act}$ the activation energy, $D_0$ is the prefactor of diffusion and $k_B$ the Boltzmann constant. Combining equations (1) and (2) yields

$$ln(d) = \frac{1}{2} ln(2t_{eq}D_0) - \frac{E_{act}}{2k_B}\frac{1}{T_{weld}}. \tag{3}$$

Here, $T_{weld}$ is the controlled welding temperature and $t_{eq}$ is the equivalent time interval around the peak of the temperature, which occurs during welding.
The slope of the linear fit from Figure 4, results in an activation energy of $E_{act} = 0.84 \pm 0.05\ eV$ for the growth of the double layer. The intermetallic layer growth by interdiffusion between Al and Cu was investigated by [20]. The resulting activation energies were specified to range from $0.85\ eV$ to $2.65\ eV$. An activation energy of $E_{act} = 1.27\ eV$ for the growth of $Al_2Cu$ and $E_{act} = 1.37\ eV$ for the growth of $Al_4Cu_9$ was determined. The experiments of [20] were conducted using annealed material. Accordingly, it has to be considered that the activation energy can be reduced by about one third. This is due to vacancy supersaturation or short circuit diffusion along dislocations or grain boundaries in highly deformed friction-stir-welded nugget material. An activation energy of $2.33\ eV$ could be measured in [14] for a comparable welding setup considering the investigated temperatures obtained in the heat affected zone (HAZ). Hence, the nugget temperatures in this study are more than $150\ K$ higher than the temperatures in [14]. The temperatures measured in the HAZ seem to be suitable for both process observation and identifying a suitable set of parameters. For a detailed understanding, a profound knowledge of the temperature of the nugget as recorded in this study is needed. The intercept of the straight line fitted in Figure 4 results in $t_{eq} * D_0 = 1.57 \cdot 10^{-4}\ cm^2$. Assuming $t_{eq}$ to be in the order of a few seconds (as done in [14]), this corresponds to $D_0$ of about $10^{-5}\ cm^2/s$, which is in good agreement to the findings in [20].

## 4. Conclusions and outlook

Dissimilar aluminium-copper lap joints were welded using temperature-controlled FSW. The thickness of the IMC layer at the interface was analysed using electron microscopy. Regarding the discussed results, the following conclusion can be drawn:

- The observation of an Arrhenius behaviour of the growth of the intermetallic layer was found, which is in agreement with previous studies by [14].
- An improved measurement system compared to [14] enabled to measure and control the welding temperature precisely. The interface temperature is assumed to almost equal the welding temperature. Therefore, an adapted Arrhenius analysis provides results, which are comparable and consistent to data of diffusion processes known from literature.
- A temperature control during FSW based on an in-tool temperature measurement combined with the derived Arrhenius correlation of welding temperature and IMC layer thickness offers new potential for joining dissimilar materials. The thickness of the IMC layer can be controlled precisely even for complex geometries and tailored to individual requirements.

## 5. References

[1] Murr L E 2010 A Review of FSW Research on Dissimilar Metal and Alloy Systems *J. Mat. Eng. and Perform.* **19** 1071–89

[2] Mubiayi M P and Akinlabi E T 2013 Friction Stir Welding of Dissimilar Materials between Aluminium Alloys and Copper - An Overview *World Acad. Sci., Eng. and Techn.* **76** 633–8
[3] Mehta K P and Badheka V J 2015 A Review on Dissimilar Friction Stir Welding of Copper to Aluminum: Process, Properties, and Variants *Mat. and Manuf. Proc.* **31** 233–54
[4] Galvão I, Loureiro A and Rodrigues D M 2016 Critical review on friction stir welding of aluminium to copper *Sci. and Tech. Weld. and Join.* **21** 523–46
[5] Abdollah-zadeh A, Saeid T and Sazgari B 2008 Microstructural and mechanical properties of friction stir welded aluminum/copper lap joints *J. All. and Comp.* **460** 535–8
[6] Bisadi H, Tavakoli A, Tour Sangsaraki M and Tour Sangsaraki K 2013 The influences of rotational and welding speeds on microstructures and mechanical properties of friction stir welded Al5083 and commercially pure copper sheets lap joints *Mat. & Des.* **43** 80–8
[7] Carlone P, Astarita A, Palazzo G S, Paradiso V and Squillace A 2015 Microstructural aspects in Al–Cu dissimilar joining by FSW *Int. J. Adv. Manuf. Technol.* **79** 1109–16
[8] Xue P, Xiao B L, Ni D R and Ma Z Y 2010 Enhanced mechanical properties of friction stir welded dissimilar Al–Cu joint by intermetallic compounds *Mat. Sci. and Eng.:* A **527** 5723–7
[9] Xue P, Xiao B L and Ma Z Y 2015 Effect of Interfacial Microstructure Evolution on Mechanical Properties and Fracture Behavior of Friction Stir-Welded Al-Cu Joints *Met. and Mat. Trans.* A **46** 3091–103
[10] Tan C W, Jiang Z G, Li L Q, Chen Y B and Chen X Y 2013 Microstructural evolution and mechanical properties of dissimilar Al–Cu joints produced by friction stir welding *Mat. & Des.* **51** 466–73
[11] Muthu M F X and Jayabalan V 2015 Tool travel speed effects on the microstructure of friction stir welded aluminum–copper joints *J. Mat. Proc. Tech.* **217** 105–13
[12] Genevois C, Girard M, Huneau B, Sauvage X and Racineux G 2011 Interfacial Reaction during Friction Stir Welding of Al and Cu *Met. and Mat. Trans.* A **42** 2290–5
[13] Avettand-Fenoël M N, Taillard R, Ji G and Goran D 2012 Multiscale Study of Interfacial Intermetallic Compounds in a Dissimilar Al 6082-T6/Cu Friction-Stir Weld *Met. and Mat. Trans.* A **43** 4655–66
[14] Marstatt R, Krutzlinger M, Luderschmid J, Zaeh M F and Haider F 2017 Formation of a diffusion-based intermetallic interface layer in friction stir welded dissimilar Al-Cu lap joints *IOP Conf. Ser.: Mater. Sci. Eng.* **181** 12002
[15] Akinlabi E T, Madyira D M and Akinlabi S A 2011 Effect of Heat Input on the Electrical Resistivity of Dissimilar Friction Stir Welded Joints of Aluminium and Copper *IEEE Africon 2011 (Livingston, Zambia)* (New York: IEEE)
[16] Wiedenhoft A G, Amorim H J, Rosendo T S and Strohaecker T R 2016Friction Stir Welding of Dissimilar Al-Cu Lap Joints with Copper on Top *KEM* **724** 71–6
[17] Galvão I, Loureiro A, Verdera D, Gesto D and Rodrigues D M 2012 Influence of Tool Offsetting on the Structure and Morphology of Dissimilar Aluminum to Copper Friction-Stir Welds *Met. and Mat. Trans.* A **43** 5096–105
[18] Bachmann A, Gamper J, Krutzlinger M, Zens A and Zaeh M F 2017 Adaptive model-based temperature control in friction stir welding *Int. J. Adv. Manuf. Technol.* **93** 1157–71
[19] Costanzi G, Bachmann A and Zaeh M F 2017 Entwicklung eines FSW-Spezialwerkzeugs zur Messung der Schweißtemperatur (‚Development of a FSW tool to measure the welding temperature') *DVS Berichte* vol 337 (Duesseldorf: DVS Media GmbH) 119–25
[20] Funamizu Y and Watanabe K 1971 Interdiffusion in the Al-Cu System *Trans. JIM* **12** 147–52

**Acknowledgements**
The research project "Mechanisms in joining of dissimilar materials by friction stir welding" is part of the priority program 1640 (SPP1640) "Joining by plastic deformation", which is funded by the German Research Foundation (DFG). The authors would like to thank for the funding and the support. The authors also thank Andreas Bachmann from the Institute for Machine Tools and Industrial Management (*iwb*) of the Technical University of Munich for technical assistance concerning the temperature control of FSW.

# The effect of additive geometry on the integration of secondary elements during Friction Stir Processing

**A Zens[1,*], M Gnedel[2], M F Zaeh[1] and F Haider[2]**

[1] Institute for Machine Tools and Industrial Management (*iwb*), Technical University of Munich, Boltzmannstr. 15, 85748 Garching, Germany
[2] Chair for Experimental Physics 1, University of Augsburg, Universitaetsstr. 1, 86159 Augsburg, Germany

[*] e-mail: amanda.zens@iwb.mw.tum.de

**Abstract.** Friction Stir Processing (FSP) can be used to locally modify properties in materials such as aluminium. This may be used, for example, to produce a fine microstructure or to integrate secondary elements into the base material. The purpose of this work is to examine the effect of the properties of the metal additives on the resulting material distribution in the processed region. For this, commercially pure iron and copper were integrated into an EN AW-1050 aluminium base material using FSP. Iron in the form of powder, wire and foil as well as copper in powder form were assessed. The various additive forms represent materials with differing surface-to-volume ratios as well as varying dispersion characteristics in the processing zone. The processing parameters for each additive form remained constant; however, two- and four-pass FSP processes were conducted. The results of CT analysis proved especially insightful regarding the spatial distribution of the various additive form within the workpiece. As expected, the powder additive was most widely distributed within the welding zone. Micro-hardness mappings showed that the powder additive contributed to the hardness within the weld nugget in comparison to the processed material without secondary elements.

## 1. Introduction and motivation

Friction Stir Processing (FSP) is an offshoot of Friction Stir Welding (FSW) and is used to create a locally refined microstructure in metal materials, most commonly aluminium. A schematic of the process, which is analogous to FSW, is shown in Figure 1a. First, a tool with rotational speed $n$, comprised of a shoulder and a probe, is plunged into the workpiece. During the plunge phase and processing, the material is softened due to the heat generated by friction between the workpiece and tool. After a short dwelling time, the tool is moved along the welding line at a feed rate $v$ and with a force $F_z$. Once the process is complete, the tool is extracted. The microstructure of the area in which processing occurred can be divided into several zones (Figure 1b) [1]. In the nugget zone (NZ), through which the tool passes, intensive material deformation and dynamic recrystallization occurs. The microstructure here is characterized by very fine and homogeneous grains. In the directly adjacent thermo-mechanically affected zone (TMAZ), plastic deformation still occurs; however, grains are larger and aligned along a certain direction. In the heat-affected zone (HAZ), no deformation occurs, but the material is subject to elevated temperatures, which leads to enlarged grains compared to the base material (BM). With regards to the material flow during FSP, a distinction must be made between the advancing side, in which the tangential velocity of the tool is in the same direction as the feed rate, and the retreating side, for which the opposite is true. It has been reported [2, 3] that the material flow in the nugget and the temperature around the nugget varies depending on which side is being considered; higher temperatures are reached on the advancing side compared to the retreating side.

FSP varies from FSW in that its primary purpose is not to weld two parts together, but rather to locally adapt the material properties. Mishra et al. [4] were the first to employ the FSP process in order to create surface composites by integrating an additive material (SiC) into the aluminium matrix in order to increase the surface hardness. Here, metal matrix composites (MMCs) were produced. Multi-pass welding is often necessary in order to assure a complete and homogeneous dispersion of the additive into the nugget. Various methods for depositing the additive material into the matrix via FSP can be found in literature. These include an application of the additive by spreading it over the material surface [4], inserting it into a groove machined out of the base material [5, 6], introducing it between plates in butt-weld configuration [7], and by integrating it in-situ via an adapted tool design [8]. Despite a variety of material deposition methods being reported, the form of the additive and its effect on the practicality of the process set-up and the resulting material properties has yet to be assessed.

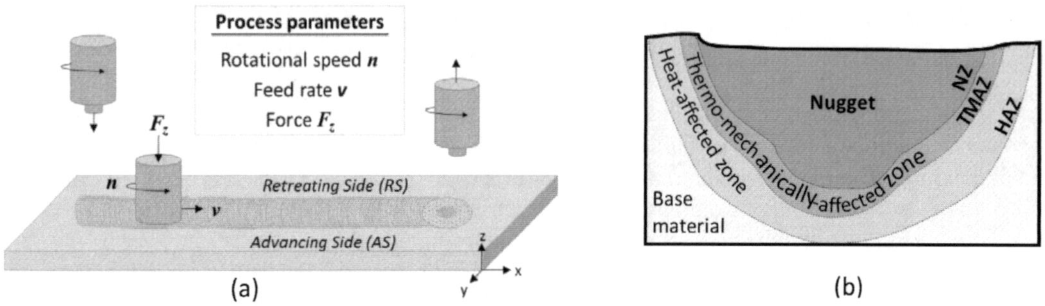


**Figure 1**. Schematic of the (a) FSP process and (b) areas of the process zone.

Researchers such as Hsu et al. [9] have reported using FSP to combine metal particles into the aluminium matrix in order to create MMCs. Here, the samples were first produced via powder metallurgy, which initiated the diffusion processes. Once multiple FSP passes were performed, it was reported that the additive material had completely reacted with the aluminium matrix to form Al-Cu intermetallic phases. This provided meaningful insights into the potential of FSP in order to process material via mechanical alloying; however, this method can only be applied to bulk materials: it cannot be used to locally change properties (i.e. chemical composition) in larger structures. The addition of copper to aluminium alloys via FSP is especially interesting because of the relatively high solubility of the Cu in Al and the potential to strengthen the material via the precipitation of secondary phases.

When powder is introduced as the additive material, the process is often referred to as Friction Powder Processing (FPP). Fujii et al. [7] performed FSP with iron powder additives in a 1050-H24 aluminium base and found that the mechanical properties of the material were enhanced by the formation of precipitates in the process zone. Abnar et al. [10] added copper powder to FSW welds of 3003 aluminium in order to counter the reduction in hardness and strength that typically occurs during FSW. They found that running the first and second passes in opposite directions (thus reversing the advancing and retreating sides) encouraged the creation of a homogeneous distribution of copper particles. In later works [11], the use of an aluminium-copper powder mixture was shown to improve the properties of the processing zone. Mahmoud and Tash [12] processed aluminium samples with additives of Fe and $Fe_2O_3$ powders in an aluminium matrix and analysed the magnetic properties.

The form of the additive is an important aspect to consider in designing a homogeneous processing zone and robust FSP process. The purpose of this work was to assess various aspects of the process set-up in order to integrate metal additives into an aluminium base using FSP. To begin, iron additives with a range of geometries were processed into an aluminium matrix and the most suitable additive form was determined. Then, copper and iron additives were integrated into an aluminium base via FSP and compared with respect to their metallurgical and mechanical properties. Copper and iron were chosen due to their distinctly different solubility properties: copper has a relatively high solubility in aluminium, whereas iron is almost insoluble.

## 2. Experimental methods

### *2.1. Experimental set-up and materials*

The base materials used in this work were sheets of 8 mm thick EN AW-1050 aluminium. Samples consisted of two 100 mm x 300 mm plates arranged in a butt-joint configuration. In order to insert the additive elements into the welding zone, a 1 mm wide groove was machined out of the (future) contact interface in one of the samples. The second plate was then placed over the additive-filled groove, thus containing the additive within the processing zone. The depth of the groove was dependent on the form of the additive. A schematic of the set-up can be found in Figure 2.

Experiments were performed on a Heller MCH 250 horizontal machining center. The employed tool consisted of a concave shoulder with diameter $d_s = 12$ mm as well as a probe with length $l_p = 3.5$ mm and diameter $d_p = 5$ mm. Temperature measurements were performed for selected trials. For this, K-type thermocouples with a diameter of 0.5 mm were placed into holes at various points along the welding line. The holes for the thermocouples were 0.6 mm in diameter and 2 mm deep. The locations of the thermocouples are given in Figure 2. The variation in employed materials and experimental set-up are as follows:

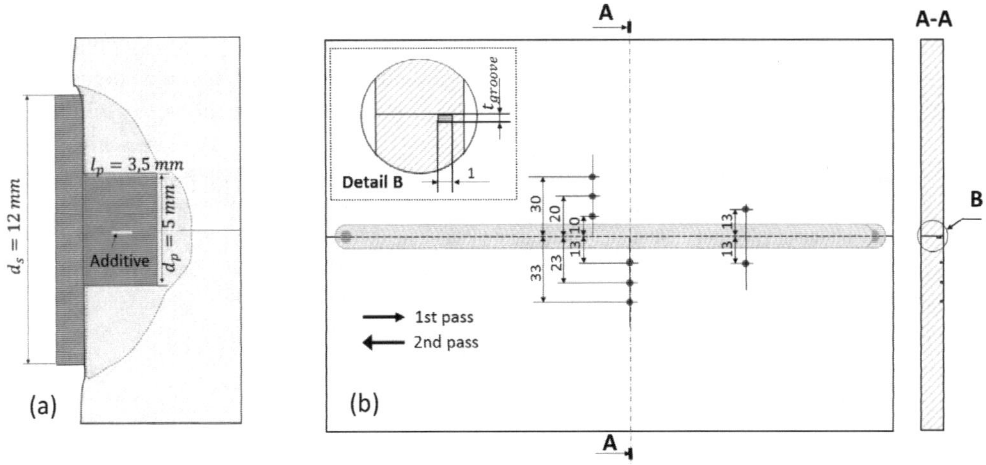


**Figure 2.** (a) Schematic of the probe geometry within the process zone and (b) schematic showing the location of the thermocouples.

#### *2.1.1. Comparison of the additive geometry (Trial A)*

Commercially pure iron in the form of powder, wire and foil was used as additive material. For foil and wire samples, a 50 µm deep groove was machined out of the aluminium sample ($t_{groove} = 50$ µm). Foil additives had a thickness of 50 µm and were cut to a width of 1 mm in order to fit into the groove. Several pieces of wire, each with a diameter of 50 µm, were placed into the groove until the amount iron present was comparable to experiments with foil additive. In order to assure a similar amount of iron in powder form (-200 mesh), the fill density was taken into account and the depth of the groove for powder additives was set to $t_{groove} = 130$ µm. Additives were located at a depth of 1.5 mm from the sample surface. The probe of the FSP-tool used in these experiments featured four flats machined at a 7° angle.

### 2.1.2. Comparison of iron and copper elements (Trial B)

Commercially pure iron and copper powders were used in order to compare the effect of the additive material on the properties of the processed zone. Iron powder with a size of -200 mesh and copper powder with diameter $d < 50$ µm were employed. The machined groove was 100 µm deep and was located at a distance of 1.0 mm from the sample surface. Additionally, a reference sample without additive material was processed for comparison. The probe of the FSP tool used in these experiments featured a screw profile with four flats machined at a 7° angle. The probe was rotated in a counterclockwise manner in order to drive the material into the processing zone and densify the material.

## 2.2. Processing parameters

The processing parameters remained constant for all experiments. The tool and additive were set up so that the additive was located at the center of the probe as it travelled through the processing zone. A rotational speed of $n = 1000$ rpm and feed rate of $v = 75$ mm/min were employed. The tool tilt angle was set to 2°. Experiments were run using position control with the plunge depth of 0.1 mm being held throughout the processing time. Samples from Trial A underwent a four-pass welding process with the feed rate direction being reversed for each pass. Samples from Trial B underwent two-pass processing with the second pass moving in the opposite direction of the first pass. The total weld line length was 265 mm.

## 2.3. Material characterization

The metallurgical and mechanical properties of the processed samples were characterized using a range of analysis methods. Computed tomography (CT) scans were performed in order to fully reconstruct and characterize the distribution of the additive elements throughout the processing zone. A voxel size of 5 µm was sufficient to get detailed information about the 3D shape of the intermixed material. Cross-sections at fixed positions revealed the spatial distribution in the given dimension. In order to assess the distribution of the additives within the processing zone, samples were ground and polished.

Samples from Trial B were etched using the Barker technique in order to increase the definition of the processing zone and to identify the dimensions of the nugget zone and the grain size. Also, hardness mappings with a spatial resolution of 300 $\mu m$ in the x- and y-directions and a peak force of 0.5 N were carried out using a Berkovich nano-indenter in order to assess the influence of additive particles and grain structure on the hardness and to evaluate the potential dissolution of the additive metal into the aluminium matrix.

# 3. Results and discussion

## 3.1. Comparison of the initial additive geometry with iron components (Trial A)

The CT images, shown in Figure 3, provide information that cannot be extracted from optical images: most importantly, the distribution of the additive along the processing line. The extent of material distribution varied greatly depending on which additive type was used. For instance, the powder additive resulted in an hourglass shape and somewhat smeared appearance with various areas of clumping along the processing line. It is unclear whether the pattern of powder distribution along the welding line is a result of the probe profile or due to an inhomogeneous distribution during set-up. An explanation for this could be cold welding of the deposited powder before the tool reaches it; the heat transfer ahead of the tool, especially at relatively low welding speeds, may have cause particles to agglomerate or weld before the tool reached them. The powder may also have compacted while being deposited into the groove. Despite the inhomogeneous deposition along the welding line, the copper powder was the additive that was most widely distributed throughout the depth of the processing zone.

The original position of the wire additive was parallel to the feed direction, almost perpendicular to the position in which the wire is seen in the CT scans. During FSP, the wire was broken up and to a large extent aligned perpendicular to the tool path (tangent to the tool rotation direction). The wire pieces are located in a band about 1 mm from the aluminium surface in a width of about 1 mm, which correlates well with the 1 mm width of the machined groove in which it was originally placed. Hence, the wire was not distributed through the depth of the processing zone.

IOP Conf. Series: Materials Science and Engineering **373** (2018) 012018 doi:10.1088/1757-899X/373/1/012018

The trial with iron foil was more difficult to assess, as very few foil remnants were able to be found in the sample. Due to the very low solubility of iron in aluminium and the lack of dissolution of the powder and wire pieces within the aluminium matrix, it is highly unlikely that the foil was dissolved into the aluminium matrix. Rather, it is expected that the foil formed large clumps, which are located sporadically along the welding line. Composite images formed by overlapping all CT images of the xy-plane over the depth of the z-axis showed several areas with iron remnants.

Each additive geometry displayed various positive and negative aspects during the experimental set-up and the integration during FSP. For example, a powder additive has a comparatively high surface-to-volume ratio and thus would theoretically allow for a more rapid integration of additives into the matrix through diffusion processes. However, as powder sizes decrease, the handling of the powder becomes challenging as particles tend to clump together due to increased surface electrostatic forces. Additionally, bringing a specified amount of powder into a given area requires suitable application strategies and the amount deposited is essentially an estimation based on the measured fill density of the powder. A more practical option from an experimental set-up and a potential industrialization viewpoint it to use foil additives. Here, the amount of material given can be easily calculated and integrated into the processing zone using a machined groove or simply placing the material between two plates in a butt joint or overlap configuration. However, foils must first be torn apart and then distributed along the process zone during processing. The possibility to completely break apart and homogeneously distribute or dissolve foil additives is questionable. The additive in wire form was easy to deposit and the theoretical amount being integrated was relatively easy to calculate; however, the wire material did not spread throughout the depth of the processing zone as completely as the powder, and it remained present in the material as large pieces. Due to the larger scattering area of the powder within the processing zone, additives in powder form were selected for further analysis.

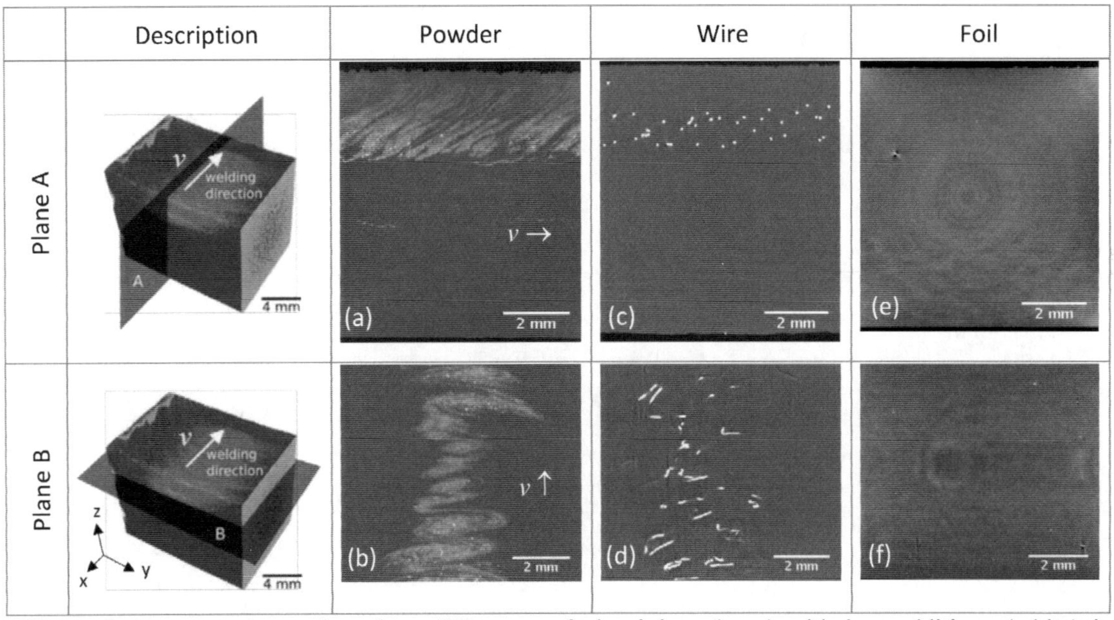


**Figure 3.** Cross sections taken from CT scans of aluminium (grey) with iron additives (white) in powder (a, b), wire (c, d) and foil (e, f) form. Images were taken from the xz-plane (plane A) and the xy-plane (plane B).

*3.2. Comparison of iron and copper powder additives (Trial B)*

A sample of the temperature measurements taken during the second pass of the FSP trial with iron powder additive can be seen in Figure 4. All trials yielded similar temperature profiles and the peak values varied only slightly, as demonstrated by the plot of the maximum temperature for each acquired data set (Figure 5).

As seen in Figure 5, the maximum temperature at a distance of 10 mm from the weld center (4 mm from the edge of the tool shoulder) ranged from 290-294 °C. Given the steep, nonlinear trend towards rising temperatures when moving closer to the process zone, it is expected that the temperature in the nugget zone was higher than the recorded temperatures. Increasing the rotational speed would also presumably increase the temperature and accelerate the diffusion processes.

There was no discernable difference in temperatures between the advancing and retreating side; however, the measurements taken from points 1 and 5 were systematically lower than those taken at point 2, although they were at the same distance from the center line. The reason for this may be the difference in proximity to the edge of the workpiece. Due to technical problems which occurred during processing, some measurement points were not able to be recorded (i.e. point 4 for the experiment with copper additives). Figure 5 shows that the thermal cycles for samples with iron and copper were very similar.

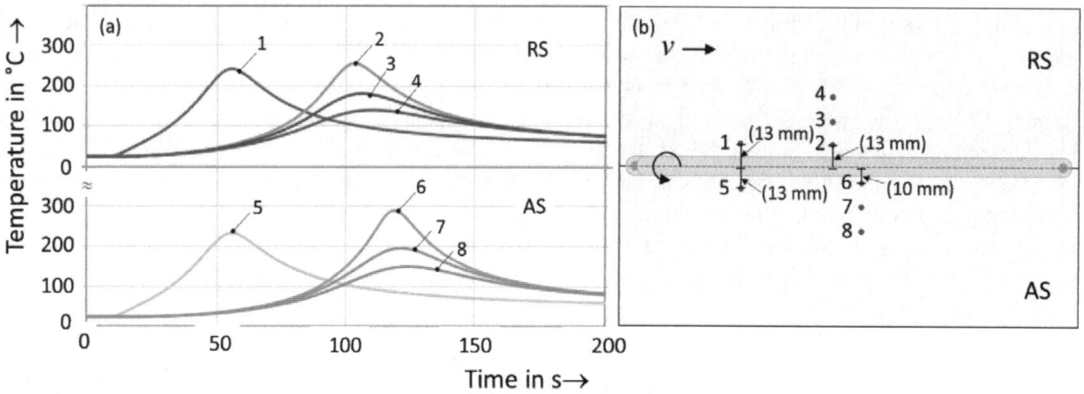


**Figure 4.** Temperature curves taken during FSP processing (a) and a schematic (b) showing the positions of the thermocouples. The numbering of the curves shows the thermocouple positions.

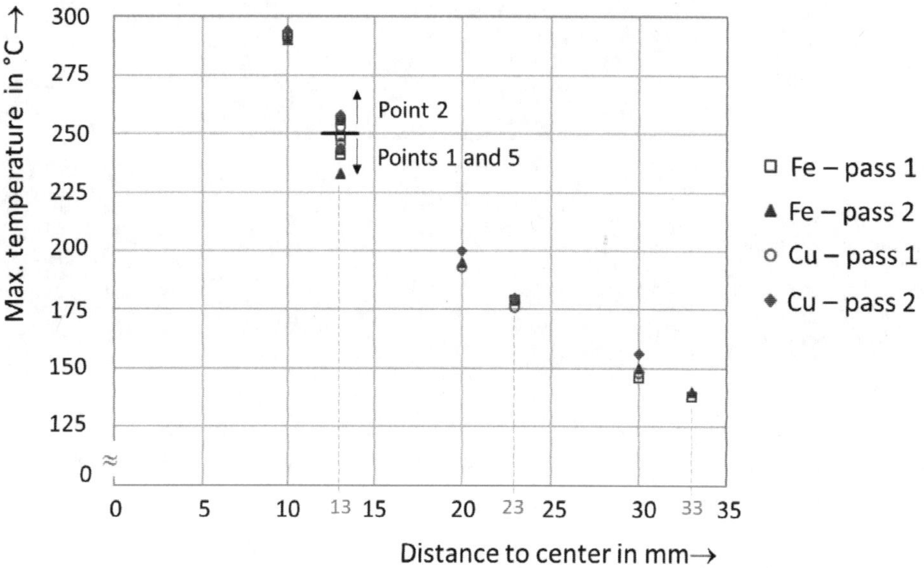


**Figure 5.** Plot of the maximum temperature for each thermocouple.

Regarding the grain structure, no significant differences between the trials with additives and the reference trial could be found (Figure 6). A higher influence of the additive metal on the dynamic recrystallization behavior during FSP is expected when the amount of additive is increased and when it is more homogeneously distributed. Within stir zone, no individual grains are visible due to lack of resolution. A significant difference in the advancing side (AS) and retreating side (RS) of the second pass was not observed; both sides appear to be symmetric in terms of grain size.

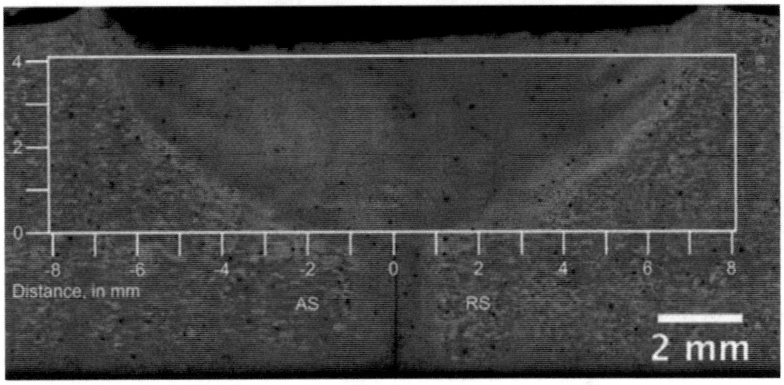


**Figure 6.** A cross section of the reference sample showing the difference in grain size between the nugget and base material. The sample was taken in the welding direction (AS and RS refer to second pass) and color-etched using the Barker-technique. The area use for hardness mappings, shown in Figure 7, is indicated.

The hardness mappings shown in Figure 7 indicate hardness up to 48 HV in the area where the copper powder was originally placed, which is up to 10 HV more than the corresponding area in on the reference sample. The iron particles led to a hardness of around 40 HV in this area. The main reason for the high peak hardness values in the area of the additive particles may have several factors; for one, intermetallic multilayers between the particles and the aluminium matrix are formed during the process, which typically exhibit substantially higher hardnesses than the base metals. An additional factor is the distribution of the powder within the nugget; if the additive material clumped in the area of the groove, rather than being distributed throughout the nugget, it would also lead to a rise in hardness.

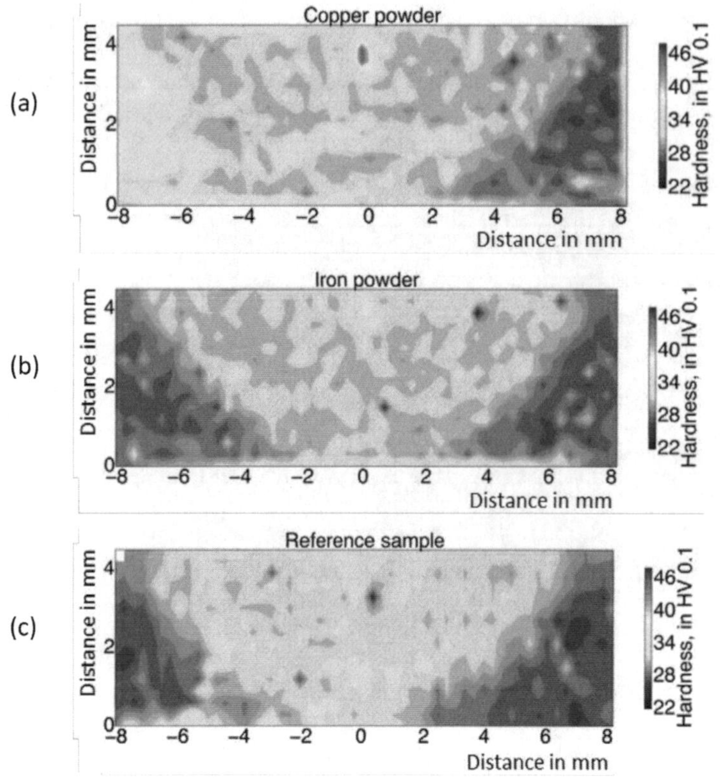


**Figure 7.** Hardness mappings with a resolution of 300 x 300 µm carried out with a Berkovich indenter of the cross section in welding direction of EN-AW 1050 with (a) copper particles, (b) iron particles, and (c) without additive particles.

Apart from the influence of the additive particles, the hardness maps reflect the grain structure in Figure 6 to the extent that the micrometer-sized dynamically recrystallized grains in the stir zone (nugget) lead to an increase in hardness of up to 10 HV. The hardness profile of the reference sample and the sample with iron powder appeared to be relatively symmetric within the processing zone; however, higher hardness values were found on the AS of the sample with copper powder, outside of the nugget zone. This occurrence cannot be explained in the framework of this paper.

## 4. Summary and outlook

In this work, the effect of the additive material and geometry in the creation of MMCs using FSP was assessed. Commercially pure iron additives in powder, wire, and foil form were integrated into an aluminium matrix using FSP. Foil and wire additives proved easy to deposit and also offered an easy way to control how much material was being added. However, based on the distribution and greater potential for powder additives to dissolve into the matrix, they were selected as the most suitable for the FSP process and were selected for further processing. Next, copper and iron additives in powder form were processed into the aluminium sheets. The hardness of the powder samples in the area of additive deposition was higher than the reference sample by about 10 HV. Despite these improvements, more suitable parameters must be found to deposit the powder material throughout the nugget and create a more homogeneous processing zone.

## 5. References

[1] Mishra R S and Mahoney M W (eds) 2007 *Friction Stir Welding and Processing* (Materials Park, Ohio: ASM International)

[2] Nandan R, Roy G G, Lienert T J and Debroy T 2007 Three-dimensional heat and material flow during friction stir welding of mild steel *Acta Materialia* **55** 883–95

[3] Shercliff H R and Colegrove P A 2007 Process Modeling *Friction Stir Welding and Processing* ed R S Mishra and M W Mahoney (Materials Park, Ohio: ASM International) 187–217

[4] Mishra RS, Ma ZY and Charit I 2003 Friction stir processing: A novel technique for fabrication of surface composite *Materials Science and Engineering: A* **341** 307–10

[5] Gandra J, Miranda R, Vilaça P, Velhinho A and Teixeira J P 2011 Functionally graded materials produced by friction stir processing *Journal of Materials Processing Technology* **211** 1659–68

[6] Lee C J, Huang J C and Hsieh P J 2006 Mg based nano-composites fabricated by friction stir processing *Scripta Materialia* **54** 1415–20

[7] Fujii H, Sun Y, Inada K, Ji Y, Yokoyama Y, Kimura H and Inoue A 2011 Fabrication of Fe-Based Metallic Glass Particle Reinforced Al-Based Composite Materials by Friction Stir Processing *Mater. Trans.* **52** 1634–40

[8] Huang Y, Wang T, Guo W, Wan L and Lv S 2014 Microstructure and surface mechanical property of AZ31 Mg/SiCp surface composite fabricated by Direct Friction Stir Processing *Materials & Design* **59** 274–8

[9] Hsu C J, Kao P W and Ho N J 2007 Intermetallic-reinforced aluminium matrix composites produced in situ by friction stir processing *Materials Letters* **61** 1315–8

[10] Abnar B, Kazeminezhad M and Kokabi A H 2014 The Effect of Cu Powder During Friction Stir Welding on Microstructure and Mechanical Properties of AA3003-H18 *Metall and Mat Trans A* **45** 3882–91

[11] Abnar B, Kazeminezhad M and Kokabi A H 2015 The Effect of Premixed Al-Cu Powder on the Stir Zone in Friction Stir Welding of AA3003-H18 *J. of Materi Eng and Perform* **24** 1086–93

[12] Mahmoud E R I and Tash M M 2016 Characterization of Aluminium-Based-Surface Matrix Composites with Iron and Iron Oxide Fabricated by Friction Stir Processing *Materials (Basel, Switzerland)* **9**

**Acknowledgements**

The results presented here were produced within the scope of the Project "Synthesis of non-equilibrium alloys using friction stir welding processes" (FSWLeg), which is funded by the German Research Foundation (DFG). The authors would like to thank the DFG for their support.

IOP Conf. Series: Materials Science and Engineering **373** (2018) 012019 doi:10.1088/1757-899X/373/1/012019

# Investigations on the Influence of Parameters During Electron Beam Surface Hardening Using the Flash Technique

**S Grafe[1], P Hengst[1], A Buchwalder[1] and R Zenker[1,2]**

[1] Institute of Materials Engineering, TU Bergakademie Freiberg, 09599 Freiberg, Germany
[2] Zenker Consult, 09648 Mittweida, Germany

**Abstract.** The electron beam hardening (EBH) process is one of today's most innovative industrial technologies. Due to the almost inertia-free deflection of the EB (up to 100 kHz), the energy transfer function can be adapted locally to the component geometry and/or loading conditions. The current state-of-the-art technology is that of EBH with continuous workpiece feed. Due to the large range of parameters, the potentials and limitations of EBH using the flash technique (without workpiece feed) have not been investigated sufficiently to date. The aim of this research was to generate surface isothermal energy transfer within the flash field. This paper examines the effects of selected process parameters on the EBH surface layer microstructure and the properties achieved when treating hardened and tempered C45E steel. When using constant point distribution within the flash field and a constant beam current, surface isothermal energy input was not generated. However, by increasing the deflection frequency, point density and beam current, a more homogeneous EBH surface layer microstructure could be achieved, along with higher surface hardness and greater surface hardening depths. Furthermore, using temperature-controlled power regulation, surface isothermal energy transfer could be realised over a larger area in the centre of the sample.

## 1. Introduction

Numerous mechanical engineering components – such as gears, camshafts and crankshafts – are subject to high levels of stress in terms of wear, impact and vibration. Such conditions require excellent component properties. Hard surface layers are essential in ensuring sufficient wear resistance. Residual compressive stresses in the surface layer increase fatigue strength. At the same time, a tough core material is necessary to avoid component failure due to brittle fracture [1, 2]. Such component properties can be achieved through thermal surface treatment. One of the most innovative thermal surface treatment technologies used today is that of electron beam hardening (EBH), which generates a hard and wear-resistant martensitic surface layer on the base material [3]. The use of electron beam (EB) as an energy source is characterised by high-speed 2D/3D beam deflection (up to 100 kHz), which provides almost inertia-free movement of the EB. The result is that almost any deflection figure can be generated with defined point and line patterns within a predetermined energy transfer field. The intensity of the energy input within the deflection field is determined by the spot interaction time of the EB and the numbers of energy inputs per spot. This means that the energy transfer function can be adapted to the specific component geometry [2, 4, 5].

Depending on the relative movement between the component and the EB during the interaction process, two beam deflection techniques may be applied – the CI (Continuous Interacting) technique and the flash technique. The CI technique – with the component moving relative to the EB – is mostly used for surface hardening of large components, e.g. camshafts [2, 6]. The flash technique is characterised by a fixed component position under the EB, and is frequently used for components where the treatment area

IOP Conf. Series: Materials Science and Engineering **373** (2018) 012019 doi:10.1088/1757-899X/373/1/012019

is not larger than the maximum usable EB energy transfer field. Thus, calottes and the valve seats of cylinders, for example, can be EB hardened using the flash technique [2, 6].

In terms of EBH using the flash technique, the potentials and limitations have not been investigated sufficiently to date. To achieve an optimum result in the hardening process, each point of the energy transfer field must be heated up rapidly to the required austenitising temperature and kept at that temperature over the entire energy impact duration. This process is termed surface isothermal energy transfer. Subsequently, rapid cooling is obtained by self-quenching [5, 7, 8]. Therefore, the goal of these investigations was to generate surface isothermal energy transfer within the energy transfer field applied. To this end, the influence of selected parameters was examined by means of the microstructure and hardness of the EBH layer. The experiments were carried out on the hardened and tempered C45E steel. Surface profile measurements and macroscopic examinations were used to characterise the EBH surface characteristics as a function of the surface hardening depth. Microscopic examinations and hardness-depth measurements were carried out to assess the EBH layer properties. The temperature distribution within the energy transfer field during the process was recorded with a thermographic camera.

## 2. Experimental details

### *2.1. Base material and sample preparation*

The material used for the investigations was C45E steel in its hardened and tempered pre-heat-treated state, and with the chemical composition given in Table 1.

**Table 1.** Chemical composition (GDOES analysis).

| Element concentration [wt.%] | | | | | | | | | |
|---|---|---|---|---|---|---|---|---|---|
| Fe | C | Si | Mn | P | S | Cr | Mo | Ni | V |
| Bal. | 0.51 | 0.27 | 0.76 | 0.01 | 0.01 | 0.14 | 0.06 | 0.22 | 0.02 |

The C45E steel was delivered as rolled square-section bar and was cut into standard samples with the dimensions 30 x 30 x 20 mm$^3$. The sample surface to be treated was ground with 500 grit sandpaper.

The hardened and tempered C45E steel exhibited a banded microstructure due to the rolling process (Figure 1a). The matrix microstructure consisted of tempered martensite and ferrite (Figure 1b). The ferrite indicated incomplete austenitisation during the volume hardening process.

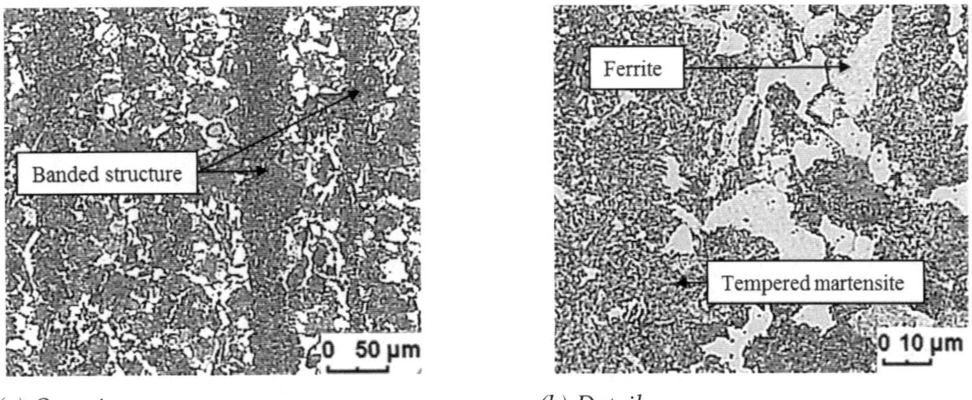


*(a) Overview* *(b) Detail*

**Figure 1.** Light optical micrographs of the hardened and tempered C45E steel.

### *2.2. Experimental setup*

Electron beam surface hardening (EBH) tests using the flash technique were carried out in a universal electron beam facility (K26-15/80, pro-beam, Germany) at the Institute of Materials Engineering, TU Bergakademie Freiberg. Figure 2a shows a schematic diagram of the experimental setup. The sample to be treated was positioned centrally under the electron beam in the working chamber. During the energy input, a flash field (20 x 20 mm$^2$) interacted with the fixed sample. The flash field applied was programmed over several points (spots) (Figure 2b). For recording the temperature distribution within the entire energy transfer field, a PI 1M thermographic camera from Optris (measurement range: 500 °C–1800 °C) was situated at an angle of 61° and at a working distance of 35 mm from the sample (Figure 2a). The emissivity value used for the thermographic camera was set at 0.24, which conformed to the emissivity value exhibited in general by freshly ground steels and ferrous metals at room temperature [9]. For the temperature-controlled power regulation, a Metis HL-22 ratio pyrometer from SensorTherm (measurement range: 550 °C–1400 °C) was placed in a position similar to that of the thermographic camera. The measurement point was located in the centre of the sample surface. The emissivity ratio for the pyrometer was 1.007. This value was determined in a previous work and adopted for the investigations carried out in this study, cf. [10].

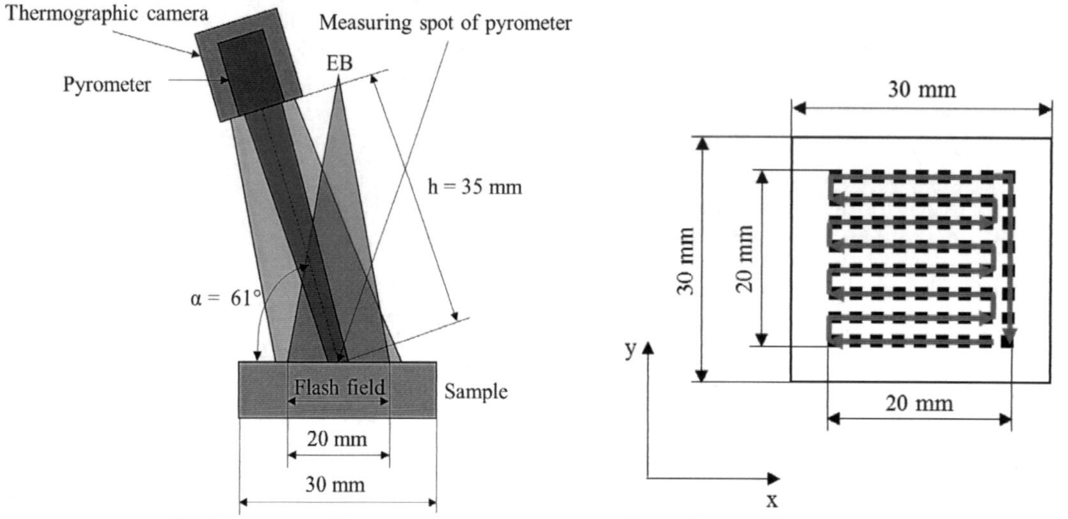


*(a) Experimental setup* *(b) Deflection figure within the flash field*

**Figure 2.** Schematic diagrams of the experimental setup and programmed flash field.

The experimental plan was divided into two work sections. Table 2 shows the parameter sets for each work section.

**Table 2.** Parameter sets.

| Work section | Test series | Beam current $I_B$ (mA) | Deflection frequency f (Hz) | Point density (points/mm$^2$) | Set temperature $T_S$ (°C) |
|---|---|---|---|---|---|
| 1 | 1 | 50 | 100–10000 | 25 | – |
| | 2 | 50 | 4000 | 6.25 | – |
| | | | 1000 | 25 | |
| | | | 444 | 56.25 | |
| | 3 | 50 | 1000 | 25 | – |
| | | 58.4 | | | |
| | | 66.8 | | | |
| 2 | 1 | 20–90 | 250–1000 | 25 | 1000 |
| | 2 | 20–120 | 1000 | 25 | 1000 |

In the first work section, the experiments were carried out without temperature-controlled power regulation ($I_B$ = constant) and the frequency, point density and beam current were varied. In the second

work section, the tests were performed using temperature-controlled power regulation, while the influence of the deflection frequency and control range of the beam current was investigated (Table 2). As described in [10], the ratio pyrometer with integrated PID controller measured the actual surface temperature and compared it to the set temperature. The manipulated variable was then sent to the CNC of the EB system. The CNC regulated the beam current according to the manipulated variable to achieve the set temperature [10].
For all of the point figures used in these experiments, the deflection frequency was programmed in the software by means of the point delay. A point delay may be defined as the time delay of the electron beam at one point [11]. Therefore, it functions indirectly as the EB spot interaction time. This provided the following correlation: The EB spot interaction time decreased with increasing deflection frequency, whereby the EB plots the deflection figure in a shorter period of time. For all tests, an acceleration voltage of 60 kV was used and the working distance, focus (surface-sharp) and process time ($t_H$ = 4 s) were kept constant.

*2.3. Test methods*
The characterisation of the EBH surface was carried out by means of macroscopic images and surface profile measurements (Figure 3). The macroscopic images were used to evaluate the geometry and to determine the area of the EBH surface $A_{EBH}$ (Figure 3a). The surface profiles were recorded with the non-contact surface measurement system SMS from Breitmeier Messtechnik GmbH. The surface profiles facilitated the determination of the volume expansion (Figure 3b) caused by the formation of martensite [12]. To this end, the line measurements were performed at a measurement point density of 100 points/mm and at a sampling frequency of 50 Hz.

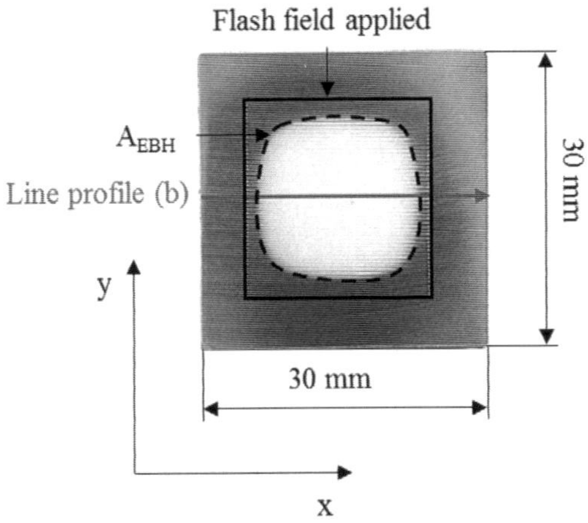


*(a) Sample with EBH surface (white area)*

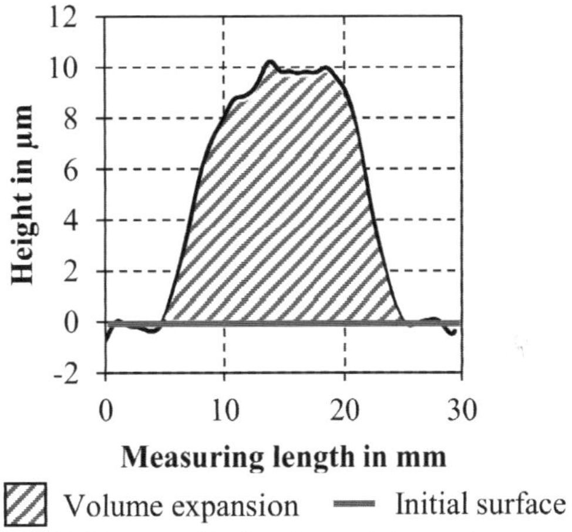


*(b) Line profile for determining volume expansion*

**Figure 3.** Characterisation of the EBH surface.

The line profiles were recorded in the x-direction (Figure 3a). The area integral between the graph and the initial surface represented the volume expansion that resulted from the martensitic transformation (Figure 3b). The light-microscopic microstructural examination of the EBH layer and determination of the hardness-depth profiles were carried out by means of sample cross sections. Hardness-depth profiles were measured at HV0.3 with a Vickers indenter. The hardness indentations were set as a triple row at a distance of 0.05 mm, with the first indentation located 0.05 mm beneath the sample surface. Depending on the surface hardening depth (SHD), 30 to 60 hardness indentation measurements were carried out per hardness-depth profile. Due to the banded microstructure of the C45E steel, three hardness-depth profiles were recorded per sample. The minimum hardness value ($HV_{min}$) and the maximum hardness value ($HV_{max}$) were determined for each measurement point of the three hardness-depth profiles. The averaging of $HV_{min}$ and $HV_{max}$ per measurement point resulted in an averaged hardness-depth profile, which was used to determine the surface hardening depth (SHD) according to DIN EN 10328. To determine the limit hardness, the minimum surface hardness value was calculated from the average of the first three hardness values of the mean hardness-depth profile.

## 3. Results

### *3.1. Work section 1: Without temperature-controlled power regulation ($I_B$ = constant)*

#### *3.1.1. Influence of the deflection frequency*

Figure 4 shows macroscopic images of the EBH surface generated with different deflection frequencies. It was found that for lower frequencies of up to 250 Hz, surface melting could occur as a result of the relatively long EB spot interaction times (Figure 4a, b). Deflection frequencies higher than 250 Hz did not lead to surface melting due to the shorter EB spot interaction times (Figure 4c–f). In addition, it could be seen that the applied EBH flash field area of 20 x 20 mm$^2$ was only achieved at a low deflection frequency of 100 Hz (Figure 4a). The higher the deflection frequency (up to 1000 Hz), the smaller the EBH surface area formed (Figure 4a–d). This could be explained by the shorter EB spot interaction time associated with increased deflection frequency, which meant that the required hardening temperature in the peripheral areas was no longer achieved or that the lateral heat flow was greater.

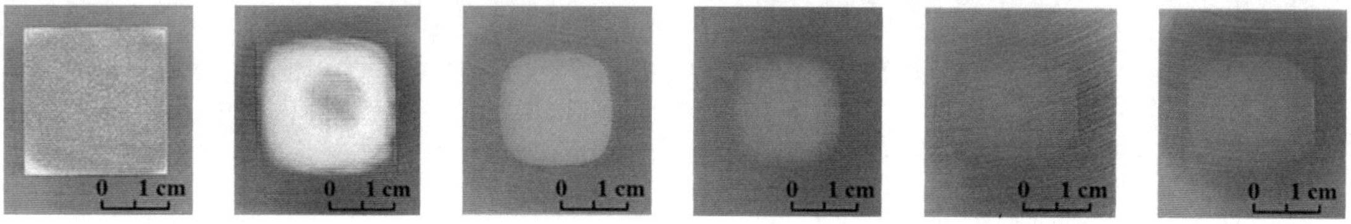


*(a) 100 Hz* $A_{EBH} = 4.0\ cm^2$ *(b) 250 Hz* $A_{EBH} = 3.3\ cm^2$ *(c) 500 Hz* $A_{EBH} = 2.5\ cm^2$ *(d) 1000 Hz* $A_{EBH} = 2.2\ cm^2$ *(e) 5000 Hz* $A_{EBH} = 2.3\ cm^2$ *(f) 10000 Hz* $A_{EBH} = 2.3\ cm^2$

**Figure 4.** Influence of the deflection frequency on the shape and size of the EBH surface.

At higher deflection frequencies ($f \geq 5000$ Hz), the applied EBH field length of 20 mm in the x-direction was almost achieved (Figure 4e, f). The higher number of energy inputs at each spot during the process seemed to compensate for the shorter EB spot interaction time. Nevertheless, the EBH surface appeared to be shortened in the y-direction. It was suspected that the electron beam moved over a longer period of time in the x-direction than in the y-direction, which meant that less energy was introduced in the y-direction. Due to the phenomenon of surface melting at $f \leq 250$ Hz, further test results are presented for deflection frequencies of $f \geq 500$ Hz.

The EBH layer cross sections (Figure 5) exhibited a lens-shaped geometry due to the more rapid heat dissipation in the near-edge areas of the samples compared to that of the sample centre.

(a) (b) (c) (d)

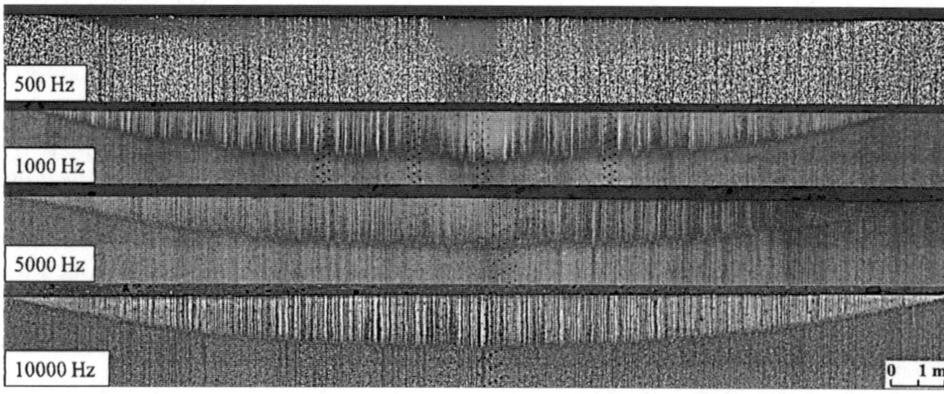


**Figure 5.** Influence of the deflection frequency on the EBH layer geometry in cross section.

For a constant point density, the number of energy inputs per spot influenced the results significantly. The higher the deflection frequency, the higher the number of energy inputs per spot. This resulted in both a wider EBH layer and a more homogeneous martensitic EBH microstructure at deflection frequencies of f ≥ 5000 Hz (Figure 5c, d).
Furthermore, higher deflection frequencies (f ≥ 5000 Hz) had a positive effect on the hardness-depth profiles measured (Figure 6a). In comparison to lower frequencies (f ≤ 1000 Hz), the average maximum EBH surface hardness was improved from approx. 509 HV0.3 to 582 HV0.3 (Figure 6a), while the average surface hardening depth (SHD) increased from approx. 0.45 mm to 0.5 mm (Figure 6b).
The higher the hardening depth, the higher was the volume expansion after the martensitic transformation (Figure 6b). This could be explained by the higher proportion of martensite formed with increasing hardening depth, which also resulted in higher surface hardness values.

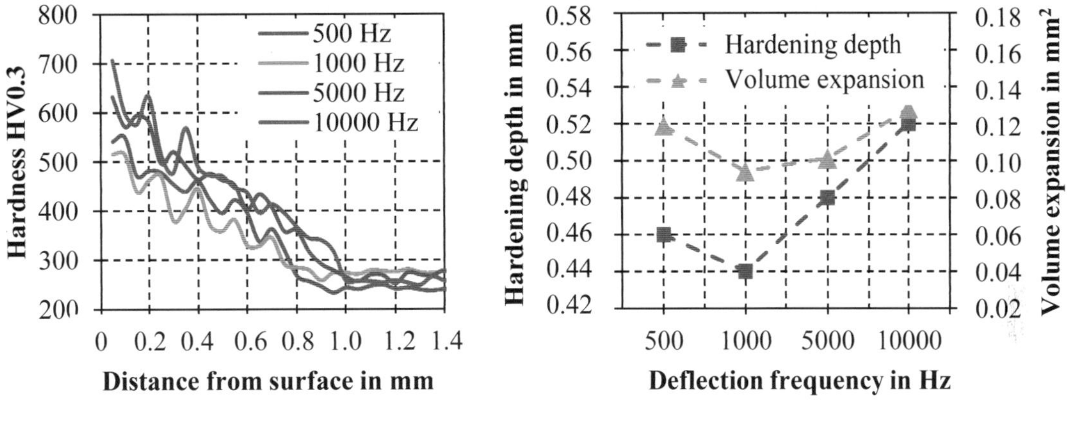


*(a) Hardness-depth profiles* *(b) Surface hardening depth and volume expansion*

**Figure 6.** Influence of the deflection frequency on the hardness-depth profiles and the characteristic parameters resulting from the martensitic transformation.

*3.1.2. Influence of the point density*

As illustrated in Figure 7 by means of etched macroscopic images, the higher the point density, the larger the hardness field formed. Compared to the energy transfer field programmed with the dimensions 20 x 20 $mm^2$, the hardness field achieved was smaller. Due to heat conduction at the field edges, the process of martensitic transformation was not completed and a heat-affected zone was formed (the white area in Figure 7). It was found that the higher the point density, the smaller the heat-affected zone generated and, conversely, the larger the EBH area (Figure 7).

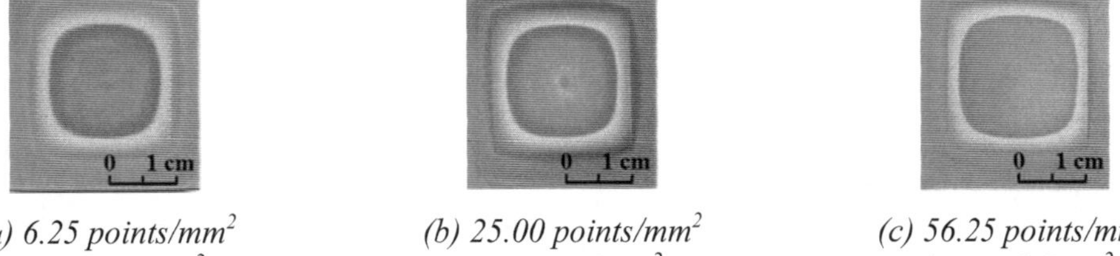


*(a) 6.25 points/mm²* $A_{EBH}$ *= 2.1 cm²* *(b) 25.00 points/mm²* $A_{EBH}$ *= 2.2 cm²* *(c) 56.25 points/mm²* $A_{EBH}$ *= 2.5 cm²*

**Figure 7.** Influence of point density on the shape and size of the EBH surface (white area: heat-affected zone, dark area: hardened zone).

The EBH layers exhibited a lenticular geometry in their cross sections (Figure 8), which corresponded to the heat conduction described above. Furthermore, the overview images show a slight increase in the EBH layer width for a higher point density of 56.25 points/mm² (Figure 8c). At low point densities of 6.25 points/mm² and 25 points/mm², the EBH layer exhibited a banded microstructure that resulted from the rolling process of the base material, and which could not be eliminated by EBH due to the short process times (Figure 8a, b). In comparison, the EBH layer microstructure appeared to be more homogeneous for the highest point density of 56.25 points/mm$^2$ (Figure 8c). The higher energy input per area suggested higher treatment temperatures, which led to more uniform carbon distribution with a more homogeneous distribution of the transformation products.

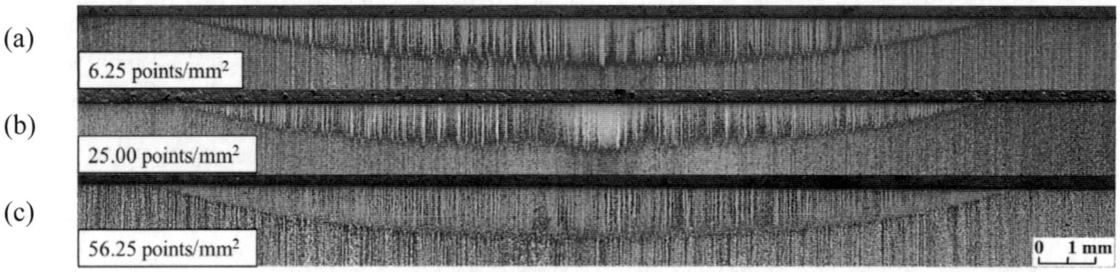


**Figure 8.** Influence of the point density on the EBH layer geometry in cross section.

In addition, an increase in EBH surface hardness values and surface hardening depths suggested a higher degree of carbon solution in austenite during the austenitisation process. As a result, the highest point density of 56.25 points/mm$^2$ led to both the highest EBH layer hardness value of 669 HV0.3 (Figure 9a) and a surface hardening depth of 0.57 mm (Figure 9b). Similar to the influence of the deflection frequency, there was a correlation between the surface hardening depth and the volume expansion (Figure 9b).

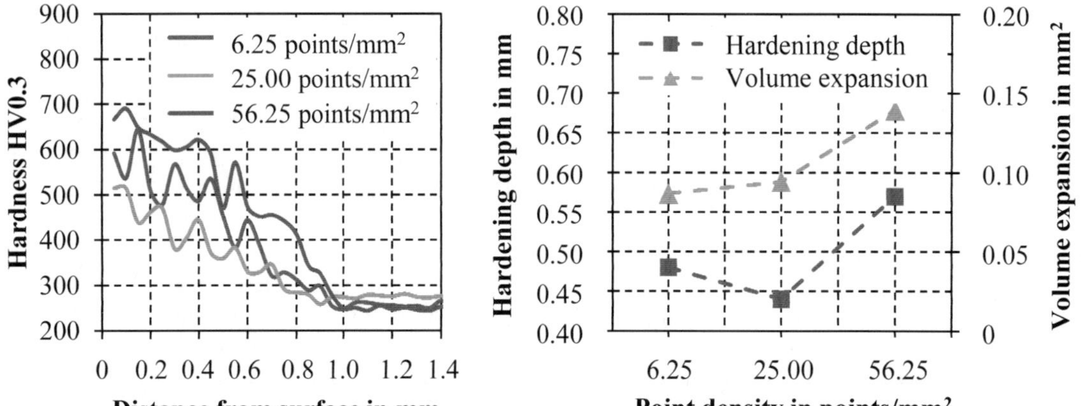


*(a) Hardness-depth profiles* *(b) Surface hardening depth and volume expansion*

**Figure 9.** Influence of the point density on the hardness-depth profiles and the characteristic parameters resulting from the martensitic transformation.

*3.1.3. Influence of the beam current*

Figure 10 shows the influence of the beam current on the macroscopic appearance of the EBH surface with respect to the field area.

*(a)* $I_B = 50.0\ mA$ $A_{EBH} = 2.2\ cm^2$ *(b)* $I_B = 58.4\ mA$ $A_{EBH} = 3.2\ cm^2$ *(c)* $I_B = 66.8\ mA$ $A_{EBH} = 3.6\ cm^2$

**Figure 10.** Influence of the beam current on the shape and size of the EBH surface.

IOP Conf. Series: Materials Science and Engineering **373** (2018) 012019 doi:10.1088/1757-899X/373/1/012019

The hardened field area increased almost symmetrically with increasing beam current (Figure 10a–c). The reason for this was the higher energy input per spot. As a result, the required transformation temperature for EBH was even achieved at the edges of the EB deflection field – despite more rapid heat dissipation in comparison to that at the sample centre. Nevertheless, the surface exhibited local surface melting (SM) in the centre of the hardened field at a beam current of 66.8 mA (Figure 10c).
The light microscopy images of the EBH layers show lenticular edge layer geometries for the above-mentioned reasons (Figure 11). In addition, it can be seen that the EBH layer width increased with increasing beam current from 15 mm (50 mA) to 19 mm (58.4 mA), cf. Figure 11a, b.

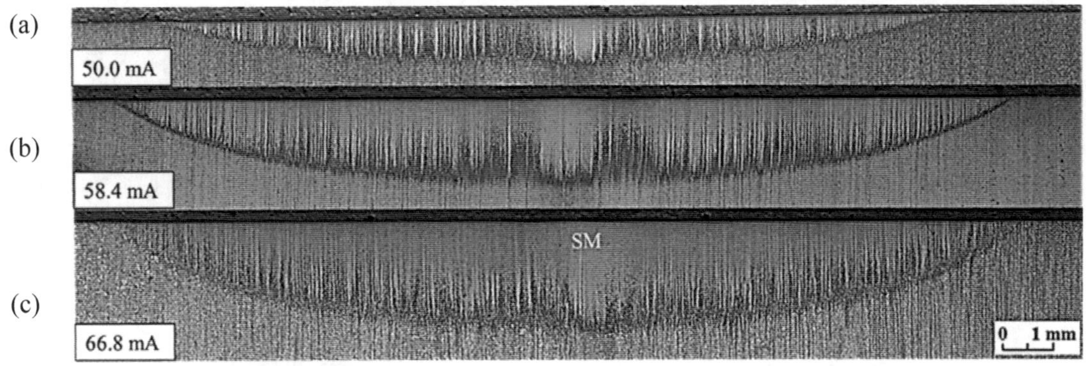


**Figure 11.** Influence of the beam current on the EBH layer geometry in cross section (SM: surface melting).

The maximum surface hardness values achieved in the EBH layer increased with increasing beam current from approx. 490 HV0.3 (50 mA) to 734 HV0.3 (58.4 mA) (Figure 12a). For $I_B$ = 50 mA, the austenitisation was incomplete, while for $I_B$ = 58.4 mA, the austenitising temperature was high enough to ensure complete carbon dissolution. As mentioned before, surface melting occurred for $I_B$ = 66.8 mA, which led to homogenised element distributions, and thus equalised hardness values up to a depth of 0.5 mm as a result of dissolved segregations (cf. Figure 11c, Figure 12a).

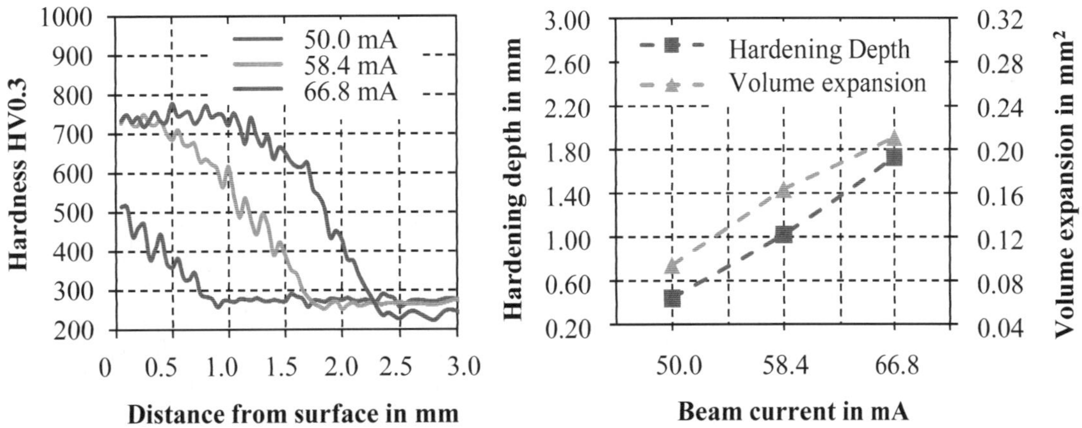


*(a) Hardness-depth profiles* *(b) Surface hardening depth and volume expansion*

**Figure 12.** Influence of beam current on the hardness-depth profiles and the characteristic parameters resulting from the martensitic transformation.

The hardening depth improved from 0.44 mm (50.0 mA) to 1.02 mm (58.4 mA) for the criterion free of melting (Figure 12b). Similarly to the other test series, the volume expansion increased with increasing surface hardening depth (Figure 12b).

### *3.2. Work section 2: Temperature-controlled power regulation*

The advantage of the temperature-controlled power regulation was the avoidance of surface melting when the conditions of heat conduction proved more difficult to influence. Nevertheless, surface melting occurred at a low deflection frequency of 250 Hz. It was assumed that the P-term selected for the PID controller integrated in the pyrometer was too small for this deflection frequency, such that surface temperatures close to the melting point were reached. No surface melting was detected for deflection frequencies higher than 250 Hz. In addition, the treatment results were largely independent of the deflection frequency and the beam current control range. Almost the same geometries and areas were achieved for the EBH surfaces and EBH layers. However, the area of the EBH surface was still smaller than the energy transfer field area applied. The average maximum surface layer hardness determined from the hardness-depth profiles was approx. 686 ± 10 HV0.3 regardless of the deflection frequency and the beam current control range. The surface hardening depths achieved mean values of approx. 0.82 mm.

When comparing the macroscopic images for the tests without (Figure 13a) and with temperature-controlled power regulation (Figure 13b), it can be established that a more homogeneous hardening field was achieved with temperature-controlled power regulation.

*(a) $I_{B,\,constant}$ = 58.4 mA, f = 1000 Hz (without temperature-controlled power regulation) $A_{EBH}$ =3.2 cm$^2$*

*(b) $I_B$ = 20–90 mA, f = 1000 Hz (with temperature-controlled power regulation) $A_{EBH}$ = 3.0 cm$^2$*

**Figure 13.** Comparison of the macroscopic appearance of the EBH surface without (a) and with (b) temperature-controlled power regulation.

The reason for this becomes clear upon examination of the images of the thermographic measurements (Figure 14), which were recorded as an example for a selected parameter set without and with temperature-controlled power regulation for certain process times (cf. Figure 13).

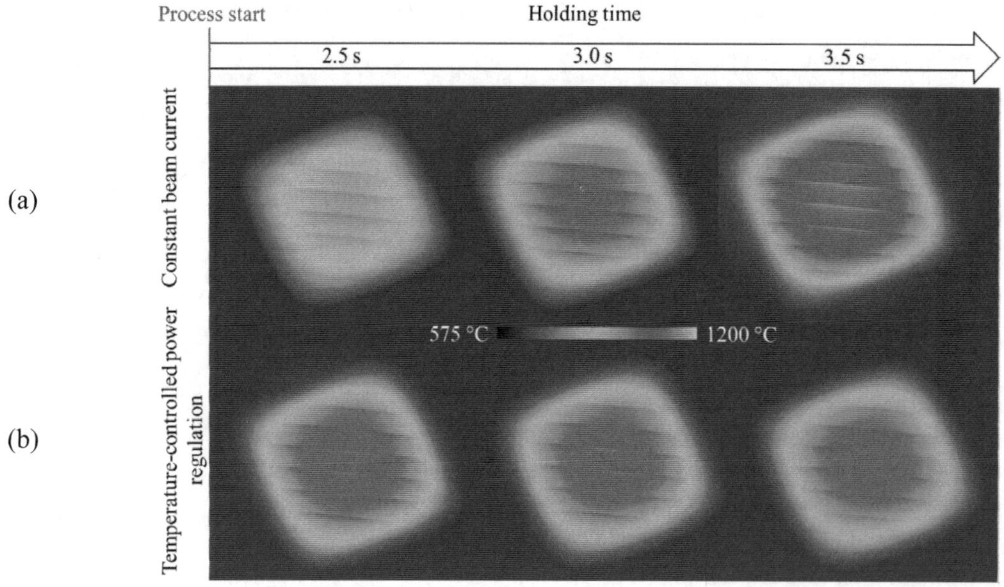


**Figure 14.** Stills of thermographic measurements at defined holding times during the EBH process without and with temperature-controlled power regulation.

It can be seen that with temperature-controlled power regulation, surface isothermal energy transfer could be achieved over a larger area in the centre of the sample (Figure 14b), which resulted in more homogeneous hardening within the flash field. In contrast, the temperature increased continuously with a constant beam current during the process without regulation (Figure 14a).

## 4. Conclusion

For EBH using the flash technique without temperature-controlled power regulation ($I_B$ = constant), low deflection frequencies ($f \leq 250$ Hz) were not found to be suitable due to local surface melting. In terms of EBH layer size and surface layer hardness, the best results were achieved at higher deflection frequencies ($f \geq 5000$ Hz). By increasing the point density and the beam current, it was possible to increase the area of the EBH surface as well as the surface layer hardness values and the surface hardening depth.

Despite temperature-controlled power regulation, surface melting could not be avoided at a low deflection frequency of 250 Hz. No surface melting was detected for deflection frequencies of $f \geq 500$ Hz, whereby the treatment result appeared to be largely independent of the beam current control range and the deflection frequency. Furthermore, surface isothermal energy transfer was achieved in the centre of the sample over a larger area than in the tests without temperature-controlled power regulation. This resulted in more homogeneous surface hardening within the energy transfer field.

## 5. References

[1] Ruge J, Wohlfahrt H 2001 Technologie der Werkstoffe: Für Studenten des Maschinenbaus und Bauingenieurwesens, der Verfahrenstechnik und der Werkstoffkunde, 6., vollständig neubearbeitete Auflage. Studium Technik. Vieweg+Teubner Verlag, Wiesbaden

[2] Zenker R, Buchwalder A 2010 Elektronenstrahl-Randschichtbehandlung: Innovative Technologien für höchste industrielle Ansprüche, 2nd edn. pro beam AG & Co. KGaA

[3] Zenker R 2013 IFHTSE Global 21: Heat treatment and surface engineering in the twenty-first century Part 15 – Progress in surface heat treatment using electron beam surface hardening. International Heat Treatment and Surface Engineering 5(2) pp 50–56. doi: 10.1179/174951411X12956207253465

[4] Zenker R 1992 Härten mit dem Elektronenstrahl. Taschenbuch der Stahl-Eisen-Werkstoffblätter pp 57–60

[5] Schiller S, Panzer S 1987 Härten von Oberflächenbahnen mit Elektronenstrahlen: Teil I: Verfahrenstechnische Grundlagen. HTM 42(5) pp 293–300

[6] Zenker R, Buchwalder A 2013 Electron Beam Surface Hardening. ASM Handbook - Steel Heat Treating Fundamentals and Processes(volume 4A) pp 462–475

[7] Rödel J 1997 Beitrag zur Modellierung des Elektronenstrahlhärtens von Stahl. Zugl.: Freiberg, Techn. Univ.,- Bergakademie, Diss., 1996. FLUX-Verl., Chemnitz

[8] Panzer S, Müller M 1988 Härten von Oberflächen mit Elektronenstrahlen: Teil II: Experimentelle Ergebnisse und Anwendung. HTM 43(2) pp 103–111

[9] Optris GmbH optris infrared thermometers: Grundlagen der berührungslosen Temperaturmessung. Accessed 01 Dec 2017

[10] Halbauer L, Hofner M, Proksch P et al. 2017 Prospects and limitations of a temperature controlled power regulation system for electron beam technologies. In: International Conference on Joining Materials - JOM 19, vol 1

[11] Kahnert M 2015 Scanstrategien zur verbesserten Prozessführung beim Elektronenstrahlschmelzen (EBM). Zugl.: München, Techn. Univ., Diss., 2014. Forschungsberichte / IWB, vol 293. Utz, München

[12] Zenker R, Müller M 1987 Randschichthärten mit Elektronenstrahlen - Verfahrenstechnische Möglichkeiten und werkstofftechnische Effekte. Neue Hütte 32(4) pp 127–134

**Acknowledgments**

The authors gratefully acknowledge the Dobeneck Technology Foundation for its financial support of the “EB-Flash” research project.

# Age-hardening behaviour, microstructure and corrosion resistance of the copper alloyed stainless steel 1.4542

**N Kauss[1], T Halle[1] and P Rosemann[2]**

[1] Institut für Werkstoff- und Fügetechnik, Otto-von-Guericke University, Universitätsplatz 2, 39106 Magdeburg, Germany
[2] BAM - Bundesanstalt für Materialforschung und –prüfung, Fachbereich 7.6 Korrosion und Korrosionsschutz, Unter den Eichen 87, 12205 Berlin, Germany

**Abstract.** The copper alloyed stainless steel 1.4542 (X5CrNiCuNb16-4) is used in different areas due to its good mechanical properties and corrosion resistance. Strength and corrosion resistance can be adjusted by the heat treatment, which is of importance for the application of this alloy. The mechanical properties (strength and hardness) are attributed to the dispersive precipitation of the copper rich ε–Phase. The additional precipitation of chromium carbides can reduce the corrosion resistance. Different ageing states were produced to investigate the precipitation behaviour with various methods. Furthermore, the influence of cold-rolling on the precipitation behaviour was studied in comparison to a solution annealed state without deformation. The microstructure was studied by SEM and the variations of hardness and magnetic proportion were characterised. The electrochemical potentiodynamic reactivation (EPR) was used to determine the corrosion resistance and detect chromium depletion in all heat-treated states. The results show that a work hardening accelerates the precipitation rate, while ageing at 600 °C reduces the corrosion resistance due to chromium depletion.

## 1. Introduction

The X5CrNiCuNb16-4 (1.4542) is a nickel martensitic, age-hardening, stainless steel used in chemical plant construction, medical technology and aerospace. In these areas, there are high demands on mechanical properties (hardness, strength, ductility and toughness) and corrosion resistance at the same time. The stability of the passive layer is influenced by various factors. These include the distribution of the alloying elements in the microstructure, the surface roughness and the surface machining [1], as well as the passivation conditions [2]. To meet the requirements for mechanical properties and corrosion resistance, a heat treatment is necessary [4]. Chromium should be distributed homogeneous in the solid solution by solution-annealing for optimal passive layer stability [3, 4]. The age-hardening steel 1.4542 receives its hardness and strength not by the martensitic conversion but by subsequent ageing, whereby the copper-rich ε-phase is precipitated [5]. The maximum hardness is achieved at short ageing times, when the precipitates are still coherent and disperse in the steel matrix [5, 6]. Too high ageing periods result in coarsening of the ε-phase while stable austenite is formed, both effects reducing the hardness [6]. Carbon enables the undesired precipitation of chromium rich carbides at the grain boundaries. This leads to a local chromium-depletion at the interface of carbide and the surrounding steel matrix, which reduces the passive layer stability and corrosion resistance [4, 7, 8]. Niobium, alloyed in the 1.4542, binds carbon in niobium-carbides to counteract the sensitisation by chromium carbide precipitation. The formation of copper-rich precipitation and the distribution of copper in the microstructure also influence the corrosion resistance. Dissolved in the solid solution, copper lowers the hydrogen overvoltage and

IOP Conf. Series: Materials Science and Engineering **373** (2018) 012020 doi:10.1088/1757-899X/373/1/012020

thus facilitates passivation, which reduces the kinetic restrain of the cathodic reaction [9]. However, ε-precipitations, which contain up to 96 wt. % copper, do not form a passive layer. These spots act as initial defects susceptible to pitting corrosion and reduce the passive layer stability of stainless steels [10]. However, the age-hardening steel 1.4542 requires the ε-phase to achieve the required mechanical properties.

This work investigates how corrosion resistance and hardness are altered by the ageing process to give recommendations for appropriate heat treatment. To detect the sensitisation, the electrochemical potentiodynamic reactivation (EPR) is used. This method has already been standardised and used for austenitic and duplex stainless steels as well as, in modified form, on martensitic stainless steels [3, 4, 11-13]. The EPR method can detect local changes of the alloy content in the microstructure which result in a destabilised passive layer.

Additional cold rolling between solution annealing and age-hardening can be used to accelerate the precipitation of the ε-phase, which may shorten the ageing time. Dislocations, generated by cold rolling, can enhance the effect called "dislocation pipe diffusion" which reduces the activation energy for diffusion and thus increases the diffusion rated [14]. This effect was studied by comparing the ageing behaviour of a solution annealed state with and without further deformation. This paper characterises the influence of cold rolling and age-hardening on hardness and corrosion resistance.

## 2. Experimental

### *2.1. Material, heat treatment and cold forming*

The material used was a 6 mm thick sheet metal of X5CrNiCuNb16-4 (1.4542), with the chemical composition shown in **table 1**. The plate was machined into samples with dimensions 9 mm x 9 mm x 6 mm on which a heat treatment, consisting of solution annealing (homogenisation) and ageing (to increase strength), was applied to achieve optimum hardness and strength.

**Table 1:** Chemical composition (wt. %) of the investigated steel X5CrNiCuNb16-4 (1.4542)

| Alloy | C | Cr | Ni | Cu | Nb | Mn | P | S |
|---|---|---|---|---|---|---|---|---|
| 1.4542 (X5CrNiCuNb16-4) | 0.05 | 15.2 | 5.15 | 3.41 | 0.30 | 0.61 | 0.02 | 0.001 |

Thermodynamic calculations were made based on the chemical composition (**table 1**) with the software *ThermoCalc* to understand the microstructural evolution and the variation of the alloy's corrosion resistance and mechanical properties by the heat treatment. Figure 1 shows the fractions of the different phases occurring in the alloy 1.4542, depending on the temperature.

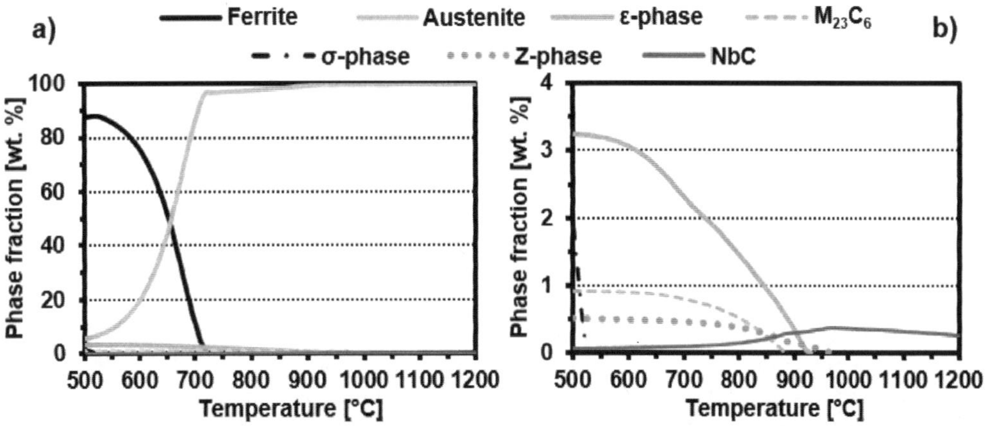


**Figure 1:** Phase fractions for the alloy 1.4542 a) 0 to 100 wt. % and b) detail from 0 to 4 wt. %

Starting from 920 °C, the copper-rich ε-phase is completely dissolved in the solid solution due to the increasing solubility for copper in the austenite. All chromium-rich phases (σ-phase, Z-phase and $M_{23}C_6$-carbides) are dissolved at > 950 °C, while carbon is mostly bound in niobium-carbides of the MC type. These are uncritical for corrosion and postpone the formation of chromium carbides during ageing treatment to longer durations.

To study the effect of age-hardening on microstructure, corrosion resistance and hardness, a precipitation-free, homogeneous initial state is required. A martensitic matrix with some remaining niobium-carbides and retained austenite (shown in figure 3 ageing duration 0 h) was achieved by solution annealing at 1050 °C for 0.5 h, followed by water quenching to room temperature. To differentiate between the effect of "pipe diffusion" and the effect of pure cold forming, solution-annealed samples were cold rolled in several steps down to 20 % of their initial height and examined concerning hardness, magnetisable proportion and corrosion resistance without further ageing.

In the next step, the influence of ageing was studied on solution annealed samples with and without deformation to 50 % of their initial height by cold rolling. The ageing temperature of 600 °C was used, and the ageing time was increased up to 8 h to precipitate copper and carbon rich phases. **Table 2** shows the used ageing parameters.

**Table 2:** Ageing parameter after solution annealing at 1050 °C for 0.5 h with water quenching

| | Temperature [°C] | Ageing time [h] | | | | | | |
|---|---|---|---|---|---|---|---|---|
| **Without deformation** | 600 | 0 | 0.25 | 0.5 | 1 | 2 | 4 | 8 |
| **With 50 % deformation** | 600 | 0 | 0.25 | 0.5 | 1 | 2 | 4 | 8 |

*2.2. Microstructure, hardness and magnetic proportion*

The microstructure was examined in a comparative way, depending on the performed heat treatment. The samples were wet ground with SiC abrasive paper until the oxidised and decarburised surface was removed (500 μm). This was followed by wet grinding with increasing grit size (80/180/360/600/1200/2400), polishing with 3 μm and 1 μm diamond suspension and etching in aged V2A stain (10 ml concentrated hydrochloric acid, 100 ml distilled water, 10 ml nitric acid) at 70 °C for 15 s. The microstructure was characterised in the scanning electron microscopy (SEM) DualBeam FEI Scios. The hardness was determined according to Vickers (HV10). In addition, the change in the magnetisable proportion was evaluated with a Feritscope, which was calibrated to 37 % and could not quantify higher levels. Therefore, the measured values show only the change of the magnetisable percentage and no absolute values.

*2.3. Electrochemical testing*

The DL-EPR method was used to detect the reduction of corrosion resistance by the formation of precipitations. For this purpose, the passivation current density ($i_{passivation}$) and the reactivation current density ($i_{reactivation}$) were determined depending on the sample condition. A classical three electrode arrangement consisting of Ag/AgCl reference electrode ($E_{NHE}$ = 197 mV), Platinum counter electrode and the sample as working electrode was used. The sample surface was wet-ground with grit 1000, cleaned with deionised water and alcohol and then dried in warm air. An area of 0,126 cm² was exposed to 0.1 mol/l $H_2SO_4$ solution with 0.001 mol/l KSCN, and polarised from -500 $mV_{Ag/AgCl}$ to +100 $mV_{Ag/AgCl}$ with 2 mV/s. In this anodic polarisation, the passive layer is formed and the necessary passivation current density ($i_{passivation}$) is determined. The maximum current density is indirect proportional to the alloying elements Cr and Mo dissolved in the solid solution [4]. The polarisation direction is reversed at +100 $mV_{Ag/AgCl}$. Weak points (chromium depletion) in the passive layer were reactivated during the cathodic polarisation from +100 $mV_{Ag/AgCl}$ back to -500 $mV_{Ag/AgCl}$. The measured reactivation current density ($i_{reactivation}$) is proportional to the dissolution of microstructural features with less corrosion resistance compared to the passive matrix. Increasing reactivation current densities show a destabilising effect of ageing treatments and microstructural alternations on the passive layer.

## 3. Results and discussion

### *3.1. Influence of cold rolling*

Figure 2 shows the effect of cold rolling from 100 % to 20 % of the initial height on hardness, magnetisable proportion, passivation and reactivation current density.

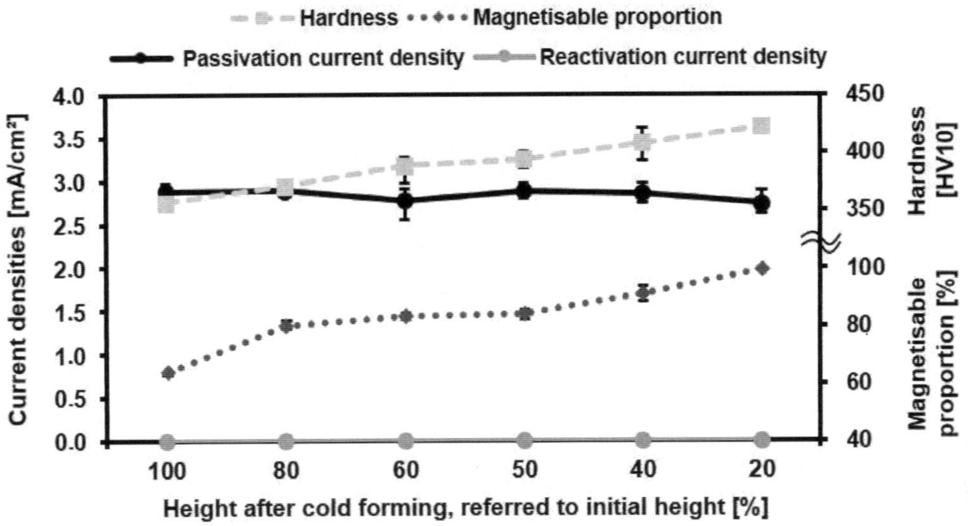


**Figure 2:** Hardness, magnetisable proportion, passivation and reactivation current density with increasing deformation due to cold forming

Hardness and the magnetisable proportion increase with the deformation. This is caused by the transformation of retained austenite into ferromagnetic deformation-martensite. At the same time, the deformation increases the dislocation density, which further increases the hardness. Furthermore, these dislocations reduce the electrical conductivity whereby the eddy current-based Feritscope does not determine a quantifiable value, but only indicates the change from the magnetisable percentage. The cold rolling has no influence on the passivation and reactivation current densities of the EPR method, and therefore it does not affect the corrosion resistance.

### *3.2. Influence of the ageing duration with and without deformation*

Figure 3 and figure 4 show the influence of ageing time at 600 °C up to 8 h on the microstructure of the deformation-free and cold rolled states. All conditions show a nickel-martensitic matrix with finely distributed niobium-carbides. Precipitations were formed as a result of ageing, whose phase proportion increases with the ageing time.

The deformed samples, which were cold rolled to 50 % of their original height, show a significant higher proportion of precipitation in the microstructure at the same ageing time. This can be attributed to the increase in diffusion velocity due the effect of "dislocation pipe diffusion". The microstructure of the rolled samples shows a dense allocation of precipitations at the grain boundaries, starting with an ageing duration of 1 h. A comparable precipitation state at the grain boundaries occurred in the samples without deformation after an ageing time of 2 h. Furthermore, the formation of precipitations ($Cr_{23}C_6$-carbides, Z-phase and NbC-carbides) within the grains starts earlier with deformation (2 h), compared to the aging without deformation (4 h).

IOP Conf. Series: Materials Science and Engineering **373** (2018) 012020 doi:10.1088/1757-899X/373/1/012020

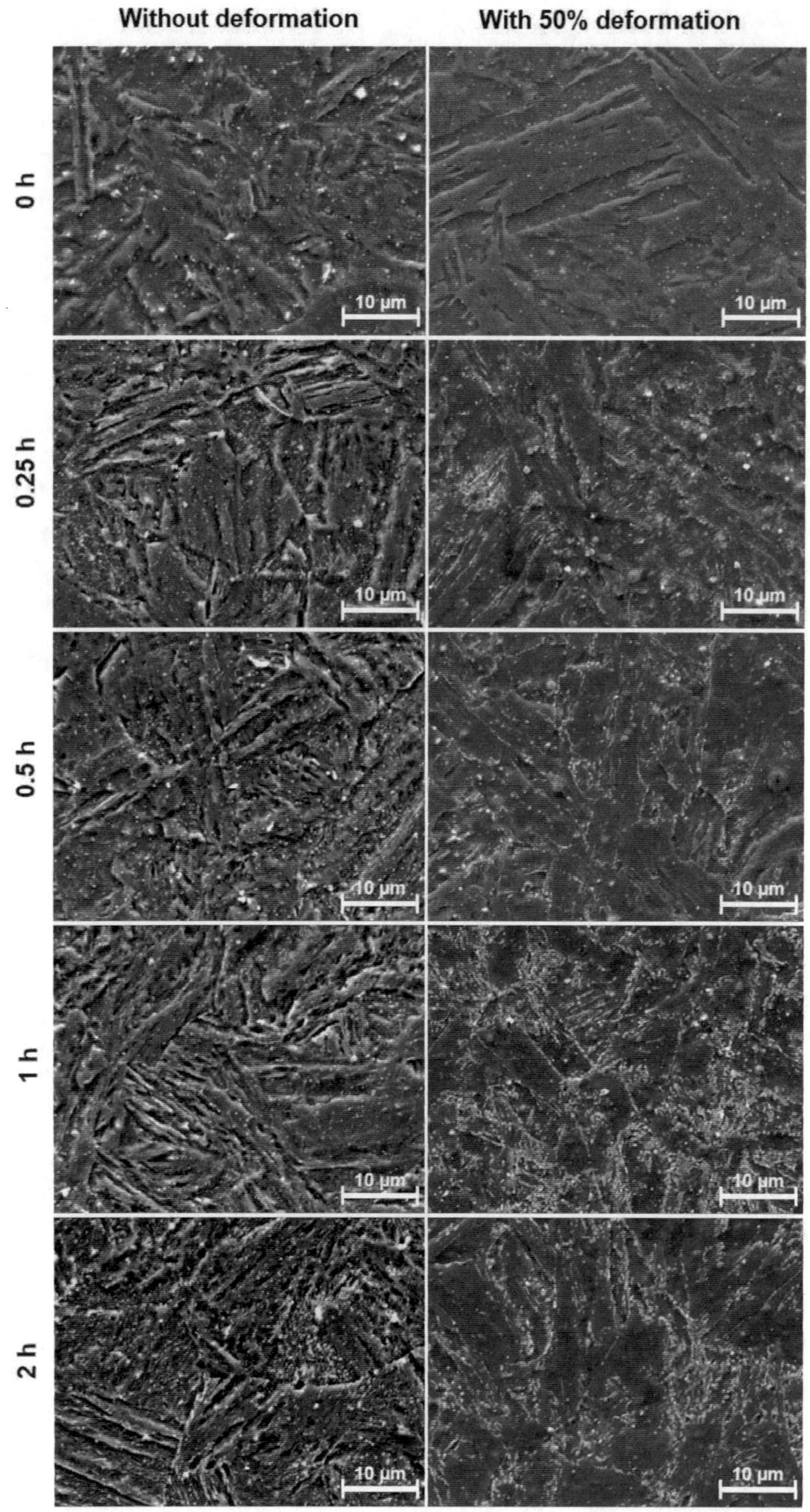


**Figure 3:** Microstructure of the steel 1.4542 after solution-annealing (1050 °C, 0.5 h, water quenched) and ageing at 600 °C for 0 h to 2 h for samples without and with 50 % deformation

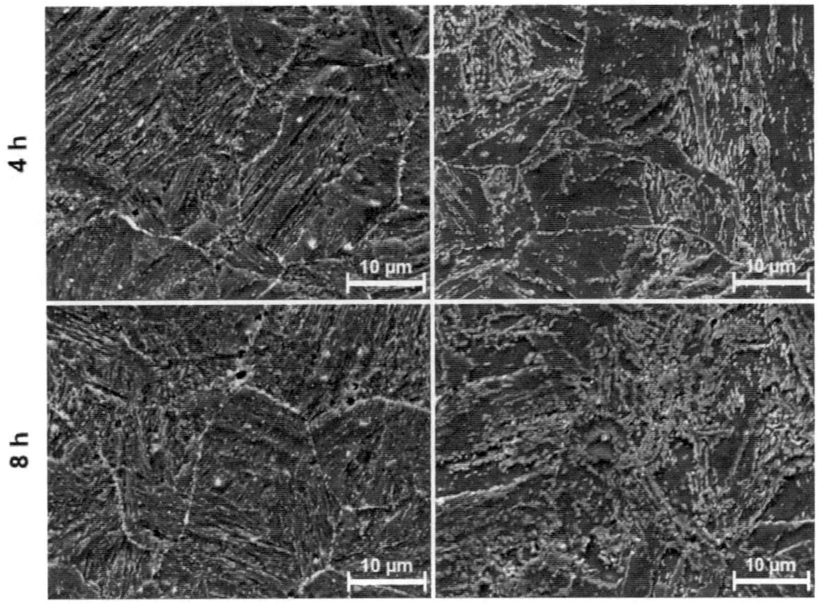


**Figure 4:** Microstructure of the steel 1.4542 after solution-annealing (1050 °C, 0.5 h, water quenched) and ageing at 600 °C for 4 h to 8 h for samples without and with 50 % deformation

Figure 5 shows the effect of ageing on hardness, magnetisable proportion, passivation and reactivation current density without deformation (figure 5 a) and with 50 % deformation (figure 5 b).

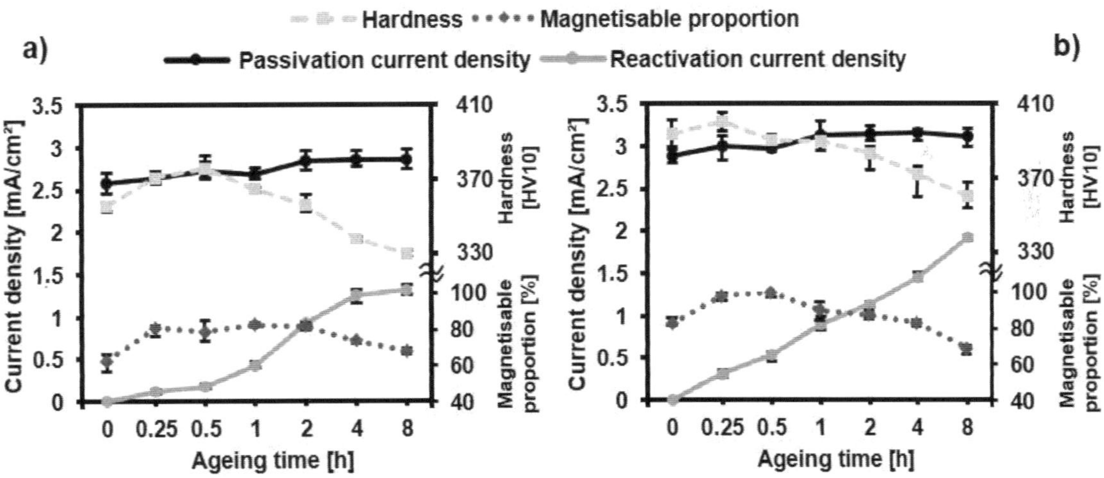


**Figure 5:** Effect of ageing duration at 600 °C on hardness, magnetisable proportion and passivation and reactivation current density, a) without deformation b) with 50 % deformation

The general magnetisable proportion is higher in the cold rolled samples than in the non-rolled samples, which is the result of the deformation induced conversion of residual austenite into martensite. The ageing increases the magnetisable proportion, and the maxima of magnetisable proportion were reached at ageing times of 1 h without deformation and 0.5 h with 50 % deformation.

Overall, the cold rolled samples showed a higher hardness as a result of the deformation induced work hardening and the transformation of retained austenite into deformation martensite. The heat treatment related hardness maxima were achieved at an ageing time of 0.5 h without deformation (figure 5a) and an ageing time of 0.25 h with 50 % deformation (figure 5b).

The earlier maxima of hardness and magnetisable proportion in the deformed state can be explained by the increased precipitation rate of the hardness-enhancing ε-phase. After reaching the maxima, both values decrease with further ageing. The deformed samples showed a stronger drop in both values, which is a result of the reduction of the dislocation density. The decrease of hardness is also reinforced by the coarsening of the now incoherent ε-phase and the reconversion from martensite to austenite.

The passivation current density ($i_{passivation}$) is almost constant, which shows that no global reduction of the alloy content of the matrix occurs. The reactivation current density ($i_{reactivation}$) increases linearly with the ageing duration. This is the result of local sensitisation in the microstructure, which reduces the passive layer stability.

The deformation of the solution-annealed samples prior to the aging causes a faster increase of the reactivation current density ($i_{reactivation}$), which also can be explained by the effect of dislocation pipe diffusion. It could not be determined which precipitation (ε-phase, $M_{23}C_6$, Z-phase) causes the reactivation, since all three phases form during ageing at 600 °C. This could be clarified with further investigations concerning the localisation of the selective corrosion during the reactivation.

## 4. Conclusion

Deformation and ageing have a significant influence on the mechanical properties and corrosion resistance of the nickel martensitic, copper alloyed, age-hardening stainless steel 1.4542 (X5CrNiCuNb16-4). The maximum hardness was achieved when the copper-rich ε-precipitations were still coherent with the matrix. The optimum corrosion resistance, on the contrary, is achieved, when all precipitations ($M_{23}C_6$, ε-phase, and Z-phase) are dissolved by the solution annealing and the alloying elements are homogeneous distributed in the solid solution.

However, ageing at 600 °C causes the formation of sensitising precipitations, which was detected based on increasing reactivation current densities with the DL-EPR method. It has also been shown that a deformation before ageing leads to higher hardness values and to the earlier attainment of the hardness maximum, which is based on the effect of "dislocation pipe diffusion". The corrosion resistance of the solution-annealed condition was not affected by the sole cold forming. Only the hardness and the magnetisable proportion increase strongly with the degree of deformation, since residual austenite is transformed into martensite and the dislocation density is increased.

The sensitisation during ageing of the 1.4542 could be caused by $M_{23}C_6$-carbides, the ε-phase or the Z-phase. Nevertheless, the sensitisation occurs even during very short ageing times (15 min for 600 °C). Carbon is tied by niobium in niobium-carbides and the formation of chromium-rich $M_{23}C_6$-carbides, and the Z-phase may need a longer ageing duration. It seems likely that the precipitation of the ε-phase is responsible for the sensitisation.

## 5. References

[1] Burkert A, Schilling K and Heyn A 2004 Einfluss der Schleifbehandlung auf das Korrosionsverhalten von Chrom-Nickel-Stählen *Materials and Corrosion* **55** 787

[2] Bierwirth M, Goellner J, Heyn A and Winkelmans M 2007 The factor time at the formation and development of the passive layer in the atmosphere *Materials and Corrosion* **58** 946

[3] Rosemann P, Kauss N, Müller C and Hall T 2015 Influence of solution annealing temperature and cooling medium on microstructure, hardness and corrosion resistance of martensitic stainless steel X46Cr13 *Materials and Corrosion* **66** 1068

[4] Rosemann P, Müller C, Kauss N and Halle T 2017 Einfluss der Wärmebehandlung auf die Korrosionsbeständigkeit von Schneidwaren *Journal of Heat Treatment and Materials* **2** 87-98

[5] Hsiao C N, Chiou C S and Yang J R 2002 Aging reactions in a 17-4 PH stainless steel *MaterialsChemistry and Physics* **74** 134

[6] Ziewiec A, Zielinska-Lipiec A and Tasak E 2014 Microstructure of Welded Joints of X5CrNiCuNb16-4 (17-4 PH) Martensitic Stainless Steel after Heat Treatment Archieves of *Metallurgy and Materials* **59** 965

[7] Bäumel A and Carius C 1961 Zusammenhang zwischen Anlaßbehandlung und Korrosionsverhalten von härtbaren nichtrostenden Chromstählen *Archiv für das Eisenhüttenwesen* **32** 237

[8] Bäumel A 1967 Korrosionsverhalten nichtrostender Vergütungsstähle mit rund 13 % Chrom *Werkstoffe und Korrosion* **18** 289

[9] Colombier L and Hochmmann L 1964 Stale odporne na korozje i stale zaroodporne *Wyd. Slask Katowice*

[10] Banas J and Mazurkiewicz A 2000 The effect of copper on passivity and corrosion behavior of ferritic and ferritic–austenitic stainless steels *Materials Science and Engineering* A**277** 183

[11] Schilling K 2005 Selektive Korrosion hochlegierter Stähle, *Dissertation, Otto-von-Guericke-Universität Magdeburg, Fakultät für Maschinenbau Magdeburg* (Shaker Verlag)

[12] Rosemann P, Müller Th, Babutzka M and Heyn A 2015 Influence of microstructure and surface treatment on the corrosion of martensitic stainless steel 1.4116, 1.4034 and 1.4021 *Materials and Corrosion* **66** 1068

[13] Schultze S 1999 Selektive Korrosion von Duplexstahl, *Dissertation, Otto-von-Guericke-Universität Magdeburg, Fakultät für Maschinenbau Magdeburg* (Shaker Verlag)

[14] Love G R 1964 Dislocation pipe diffusion *Acta Metallurgica* **12** 731

# Heat treatment condition of EN AW-7075 influencing the anodic oxidation process and coating properties

**R Morgenstern*, I Scharf and T Lampke**

Institute of Materials Science and Engineering, Chemnitz University of Technology, 09125 Chemnitz, Germany,

* e-mail: roy.morgenstern@mb.tu-chemnitz.de

**Abstract**. The age-hardenable aluminium alloy EN AW-7075 exhibits outstanding specific mechanical properties and therefore offers a high potential for lightweight construction. Anodising in aqueous oxalic acid solutions is suitable to produce a protective oxide ceramic conversion layer on this alloy. This study examines the influence of the precipitation state of the substrate alloy on microstructure and properties of anodic oxide layers. Therefore, EN AW-7075 sheets in the heat treatment conditions T4, T6 and T73 were anodized in 0.8 M oxalic acid solution at constant voltage. The current efficiency was determined on the basis of the electrical charge quantity, coating thickness and coating mass. Instrumented indentation tests were applied in order to evaluate the coating hardness. The microstructure of the anodic oxide layer was illustrated using field emission electron microscopy. It was shown that the current efficiency strongly depends on the heat treatment condition.

## 1. Introduction

Due to the highest mechanical strength among aluminum alloys, the 7xxx series are widely applied for lightweight constructions in automotive and aerospace applications. However, the low hardness and high adhesion tendency of aluminum alloys limit their applicability under tribological load. This restriction can be overcome by the generation of hard alumina conversion coatings using the anodic oxidation process [1]. Industrial anodizing processes are typically conducted in sulfuric acid electrolytes, because of the low process costs. Conversion coatings with a high maximum thickness, high hardness and a low hardness decline across the coating thickness can be obtained by reducing the bath temperature below 5 °C. This is due to the reduced chemical dissolution of the pore walls and hence the lower conicity of the pore walls. Unfortunately, additional energy and equipment costs arise from the electrolyte cooling and the increasing process voltage. Accepting higher process voltages, anodic aluminum oxide (AAO) coatings with a comparably high hardness and low hardness gradient can be obtained without extensive electrolyte cooling from oxalic acid electrolytes at room temperature, alternatively. Currently, efforts are made in order to generate wear resistant anodic oxide coatings only at tribologically stressed functional surfaces of components. This enables further savings in energy and chemicals and dimensional tolerances of other component surfaces are not affected. This can be realized, inter alia, using sealed electrochemical cells, e.g. capillary cells [2, 3] or open systems, e.g. continuous electrolyte jet anodizing [4]. Especially for the latter processes, oxalic acid electrolytes might be more suitable due to less corrosion of the plant equipment compared with sulfuric acid electrolytes.

As anodizing operates by the conversion of the substrate material, the coating properties are not only determined by the process characteristics (e.g. electrolyte, temperature, electrical parameters) but also by the chemical composition and the microstructure of the substrate alloy. Because of the less negative enthalpy of oxide formation compared with Al, the main alloying element of 7xxx alloys Zn enriches at the substrate-coating interface during anodic oxidation, while Al is preferably oxidized. Regarding model alloys containing 0.2 to 0.9 at% (0.5 to 2.2 wt%) Zn, the authors Zhou et al. observed a Zn enriched zone with a Zn concentration of at least 7.8 at% (17 wt%) and a thickness of up to 10 nm [5]. Apart from a higher Zn content of 5.5 wt%, alloy EN AW-7075 contains up to 2.9 wt% Mg, up to 2 wt% Cu and small amounts of Fe, Si, Mn, Cr und Ti. Similar to Zn, the elements Cu, Cr and Fe are enriched at substrate-alloy interface during anodizing, as well. However, being semi-conductors, the oxides of these elements allow electronic conduction, thus enabling oxygen generation at the anode. With proceeding oxide growth, the emerging gas bubbles leave small voids along the pore channels [6]. Consequences are an increased coating porosity and a reduced current efficiency of oxide formation. Investigations at different heat treatment conditions of the alloy EN AW-2024 (AlCu4Mg1) showed that the oxide formation is less negatively affected, if the element Cu is not homogeneously dispersed in the Al solid solution (Alss) or in the form of atomic clusters but bound in $Al_2CuMg$ precipitates as they are rapidly dissolved during the anodizing process [7].

It might be reasonably assumed, that the conversion and dissolution behavior of intermetallic phases determines the anodizing behavior of alloy EN AW-7075 as the elements Fe, Mn and Cr are almost insoluble in the Alss, even at elevated temperatures. They already form primary precipitates during the solidification process and may also bind Cu atoms. Mukhopadhyay et al. observed $Al_{12}(FeMn)_3Si$ precipitates containing a small amount of Cu and Cr with a size of up to 12 μm in alloy EN AW-7075 with commercial purity [8]. These coarse precipitates are not completely dissolved during anodizing in a sulfuric acid electrolyte and locally impede the oxide growth. This results in an increasing process voltage [8]. In contrast, the $Al_7Cu_2Fe$ phase, which dissolves during anodizing without impeding the process significantly, was observed in a model alloy containing smaller amounts of the impurities Si, Mn and Fe [8]. This is confirmed by Veys-Renaux et al. who observed the dissolution of the $Al_7Cu_2Fe$ phase during anodizing of EN AW-7175 (same alloy composition as EN AW-7075, lower impurity content) in a sulphuric acid electrolyte [9]. Moreover, the authors observed that the strengthening precipitates $MgZn_2$, AlCuMg and $Al_2CuMg$ are also dissolved during anodizing leaving voids with corresponding sizes between 100 and 200 nm within the oxide conversion coating [9]. Growth rate and current efficiency are similar for the anodizing of alloy EN AW-7175 and commercially pure aluminum (EN AW-1050, Al99.5) and significantly larger compared with alloy EN AW-2618 [9]. Although, the heat treatment conditions of EN AW-7175 and EN AW-2618 are not mentioned explicitly, some conclusions can be drawn on the basis of the microstructural descriptions. Presumably, alloy EN AW-7175 was used in an overaged condition because the strengthening phases were no longer present as Guinier-Preston (GP) zones, but already as stoichiometric phases. On the contrary, no strengthening phases could be detected for the EN AW-2618 alloy; thus a naturally aged condition (e.g. T3, T4) can be suggested. The results of Veys-Renaux et al. [9] broadly correspond with those of Morgenstern et al. [7] indicating that the properties of anodic oxide coatings are not only influenced by the alloy composition but also by the distribution of the alloying elements and therefore by the heat treatment state.

Up to the present, the impact of the heat treatment condition of EN AW-7075 on the properties of anodic oxide coatings is not covered systematically. Furthermore, few knowledge is available on the dissolution and conversion behavior of precipitates in EN AW-7075 during anodizing in oxalic acid electrolytes. The current article strives to close these knowledge gaps and to explain the complex correlations between substrate microstructure (precipitates), process characteristics (e.g. current efficiency), coating microstructure (porosity) and coating properties (coating thickness and hardness). Furthermore, the article is focused on the industrially important conditions T6 (highest strength) and T73 (improved resistance against stress corrosion cracking). Additionally, the naturally aged state T4 is included as a reference state.

## 2. Materials and methods

Sheets of the aluminum alloy AlZn5.5MgCu (EN AW-7075, nominal composition in wt%: Si $\leq$ 0.4, Fe $\leq$ 0.5, Cu 1.2–2.0, Mn $\leq$ 0.3, Mg 2.1–2.9, Cr 0.18–0.28, Zn 5.1–6.1, Ti $\leq$ 0.2) with the dimensions of 100 x 25 x 1.5 $mm^3$ were used as substrate material. All sheets were solution annealed (470 °C, 1 h) and quenched in water, followed by natural aging (room temperature, > 14 days, condition T4), peak aging (120 °C, 24 h, condition T6) and overaging (120 °C, 24 h and 160 °C, 30 h, condition T73) according to [10].

Three samples of each heat treatment condition were prepared for anodizing. Sample pretreatment included etching in 3 wt% NaOH solution (60 °C, 3 min) and pickling in 1:1 diluted nitric acid (room temperature, 30 s). The electrolyte for the anodic oxidation was comprised of 0.8 mol/l oxalic acid (supplied as oxalic acid dehydrate, Merck). The electrolyte temperature was 25 °C. All the used chemicals were of analytical grade. The anodizing voltage was linearly increases from 0 to 60 V within 30 s and afterwards kept constant at 60 V. The electrolyte was constantly stirred during the total anodizing duration of 15 min. The power supply was a pe1028 (Plating Electronic). The values of current and voltage were logged internally during the process with an acquisition rate of 1 sample per second. The electrical charge per area was calculated by the integration of the current density profile over the anodizing time. The coating thickness was measured by optical microscopy on the cross-sections of all samples.

The mass of all samples was determined before and after the dissolution of the conversion coating in chromic/phosphoric acid (35 ml/l phosphoric acid + 20 g/l chromium(VI)oxide) at 60 °C for 4 h with a X1003S balance (Mettler Toledo). Pretests showed that no measureable dissolution of the substrate alloy occurs during the exposition in chromic/phosphoric acid, therefore, the mass difference equals to the coating mass. The porosity $p$ of the coatings was estimated by the following formula using the coating mass $m$, the surface area $A$, the coating thickness $s$ and the density of alumina $\rho$ (3.95 $g/cm^3$):

$$p = 1 - \frac{m}{A * s * \rho} \tag{1}$$

The current efficiency $\eta$ was determined by dividing the coating mass $m$ by the theoretical mass $m_{th}$. The theoretical mass $m_{th}$ was calculated according to the Faraday´s law using the electrical charge $Q$, the molecular mass $M$ of $Al_2O_3$ (101,96 g/mol), the number of electrons involved in the generation of one $Al_2O_3$ molecule (u = 6) and the Faraday constant $F$ (96485,33289 As/mol):

$$m_{th} = \frac{Q * M}{u * F} \tag{2}$$

Metallographic cross-sections were prepared by diamond grinding accomplished by a finish using a silica oxide polishing suspension. The microstructure was investigated by field emission scanning electron microscopy (FE-SEM, Zeiss, NEON40EsB). Both secondary electron (SE, topography contrast) and backscattered electron (QBSD, element contrast) detectors were applied. Hardness profiles of the AAO cross-sections were obtained from instrumented indentation tests at the cross sections of the coatings with a Berkovich indenter (Asmec Unat) by placing the indents at different locations and measuring the distances of the indents from the substrate surface via optical microscopy. A load of 5 mN was applied (load time 10 s, hold time 5 s, unload time 4 s).

## 3. Results and discussion

### *3.1. Current efficiency*

Typical current density profiles for the anodic oxidation of different heat-treatment conditions of EN AW-7075 are shown in figure 1. During the first seconds, the current rises steeply with the increasing anodic potential. After 30 s, the constant potential of 60 V is reached and soon afterwards, a continuous and dense alumina barrier layer evolved. Because of the presence of the electrically insulating barrier layer, the anodic current is reduced to a minimum after about 50 to 100 s. With the concentration of the current flow at weaknesses of the barrier layer, the anodic dissolution becomes localized and pores evolve. The steady growth of the coating thickness is accompanied by a slight increase of the anodic current. It can be seen from figure 1 that the current density profile for the T4 condition runs at a slightly higher level compared with the T6 condition, whereas a significantly lower current level can be observed for the T7 condition. This observation is quantitatively displayed by the electrical charge density $\sigma_q$ in table 1. Accordingly, the electrical charge quantity $\sigma_q$ decreases with increasing intensity of the aging treatment from the condition T4 to T7 by about 30 %. Exactly the same proportional decline can be observed for the coating mass per area (referred to as specific mass). Consequently, the current efficiency, i.e. the amount of oxide generated from a certain amount of electrical charge, is constant. In relation to the maximum possible coating mass according to the Faradays law, the current efficiency amounts to about 46 % for all heat treatment conditions. Furthermore, the coating thickness also decreases by about 30 %. As the coating thickness is proportional to the specific coating mass, a constant overall coating density can be assumed. Considering the density of alumina, a coating porosity of about 43 % applies to all heat treatment conditions. Summarizing the above, it can be stated that the aging condition only influences the anodic charge quantity. Coating mass and coating thickness change accordingly and mass-related properties, e.g. current efficiency and porosity stay constant.

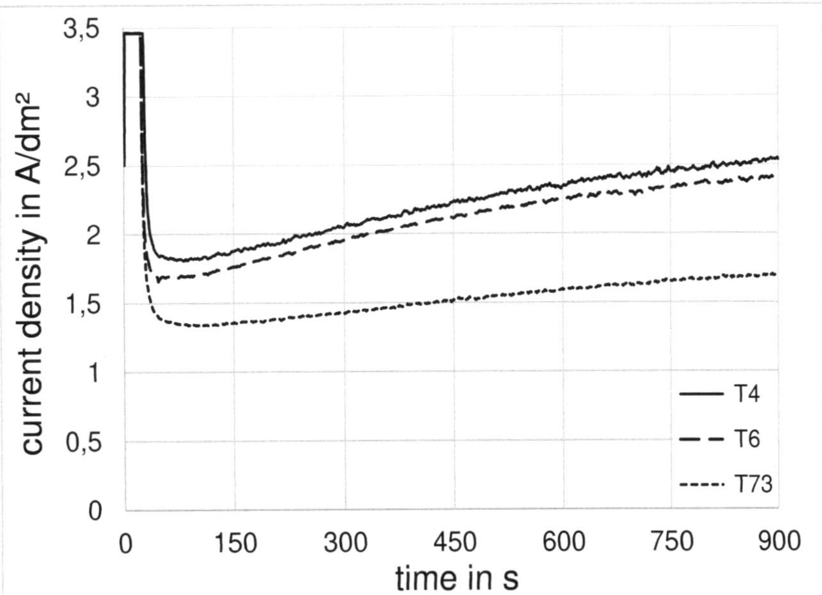


**Figure 1.** Profiles of the current density for the anodic oxidation of the conditions T4, T6 and T3 in 0.8 mol/l oxalic acid at a constant voltage of 60 V after the linear voltage ramp (first 30 s).

IOP Conf. Series: Materials Science and Engineering **373** (2018) 012021 doi:10.1088/1757-899X/373/1/012021

**Table 1.** Survey of process characteristics and coating properties.

| | symbol | unit | T4 | T6 | T73 |
|---|---|---|---|---|---|
| **charge density** | $\sigma_q$ | As/dm² | 2069±113 | 1899±39 | 1441±24 |
| **specific mass** | $\rho_A$ | mg/dm² | 169.5±5.5 | 152.9±7.0 | 118.1±0.7 |
| **current efficiency** | $\eta$ | % | 46.5±0.2 | 45.7±1.2 | 46.5±0.5 |
| **coating thickness** | $s$ | µm | 8.7 ±0.2 | 8.1±0.5 | 6.0±0.1 |
| **coating porosity** | $p$ | % | 43.5±2.6 | 45.1±5.5 | 42.6±0.8 |

### *3.2. Microstructure and hardness*

QBSD examinations of the substrate materials were used in order to visualize the precipitations. As most of them consist of heavier elements (e.g. Zn, Fe), they appear bright in comparison to the dark Al solid solution (Alss). On the maximum magnification level, which can be realized by FE-SEM under the given conditions, no systematic differences can be observed between the conditions T4 and T6.

Figure 2 shows QBSD images of the conditions T6 (a) and T7 (b). According to [11], the microstructure of the condition T6 is characterized by spherical GP zones (less than 7.5 nm in diameter) and a small amount of the η' phase (about 10 nm in diameter). These small precipitates cannot be recognized by FE-SEM using QBSD imaging. However, as can be seen from figure 2a, also larger precipitates with up to 200 nm are present in the T6 state, both along the grain boundaries and randomly distributed within the grains. This corresponds well with the observations of Singh et al. who determined the size distribution of submicron η precipitates in EN AW-7075 T651 using FIB tomography [12]. These comparatively coarse precipitates must have developed during the processing of the sheet material prior to the T6 heat treatment (e.g. during hot rolling) and they did not dissolve completely during the solution annealing. Emani et al. observed similar η precipitates in different heat treatment conditions of EN AW-7075 and attributed their stability during solution annealing to chromium enrichment [13]. Of cause, the submicron η precipitates are also present in the T73 condition. Additionally, small bright spots can be observed in the grain interior at figure 2b. These precipitates obviously developed during the second annealing step at 160 °C which was part of the T73 treatment. Similar precipitates with up to 50 nm in diameter were identified as η' precipitates by Emani et al. after annealing at 177 °C [13]. Due to the other alloying elements and impurities, additional nano and submicron scale precipitates might be present in EN AW-7075, e.g. $Mg_2Si$ [3]. However, because of the small differences in the atomic weight of Mg, Si and the Alss, this phase might be hardly recognizable at the QBSD images.

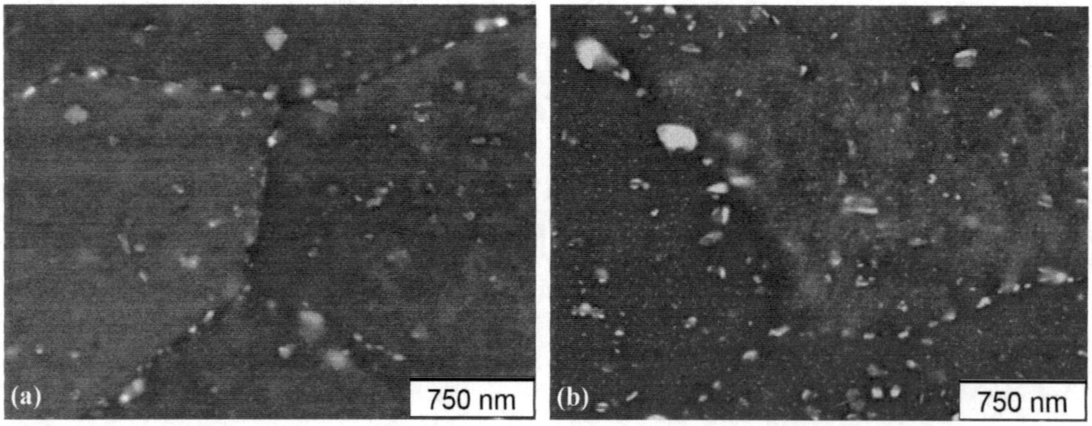


**Figure 2.** QBSD images of EN AW-7075 in the heat-treatment states (a) T6 showing submicron η precipitates and (b) T73 showing submicron η and nanoscale η' precipitates.

Figure 3 represents the microstructure of anodic oxide coatings on the T6 condition. The overview image (figure 3a) shows occasionally appearing oval areas (marked with (1)) within the coating exhibiting cracks perpendicular to the oxide growth direction (or parallel to the substrate coating interface, respectively). Their size and shape correspond to occasionally occurring micron scale precipitates within the substrate. The cracks indicate that the conversion of these precipitates is accompanied by a smaller volume expansion compared with the conversion of the surrounding Alss. As the cracks are not continuing through the surrounding oxide and because of their small number, these coating defects will probably not deteriorate the coating properties (e.g. hardness) significantly. Furthermore, areas with higher porosity occur at the exterior of the coatings (marked with (2)). This might be due to the chemical dissolution of the pore walls in the acidic electrolyte. The impression of this effect might be further pronounced by the final polishing of the coating cross section as more porous areas are removed faster.

Figure 3b shows the submicron and nano scale porosity of the anodic oxide coating on EN AW-7075 T6 in more detail. The characteristic pore channels are not aligned perfectly perpendicular to the substrate and parallel to each other. Some pore channels are branching and take over the area of other pore channels, while the growth of the latter stops. The competitive growth of pore channels occurs due to the faster dissolution at the bottom of some channels and therefore the faster growth of those channels [14]. Li et al. observed a highly non-conformal pore growth on high purity aluminum in 0.3 M oxalic acid at 0-5 °C and high voltages between 140 and 400 V [15]. However, some non-uniform pore growth might already occur at 60 V under the anodizing conditions described in chapter 2. Especially, the preferential dissolution of the submicron η precipitates might accelerate the progress of some pore channels and disturb the conformal pore growth. After rapid dissolution, these precipitates leave voids with similar size and shape within the oxide coating (marked by arrows in figure 3b).

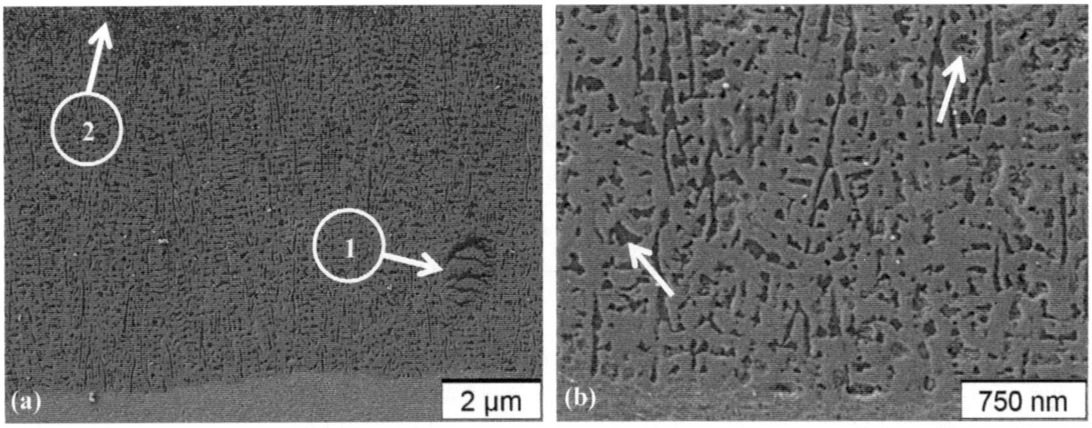


**Figure 3.** SE images of anodic oxide coatings on EN AW-7075 T6, (a) survey with (1) oxidized micron scale precipitate and (2) outer coating area with higher porosity, (b) detailed view in the vicinity of the substrate-coating interface showing voids due to the dissolution of submicron precipitates (marked by arrows).

As the submicron scale precipitates are also present in the T73 condition to a similar extent, the submicron scale voids are present in the oxide coatings as well (marked by arrows in figures 4a and b). Moreover, the pore channels are non-conformal for both heat treatment conditions to a similar extend. Solely the coating thickness is lower for the T73 condition due to the smaller electrical charge quantity, as already mentioned above. Possibly, the smaller η' precipitates (visible in figure 2b) give rise to the electrical resistance during anodizing leading to a decrease in the anodic current at constant voltage. However, as these small precipitates (< 50 nm) are smaller than the pore cells (~ 150 nm), they do not significantly influence the pore growth and therefore not affect the pore microstructure.

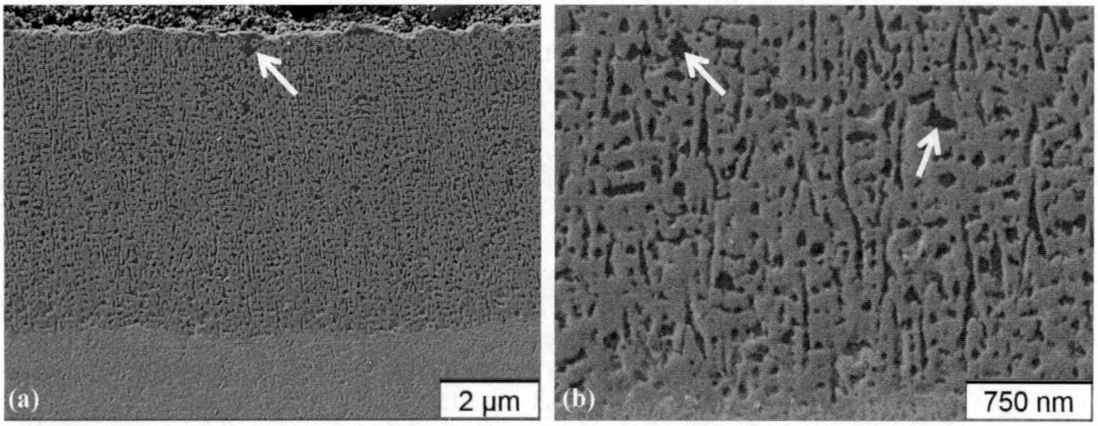


**Figure 4.** SE images of anodic oxide coatings on EN AW-7075 T73, (a) survey and (b) detailed view in the vicinity of the substrate-coating interface. Voids due to the dissolution of submicron precipitates are marked by arrows.

It is known from mild anodizing in sulfuric acid electrolytes at room temperature that the coating hardness significantly decreases with increasing distance from the substrate due to the chemical dissolution of the pore walls. This effect is less pronounced for the anodic oxide coatings obtained from the oxalic acid electrolyte. Figure 5 uses linear trend lines averaging about 25 data points per condition in order to display the hardness profiles more clearly. Additionally, all data points of the T4 condition are illustrated in figure 5 for giving an impression of the deviation between the trend line and the data points. It can be seen that the deviation of single data points amounts to up to 1 GPa, while most of the data points are located within a deviation of 0.5 GPa. The trend lines suggest an almost constant hardness of about 3.5 to 4 GPa (being also a typical hardness range for hard anodizing in sulfuric acid electrolytes) over a wide range of the coating thickness. Whereas for hard anodizing in sulfuric acid electrolytes, the coating hardness typically decreases more significantly within the oxide coating. This is due to the stronger chemical dissolution of the anodic alumina in sulfuric acid. As illustrated in figure 3a, areas with higher porosity are also existing in the outer part of anodic oxide coatings generated from the oxalic acid electrolyte. Therefore, a further decrease of the hardness can be expected at higher distances from the substrate-coating interface. However, due to the missing supporting effect of the surrounding material, reliable nanoindentation measurements cannot be performed in the close proximity of the outer edge. With respect to the typical deviations of the data points, the hardness levels of the different heat treatment conditions do not differ significantly. This corresponds with the fact that the anodic oxide coatings exhibit similar pore morphologies for all heat treatment conditions.

IOP Conf. Series: Materials Science and Engineering **373** (2018) 012021 doi:10.1088/1757-899X/373/1/012021

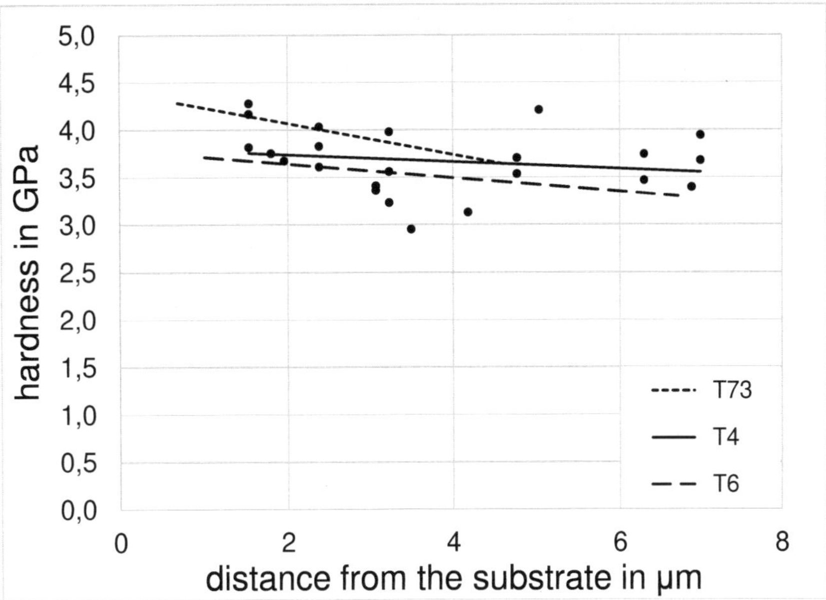


**Figure 5.** Linear trend lines of hardness profiles of anodic oxide coatings on heat treatment conditions T4, T6 and T73 and data points of the T4 state.

## 4. Conclusion

This paper describes the dependencies of coating properties (e.g. thickness, porosity, hardness) and characteristic parameters of the anodizing process in an oxalic acid electrolyte (e.g. current efficiency) on the heat treatment condition of EN AW-7075. It was found that the electrical charge density decreased by about 30 % with increasing intensity of the artificial aging from the condition T4 towards the condition T73. As coating thickness and specific mass decreased in the same proportion, it can be concluded that current efficiency and overall coating porosity were constant for all heat treatment conditions.

In order to explain the evolution of the characteristic oxide microstructure during anodizing, special attention was initially paid to the precipitates being present in the substrate materials of the industrially important conditions T6 and T73. For both conditions, a large number of submicron η precipitates were found. During anodizing, these precipitates are preferentially dissolved leading to the presence of submicron voids. Moreover, the dissolution of the precipitates disturbs the even conversion front which could be the reason for the non-conformal growth of the pore channels. The T73 condition exhibits also the presence of η' precipitates with a size of up to 50 nm. Because of their small size, they do not influence the morphology of the pore channels significantly, however they might impede the conversion process or the charge transfer and increase the electrical resistance. Thus, the electrical current and the charge quantity are decreased. The exact role of the η' precipitates during anodizing in oxalic acid electrolytes should be further investigated in the future.

Due to the similar pore morphologies of anodic oxide coatings on the different heat treatment conditions, similar hardness profiles were obtained by nanoindentation. The hardness level of about 3.5 to 4 GPa equals the hardness levels observed for hard anodizing in sulfuric acid electrolytes. However, unlike sulfuric acid, oxalic acid does not dissolve the pore walls in the inner part of the oxide coatings significantly. Therefore, the hardness remains almost constant over a wide range of the coating thickness.

It can be reasonably assumed that the electrolyte composition influences the conversion or dissolution of precipitates. Thus, future work should address the impact of other base electrolytes (e.g. sulphuric acid) and electrolyte additives on the anodizing behavior of different heat treatment conditions. Moreover, the conversion and dissolution kinetics of different phases depend on the applied potential. Consequently, different electrical regimes could be applied in order to modify the morphologies of pore channels and voids with the aim of achieving desired coating properties.

## 5. References


[1] Wielage B, Alisch G, Lampke T and Nickel D 2008 *Key Eng. Mat.* **384** 263

[2] Sieber M, Morgenstern R, Kuhn D, Hackert-Oschätzchen M, Schubert A and Lampke T 2017 *IOP Conf. Ser.-Mat. Sci.* **181** 012044

[3] Nickel D, Dietrich D, Morgenstern R, Scharf I, Podlesak H and Lampke T 2016 *Adv. Mater. Sci. Eng.* 1374897

[4] Kuhn D, Martin A, Eckart C, Sieber M, Morgenstern R, Hackert-Oschätzchen M, Lampke T and Schubert A 2017 *IOP Conf. Ser.-Mat. Sci.* **181** 012042

[5] Zhou X, Habazaki H, Shimizu K, Skeldon P, Thompson GE and Wood GC 1996 *Corros. Sci.* **38** 1563

[6] Thompson GE, Habazaki H, Shimizu K, Sakairi M, Skeldon P, Zhou X and Wood GC 1999 *Aircr. Eng. Aerosp. Tec.* **71** 228

[7] Morgenstern R, Dietrich D, Sieber M and Lampke T 2017 *IOP Conf. Ser.-Mat. Sci.* **181** 012043

[8] Mukhopadhyay AK 1998 *Metall. Mater. Trans. A* **29A** 979

[9] Veys-Renaux D, Chahbouna N and Rocca E 2016 *Elektrochim. Acta* **211** 1056

[10] Li J, Peng Z, Li C, Jia Z, Chen W and Zheng Z 2008 *Trans. Nonferrous Met. Soc. China* **18** 755

[11] Delasi R and Adler PN 1977 *Metall. Trans. A* **8A** 1177

[12] Singh SS, Loza JJ, Merkle AP and Chawla N 2016 *Mater. Charact.* **118** 102

[13] Emani SV, Benedyk J, Nash P and Chen D 2009 *J. Mater. Sci.* **44** 6384

[14] Chen S, Ling Z, Hu X, Yang H and Li Y 2010 *J. Mater. Chem.* **20** 1794

[15] Li Y, Ling Z, Hu X, Liu Y and Chang Y 2011 *J. Mater. Chem.* **21** 9661



**Acknowledgments**

The authors gratefully acknowledge funding of this work by the German Research Foundation (Deutsche Forschungsgemeinschaft, DFG, project LA1274/40-1). The support by Dagobert Spieler, Elke Benedix, Steffen Clauß, Christel Gläser and Paul Clauß (all from the Institute of Materials Science and Engineering) is gratefully acknowledged.

# Investigation on the influence of nitrogen in process atmospheres on the corrosion behavior of brazed stainless steel joints

**V Fedorov[1], T Uhlig[1], G Wagner[1], A Langohr[2] and U Holländer[2]**

[1] Institute of Materials Science and Engineering, Chemnitz University of Technology
[2] Institute of Materials Science, Leibniz University Hannover

**Abstract.** Brazing of stainless steels is commonly carried out using nickel-based brazing fillers, which provide a high corrosion and oxidation resistance of the resulting joints. These brazed stainless steel joints are mostly used for manufacturing of heat exchangers for energy and air conditioning technologies. The joints of the study were produced at temperatures of 1000 °C, 1125 °C and 1150 °C in vacuum furnaces or continuous furnaces. In both cases, the parts interact with process gases like nitrogen within the brazing process, especially during cooling. The amount of nitrogen in the braze metal as well as in the base material was determined by the carrier gas hot extraction technique. The occurring diffusion of nitrogen into the braze metal and the base material causes a shift in the corrosion potentials. In this work, the influence of the nitrogen enrichment on the corrosion behavior was investigated using a capillary microcell. The corrosion measurements were carried out on the braze metal and the base material. The results of samples, brazed with and without the influence of nitrogen, were compared.

## 1. Introduction

Brazed stainless steel components such as heat exchangers for energy and air conditioning technology are manufactured millions of times in a wide range of sizes and types. A significant part of these components is brazed with nickel-based fillers, which offer the highest corrosion and oxidation resistance of the resulting joints [1-3]. By alloying with the elements silicon, boron and phosphorus, the melting temperature of the nickel or the nickel-chromium matrix is reduced and thus brazing temperatures between 1000 °C to 1200 °C are possible. Nickel-based fillers like (NiCoCr)-B-Si amorphous brazing filler metals, Ni-Cr-Si-B-P amorphous brazing foils and new nickel-based brazing alloy Ni-29Cr-8Mo-xFe-ySi offer a great potential for joining of stainless steels. The processing, structure and properties of the resulting brazed joints have been extensively reported [4-6]. The alloying elements provide excellent flowing and wetting properties, because they are able to eliminate the oxide film on the surface of the base materials. However, the alloying elements adversely affect the microstructure and the mechanical properties of the joints when the gap size exceeds 50 µm [7]. The joints are commonly brazed in vacuum furnaces or continuous furnaces. Nitrogen ($N_2$) or mixtures of nitrogen and hydrogen ($N_2$+$H_2$) are commonly used as process gases in continuous furnaces. In vacuum furnaces, $N_2$ is often added during the cooling process. For this reason, a nitrogen enrichment takes place in the braze metal and in the near-surface area of the base material. However, the influence of the nitrogen enrichment on the properties of the brazed joints has not been investigated yet. In this context, it was supposed that nitrogen is able to influence the corrosion properties of stainless steel joints, brazed with nickel-based fillers [8]. In this

IOP Conf. Series: Materials Science and Engineering **373** (2018) 012022 doi:10.1088/1757-899X/373/1/012022

work, the influence of nitrogen in process atmospheres on the corrosion behavior of brazed stainless steel joints is investigated. The nitrogen content in the braze metal as well as in the base material is determined using the carrier gas hot extraction technique (CGHE). The influence of the nitrogen enrichment on the corrosion behavior is analyzed using a capillary microcell. The corrosion measurements are performed on the braze metal and the base material. The results of samples brazed with and without the influence of nitrogen are compared.

## 2. Experimental procedures

Austenitic stainless steel (AISI 304) was used as base material. The components to be brazed (circular and semi-circular blanks with a diameter of 30 mm) were made of finely rolled sheets by laser cutting, which leads to burr-free and distortion-free samples. The fillers Ni710, Ni60CrPSi and Ni650 were applied on the base material as powder. The thickness of the produced brazed joint was adjusted at 50 µm by means of inserted spacer foils, figure 1. The quantity of applied filler was the same for all samples and was dimensioned, such that a joint thickness of at least 50 µm was ensured on average. Binder for a brazing powder was avoided to eliminate carbon residues and their consequences, for example, the formation of chromium carbide in the braze metal.

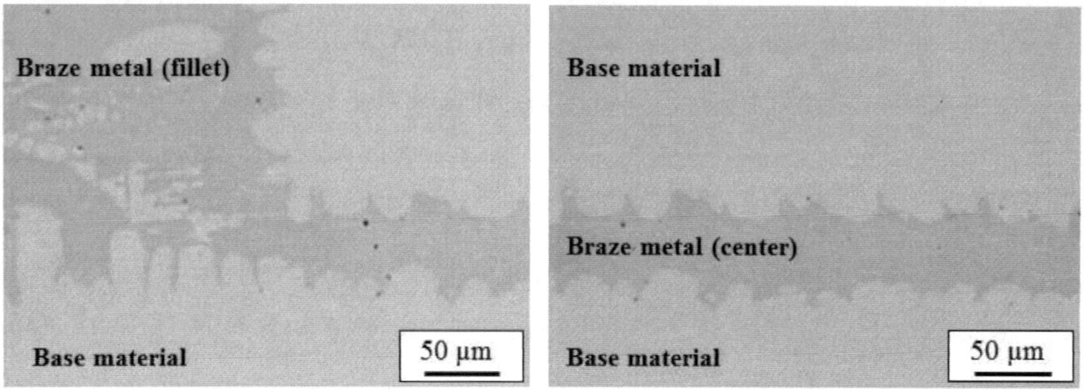


a) b)

**Figure 1.** Light microscope image of the sample, brazed with Ni710: a) area of braze fillet; b) sample center

The brazed single lap samples, used for the analysis of the nitrogen enrichment and the corrosion measurements, were produced at temperatures of 1000 °C, 1125 °C and 1150 °C in a vacuum furnace and a continuous furnace in process gas, Table 1. The heating rate and cooling rate was adjusted at 50 K/min in both furnaces. The brazing process in the continuous furnace was carried out using different holding times of 5 min and 10 min and different process gases, 100% $N_2$ and 50% $N_2$/50% $H_2$. The brazing temperature was recorded as sample temperature by a trailing thermocouple attached just below the sample surface. In the vacuum furnace, a holding time of 10 min and a cooling process in an argon atmosphere were performed. The brazing temperature was measured by a thermocouple, attached in the base material below the brazed surface.

IOP Conf. Series: Materials Science and Engineering **373** (2018) 012022 doi:10.1088/1757-899X/373/1/012022

Table 1: Overview of the brazed samples

| **Brazing filler** | **Continuous furnace** | | | | **Vacuum furnace** |
|---|---|---|---|---|---|
| | **Holding time** | | | | |
| | 5 min | | 10 min | | 10 min |
| | **Process gas** | | | | |
| | 100% $N_2$ | 50% $N_2$/50% $H_2$ | 100% $N_2$ | 50% $N_2$/50% $H_2$ | 0% $N_2$ |
| **NiCrP (Ni710)** $T_{Brazing}$ = 1000°C | 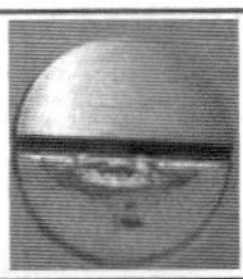 |  |  |  | 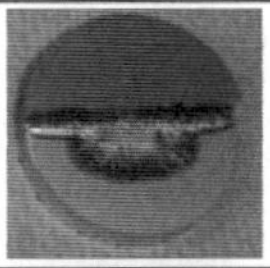 |
| **NiCrPSi** $T_{Brazing}$ = 1125°C | 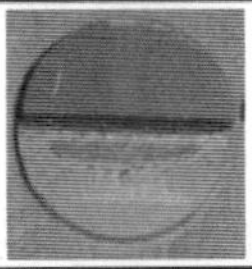 |  |  | 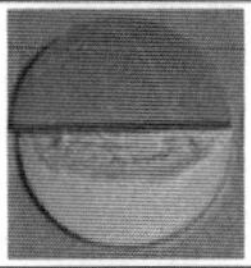 |  |
| **NiCrSi (Ni650)** $T_{Brazing}$ = 1150°C | 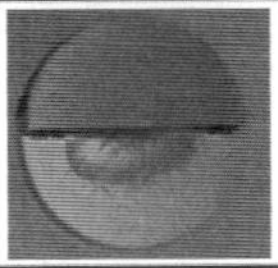 |  |  | 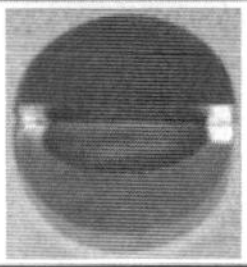 |  |

## 3. Results and discussion

### *3.1. Analysis of the nitrogen enrichment*

A quantitative determination of the nitrogen enrichment in the braze metal as well as in the base material are possible by means of CGHE. For this purpose, sample masses of about 100 mg are required. The process principle is described by Bagel [9]. Specific areas of the overlap samples are extracted by milling using a cutter with an associated cemented carbide indexable insert. The sample preparation for the analysis of the nitrogen enrichment is schematically presented in figure 2. The surface of the base material is cut off on the surface of the upper part of the sample in a depth of 100 µm. The braze metal is cut out of the braze fillet and the wetted surface without contact to the base material. The emerging material chips are used for CGHE.

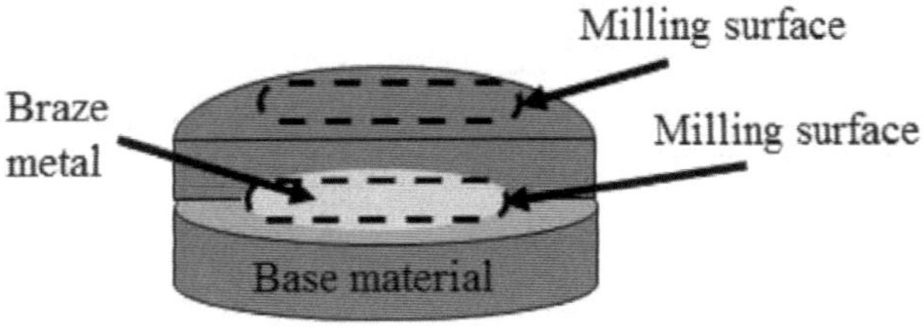


**Figure 2.** Scheme of sample preparation for the analysis of nitrogen enrichment

At first, the contents of $N_2$ in the brazing fillers and the base material were measured in the as-delivered condition. The values are 0.02 wt.% for Ni710, 0.04 wt.% for NiCrPSi and 0.05 wt.% for Ni650. The difference in the contents can be explained by the different chemical compositions of the used fillers. The $N_2$ content of the base material is 0.07 wt.%. After that, the $N_2$ contents of vacuum brazed samples were measured. The values are 0.06 to 0.07 wt.% in the base material and 0.03 to 0.06 wt.% in the braze metal. As expected, the vacuum brazing process does not influence the $N_2$ contents of base material and braze metal in comparison to the as-delivered condition. In case of samples that were brazed in the continuous furnace, the brazing process influences the $N_2$ contents of base material and braze metal significantly. In figure 3, the measured values of $N_2$ in the braze metal and in the base material are shown as a function of the brazing temperature resp. brazing fillers. With an increase of the brazing temperature and the use of a higher melting brazing filler, the $N_2$ contents in the braze metal as well as in the base material increase. This can be explained by an enhanced diffusion of nitrogen atoms into the braze metal and the base material, as it is known from nitriding processes of steels. The higher the temperature, the faster the diffusion of $N_2$ atoms takes place [10]. Additionally, it was found out that the process parameters holding time and $N_2$ content in process gas, do not have a significant influence on the $N_2$ contents of base material and braze metal. Hence, the scatter of the values is shown as standard deviation in the diagram. The braze metal is more intensely enriched with $N_2$ in comparison to the base material. The difference can be explained by the fact that liquid metals can absorb a higher amount of $N_2$ in comparison to solid metals, which is dissolved in the braze metal after solidification. During the brazing process, the braze metal is liquid while the base material is in the solid state.

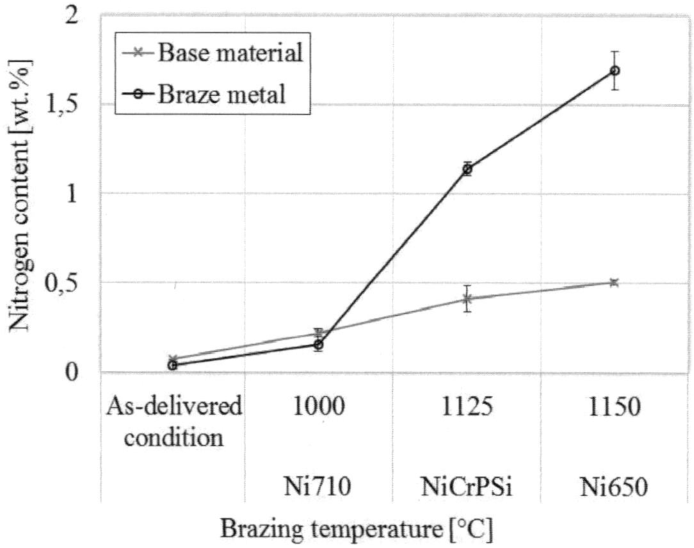


**Figure 3.** Nitrogen contents in the braze metal and in the base material as a function of the brazing temperature resp. brazing fillers, determined by CGHE

### *3.2. Corrosion measurements using a capillary microcell*

The influence of the nitrogen enrichment on the corrosion behavior is analyzed using a capillary microcell. The measurement principle of capillary microcell and the measurement positions on the brazed joints are schematically presented in figure 4. A drop of electrolyte is precisely applied on the tested sample using a capillary microcell. The aggressive model condensate K 1.2 ($H_2SO_4$ + $HNO_3$ (pH 1.2) + 1000 mg/l NaCl) electrolyte and an Ag/AgCl electrode are used [11]. For the measurements, a capillary with a diameter of 70 μm was chosen. In this case, the different microstructure components of the brazed joints can be examined. The corrosion measurements of the base material are performed at different positions: in the cross-section (point 1), in the cross-section near the braze fillet (point 2) on the surface (point 3) as well as in the cross-section on the braze metal (point 4), figure 4. It was determined that the corrosion potentials from point 1 do not differ for all the samples brazed in a vacuum and a continuous furnace. This can be explained by the fact that the base material is in the solid state during the brazing process and the nitrogen does not have driving forces to diffuse into the depth of the base material. Additionally, the corrosion potential of point 1 is also similar to that of point 2. Consequently, there is no influence of the process atmosphere on the corrosion behavior of the base material at these positions.

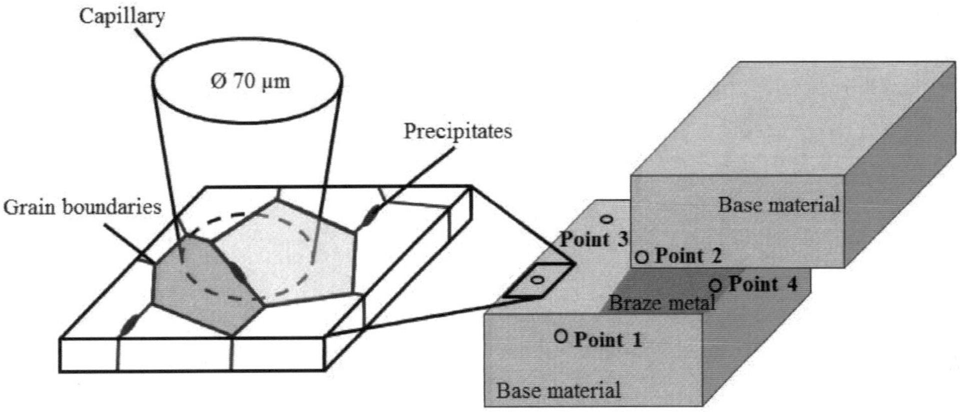


**Figure 4.** Principle of capillary microcell measurement and measurement positions on the samples

In figure 5, the corrosion potentials of the base material (point 3) and the braze metal (point 4) are presented as a function of the brazing temperature and the brazing fillers, respectively. It was found out that the used holding times and the $N_2$ contents in the process gas during continuous furnace brazing have a minor influence on the corrosion potentials of the base material and the braze metal. Hence, the scatter of the values is shown as standard deviation in the diagram. With an increase of the brazing temperature, the corrosion potentials get more negative in all cases. In comparison to the reference sample in the as-delivered condition (-300 mV), the corrosion potential of the base material is more positive for all brazed samples. During the brazing process, the oxide layer on the surface of the base material is affected. The samples that were brazed in process gas show a negative shift of the corrosion potential of -80 to -100 mV in comparison to the vacuum brazed samples dependent on the brazing temperature. That means, the base material of the samples that were brazed in process gas, becomes more ignoble than those of the vacuum brazed samples. This is caused by the enrichment of the base material with $N_2$ during the brazing in process gas.

The braze metals show less negative corrosion potentials than the base materials for both brazing processes. As already mentioned, liquid metals can absorb a higher amount of $N_2$ in comparison to solid metals. It was determined that nitrogen in the process gas causes a shift in the corrosion potentials of the braze metal. When compared to the vacuum brazed samples, the braze metals of samples, that were brazed in process gas, show a positive shift of the corrosion potential of about 10 to 30 mV dependent on the brazing temperature. It means that the braze metal of the joints is more noble than that of the vacuum brazed joints. This effect can be explained by the nitrogen enrichment, determined by means of CGHE, in the braze metal as well as in the base material.

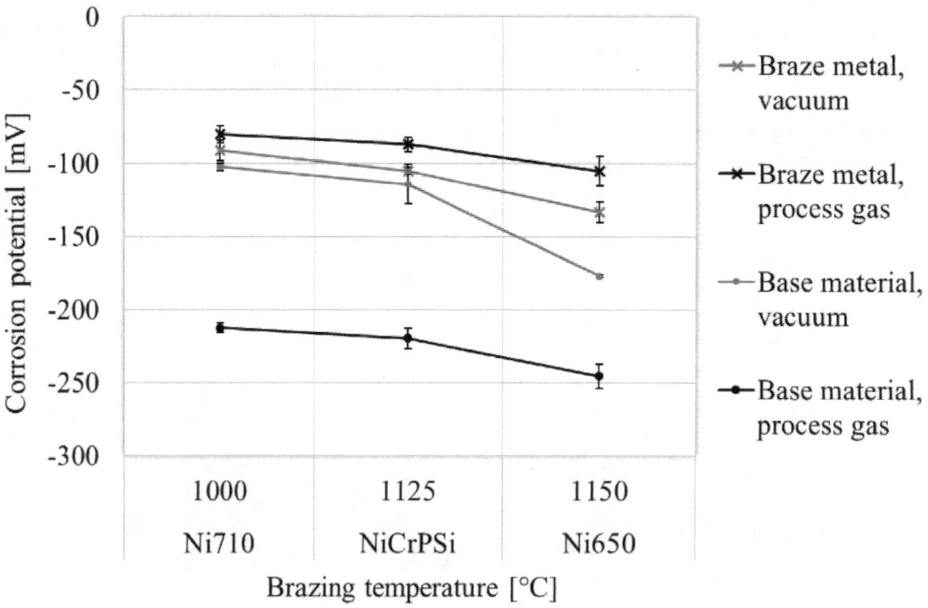


**Figure 5.** Corrosion potentials of the braze metal and the base material as a function of the brazing temperature resp. brazing fillers

All samples were observed using a light microscope after the capillary microcell measurements. The results of the corrosion measurements on the braze metal of the joints (point 4), brazed with Ni710 without influence of $N_2$ (vacuum) und with influence of $N_2$ (process gas), are exemplarily presented in figure 6. It can be seen that the braze metal of the joint, brazed in vacuum, is slightly more affected by the corrosion in comparison to the joint, brazed in process gas (black spots in the measurement point). This confirms the obtained results of corrosion measurements, shown in figure 5.

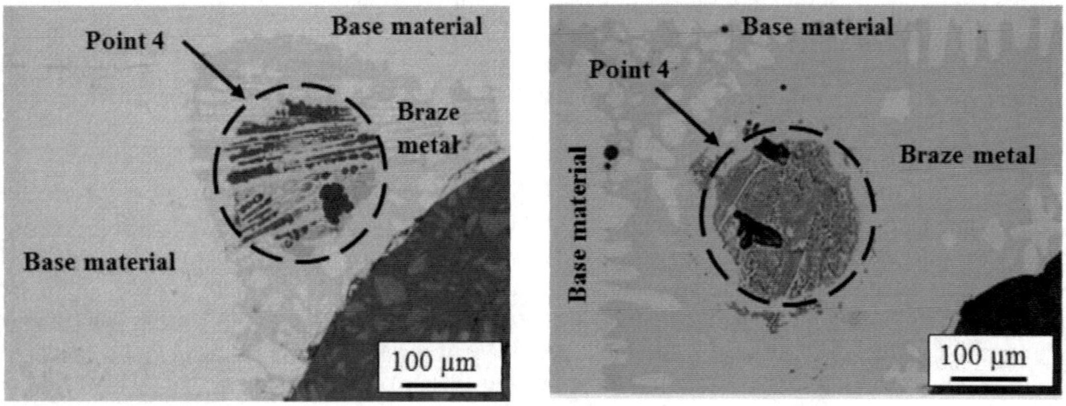


a) b)

**Figure 6.** Corrosion measurements on the braze metal of the joints (point 4), brazed with Ni710: a) vacuum; b) process gas

## 4. Conclusions

The joints were produced using nickel-based brazing fillers at temperatures of 1000 °C, 1125 °C and 1150 °C in a vacuum furnace and a continuous furnace. The $N_2$ content in the braze metal as well as in the base material was determined by CGHE. It was found out that the different process parameters (holding time and $N_2$ content in the process gas) during continuous furnace brazing do not have a significant influence on the $N_2$ contents in the base material and the braze metal. In its turn, the samples, brazed in a continuous furnace, are more intensely enriched with nitrogen in comparison to samples in the as-delivered condition and vacuum brazed samples. With an increase of the brazing temperature and the use of a higher melting brazing filler, the $N_2$ content in the braze metal as well as in the base material increases for the joints that were brazed in a continuous furnace. Additionally, it was determined that the braze metal is more intensely enriched with nitrogen in comparison to the base material. The influence of the nitrogen enrichment on the corrosion behavior is analyzed using a capillary microcell. The different braze metals show less negative corrosion potentials than the base materials for both brazing processes. With an increase of the brazing temperature, the corrosion potentials are more negative in all cases. Additionally, it was determined that nitrogen in the process gas causes a shift in the corrosion potentials of the braze metal and the base material. In case of the braze metal, the nitrogen influences the corrosion potential of the brazed joint positively. In contrast to that, the nitrogen has a negative influence on the corrosion potential of the base material. It was also supposed, that nitrogen can influence the fatigue behavior of stainless steel joints, brazed with nickel-based fillers [8]. In further investigations, the fatigue behavior of brazed stainless steel joints will be observed.

## 5. References


[1] Hoffmann E E, Patriarca P and Leitten Jr. C F 1956 An Evaluation Of The Corrosion And Oxidation Resistance Of High-Temperature Brazing Alloys Welding Journal

[2] Achete M, Lugscheider E and Pelster H 1987 Korrosionsbeständigkeit von Hochtemperaturlötverbindungen in organischen und wässrigen Lösungen Schweißen & Schneiden 49 280-284

[3] Brandl W, Minarski P and Podieschny R 1990 Korrosionsverhalten von gelöteten CrNi-Stahlverbindungen in Trinkwässern DVS-Berichte 132

[4] Rabinkin A 2004 Brazing with (NiCoCr)–B–Si amorphous brazing filler metals: alloys, processing, joint structure, properties Science and Technology of Welding and Joining 9 181-199

[5] Hartmann T and Nuetzel D 2009 New amorphous brazing foils for exhaust gas applications In Proceedings of the 4th International Brazing and Soldering Conference (IBSC) 110-117

[6] Persson U and Bornegård N 2016 New nickel based brazing alloy suitable for very corrosive environments DVS-Berichte 325 224-226

[7] Lugscheider E, Schittny T and Halmoy E 1989 Metallurgical Aspects of Additive-Aided Wide-Clearance Brazing with Nickel-Based Filler Metals Welding Research Supplement 9-13

[8] Holländer U, Weber C, Möhwald K and Maier H J 2015 Ermittlung von Versagenskriterien mechanisch-korrosiv belasteter, hartgelöteter Edelstahlblechverbindungen unter Berücksichtigung der Nickellotmetallurgie und der Fertigungsbedingungen Schweißen & Schneiden 66 174-182

[9] Bagel A 1985 Die Ermittlung des Gesamtstickstoffanteils Handbuch für das Eisenhüttenlaboratorium Verlag Stahleisen Band 2A (Düsseldorf)

[10] Liedtke D 2006 Wärmebehandlung von Eisenwerkstoffen, II: Nitrieren und Nitrocarburieren Expert Verlag

[11] VDA 230-214 2010 Beständigkeit metallischer Werkstoffe gegen Kondensat-Korrosion in abgasführenden Bauteilen Verband der Automobilindustrie



## Acknowledgments

Part of the research was carried out within the scope of the project # 19056 BR funded by the Bundesministerium für Wirtschaft und Energie (BMWi) via the Arbeitsgemeinschaft industrieller Forschungsvereinigungen "Otto von Guericke" e.V. The authors thank the BMWi for financing this project.

IOP Conf. Series: Materials Science and Engineering **373** (2018) 012023 doi:10.1088/1757-899X/373/1/012023

# Vacuum brazing of 316L stainless steel based on additively manufactured and conventional material grades

**W Tillmann, T Henning and L Wojarski**

Institute of Materials Engineering, TU Dortmund University, 44227 Dortmund, Leonhard-Euler-Straße 2, Germany

**Abstract.** Many industrial applications require components with an increasing geometric complexity and specific material properties. Furthermore, the production costs and the affordable production time have to be minimized in order to ensure competitiveness. These divergent objectives are difficult to achieve with a single manufacturing technology. Therefore, joining of selective laser melted (SLM) complex shaped parts to conventionally produced high-volume components provides a high potential. The current investigation focuses on vacuum brazing conventionally manufactured to non-hipped SLM generated AISI 316L stainless steel. Cylindrical samples (Ø 14 mm) were brazed using a B-Ni2 foil (50 µm) at 1050 °C for 30 minutes in vacuum (< $4.5 \cdot 10^{-5}$ mbar) and directly cooled down to room temperature with 4 bar overpressure to prevent the formation of chromium carbides within the base material. It could be proven that the brazing quality is extremely sensitive to even marginal porosities (< 0.2 %) and/or oxide inclusions of the SLM microstructure. Therefore, the fracture mirror in SLM/conventional steel brazements was at the joint braze/SLM steel interface, leading to a joint strength of 317.4 MPa. This corresponds to only 67.4 % of the joint strength obtained with conventional steel, where the fracture propagated through the diffusion area.

## 1. Introduction

Additive manufacturing processes offer an almost unlimited freedom of scope with respect to complex component structures. Therefore, the utilization of selected laser melting (SLM) that started as a prototyping process and has become more and more end consumer application oriented in the last decade [1]. At least until today, high manufacturing costs, comparatively long production times, and restricted construction installation spaces slow the replacement of conventional manufacturing processes like drilling, die sinking, and milling down. Thus, there is an economic intersection if highly complex SLM-components are joined to high-volume and simply shaped components, manufactured by conventional methods. Vacuum furnace brazing is a predestined process to combine the joining process of segmented components with a precise and well-defined heat treatment for the resulting hybrids [2,3]. In addition, this procedure offers the possibility to achieve an equalization of the lamellar SLM microstructure into a homogeneous microstructure of conventional material grades due to an adequate heat treatment [4-6]. Furthermore, there are other significant differences between SLM and conventional material grades that have to be considered if hybrid composites are produced. In terms of SLM process technology, the produced components contain high residual stresses which primarily occur at the outer component areas in the as-built condition [7]. As a result of the specific lamellar SLM microstructure and the size of carbide precipitations, there are also differences concerning to the hardness as well as the tensile and yield strength [2,8]. In turn, a well-defined heat treatment during the brazing cycle facilitates the production of hybrid composites with almost identical microstructures and mechanical properties. However, a combined utilization of these technologies is barely investigated.

Additively manufactured components of 316L are mainly used for heat exchangers, steam or nuclear power plants, as well as in the aviation industry [7,9]. These components are often exposed to very high mechanical, thermal, and corrosive loads which justifies a profound research, taking several manufacturing technologies into consideration. The tensile strength of SLM 316L is generally between 580-620 MPa as a function of the grain size [10]. In order to achieve a high joint strength and a reasonable oxidization and corrosion resistance, vacuum brazing of conventional 316L material grades is usually realized with nickel based brazing alloys [11]. In special, B-Ni2 ($T_{S/L}$: 969-1024 °C) or preferably B-Ni5b ($T_{S/L}$: 1030-1126 °C) brazing alloys are frequently used [12]. The temperature range recommended for solution annealing of 316L is between 1020-1120 °C [12]. Therefore, this step can be easily implemented in the brazing cycle. It is vital that the subsequent cooling rate between 900-500 °C is high enough to prevent the formation of chromium carbides in order to guarantee a sufficient corrosion resistance of the base material itself. Hence, cooling with nitrogen overpressure and an enforced gas flow to a heat exchanger is a common set up in vacuum furnaces [13].

The objective of the present work is to evaluate the influence of oxide inclusions and the porosity within the SLM microstructures, related to their negative influence of the brazing quality [14]. In case of oxide inclusions, the cause of failure can be related to the ionic-covalent atomic bond of the oxides in contrast to the metallic bond of the braze and the base material. Consequently, brazing was conducted using all combinations of SLM and conventional material grades in a single furnace run (SLM/SLM, SLM/conv., conv./conv.). This allowed to directly trace bonding defects at the interface braze/diffusion area as well as the related joint strength to occurring oxide inclusions and the porosity of the SLM microstructure. The base materials were analyzed in the as-delivered/as-build condition by means of light microscopy, scanning electron microscopy, EDS-analysis and Vickers hardness testing. Additionally, tensile testing and fracture analyses were conducted for the brazed samples.

## 2. Materials and methods

### *2.1. Reactive systems as innovative heat source*

SLM AISI 316L was manufactured at the Direct Manufacturing Research Center (DMRC) of the university of Paderborn using cylinders with 15 mm diameter that were 75 mm high [10]. In order to analyze the influence of porosity, non-hipped material was used [15]. The samples were machined to a diameter of 14 mm by turning in order to remove the inhomogeneous surface. Conventional AISI 316L, produced by continuous casting, rolling, and solution annealing, was machined identically. Table 1 shows the chemical composition of the base material according to specification DIN EN 10088-3. In addition, the base materials used (conventional/SLM) were examined by EDS-analyses on metallographically prepared and polished cross-sections of three samples each. It should be noted that EDS measurements of carbon are not suitable to obtain a quantitatively exact value. Previous investigations using optical emission spectroscopy (OES) of the used 316L powder grade showed a much lower carbon content (marked by * in table 1) [10]. In addition, the measured amount of molybdenum in the SLM material was unusually high when using EDS.

**Table 1.** Chemical composition of 316L according to DIN EN 10088-3 in comparison to the composition of the conventional and SLM manufactured steel grades.

| [wt.-%] | C | Si | Mn | P | S | Cr | Mo | Ni | N |
|---|---|---|---|---|---|---|---|---|---|
| **spec.** | <0.03 | <1.0 | <2.0 | <0.045 | <0.03 | 16.5 -18.5 | 2.0 - 2.5 | 10.0 – 13.0 | <0.1 |
| **conv. (EDS)** | (0.48) | 0.64 | 1.57 | - | - | 17.11 | 2.39 | 11.01 | - |
| **SLM (OES)*** | 0.018 | - | - | - | - | 17.03 | 2.27 | 10.52 | - |
| **SLM (EDS)** | (0.53) | 0.87 | 1.12 | - | - | 17.89 | 3.28 | 11.61 | - |

Vickers hardness measurement of these samples revealed an average hardness of 290.6 $HV_{0.3}$ (standard deviation s = 3.3) for the conventional material and 235.5 $HV_{0.3}$ (s = 2.7) for the SLM graded material longitudinal to the built-up direction as well as 229.0 $HV_{0.3}$ (s = 4.2) for the transverse direction. Hence, the conventional material showed a slightly higher hardness. The samples were subsequently etched by V2A at 70 °C. The microstructures are shown in figure 1 (a-c).

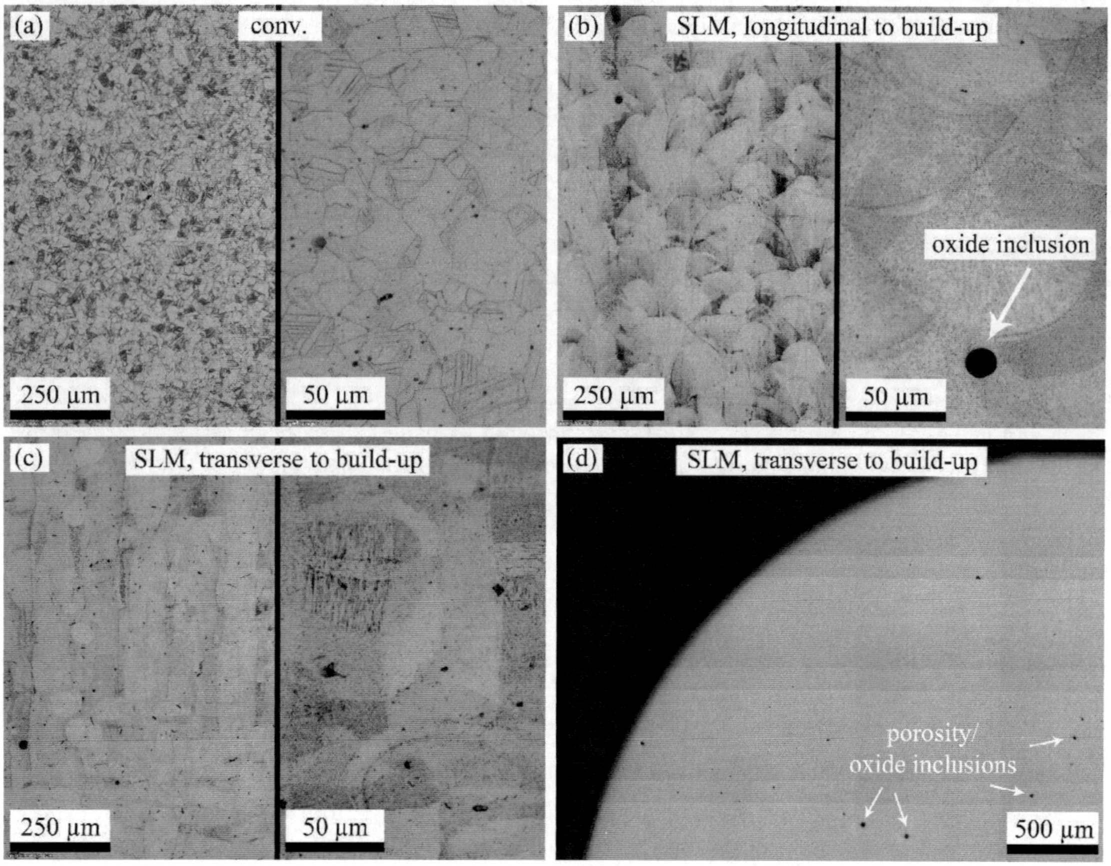


**Figure 1.** Light microscope images of the as-delivered/as-build condition of the base materials: (a) conventional, (b) SLM longitudinal to the build-up direction, (c) SLM transverse to the build-up direction, (d) Quarter of a joining area (SLM transverse to the build-up direction) with a typical proportion of porosity and oxide inclusions in SLM.

The microstructure of the conventional material featured an average grain size of 21.4 µm and was completely austenitic. Dislocation structures were clearly visible as lines within the grains. As expected, there was no difference between the longitudinal and the transverse cross-section of the rod axis detectable. Some very few inclusions of manganese sulfide were detected (figure 1(a)). In contrast to this, the microstructure of the longitudinal and the transverse cross-section related to the build-up direction, were completely different. The longitudinal cross-section showed a semi-circular superimposed and thus lamellar structure, which is the result of a local melting behavior of the laser (figure 1(b)). In transverse direction, the exposure strategy of the laser led to a reticular microstructure (figure 1(c)). For both directions, porosity and oxide inclusions were examined by grey scale value tolerance measurements, based on the light microscopic images. The ratio of comparatively large pores and oxide inclusions to the residual surface of the base material was merely 0.14 % (s = 0.03). Nevertheless, finely distributed defects were not detectable with this method and no detailed analysis was made to distinguish pores and oxide inclusions from each other because both appeared in black color. An exemplary image of one quarter of the faying surface of a sample with this ratio is shown in figure 1(d). Amongst oxygen, EDS-analyses of the oxides (average size: 10-35 µm) revealed high amounts of chromium, manganese, silicon, and aluminum. Each of these elements has an exceedingly high affinity to oxygen. It should be noted that aluminum is generally not included in AISI 316L. Therefore, aluminum should be traced back to powder residues from a previous manufacturing process. Nevertheless, no aluminum was detected within the steel.

An amorphous nickel-based foil of type AWS B-Ni2 (MBF-20, $T_{S/L}$: 969-1024 °C) with a thickness of 50 µm was used for the vacuum furnace brazing process. This brazing alloy was chosen in order to combine a high joint strength with a suitable corrosion resistance at a comparatively low brazing temperature of 1050 °C at the lower end of the solution annealing temperature [16].

### *2.2. Sample preparation*
A butt-joint geometry was used for the brazing experiments. Cylindrical specimens with a diameter of 14 mm and a height of 15 mm for metallographic inspections and a height of 50 mm for mechanical tests were selected. The transverse to the build-up direction in the case of the SLM substrates were chosen to be used as faying surfaces, which were mechanically face grinded from quality level N5 to N6. The experimental design was set to braze and analyze all combinations of the different material grades of AISI 316L: conventional/conventional, SLM/conventional, as well as SLM/SLM. Three samples each were brazed for a metallographic inspection. In case of tensile testing, five conventional/conventional and SLM/conventional samples were prepared. The SLM parts were placed on top in every case. Due to the height of the tensile samples, a fixture was used to assure an axial alignment. The samples were cleaned in an ultrasonic ethanol bath for five minutes immediately prior to the brazing and dried with hot air.

### *2.3. Vacuum furnace brazing and heat treatment*
Brazing was conducted in a vertical vacuum furnace (SCHMETZ type EU80/1H/2RV) at 1050 °C for 30 minutes and a vacuum higher than $4.5 \cdot 10^{-5}$ mbar. In order to assure a uniform temperature distribution in the molybdenum heating chamber, a heating rate of 10 K/minute and dwell times of 15 minutes each were used at 650 °C, 850 °C, and 920 °C. Simultaneously to the brazing temperature, the base material was solution annealed to dissolve already existing chromium carbides. Afterwards, the samples were rapidly cooled down to room temperature, using a directed cooling gas flow of 4 bar nitrogen and a heat exchanger. A cooling rate of approximately 4.5 °C/s was achieved in the temperature range of 900-500 °C in order to prevent the generation of chromium carbides in the base material.

### *2.4. Machining and testing of tensile samples*
The brazed cylinders with a height of 100 mm were manufactured by turning to tensile samples of type B8x40 with a M12 thread according to DIN EN 50125:2016-12. Tensile tests were carried out with the universal testing machine ZWICK 1488.100/01 with a position controlled movement speed of 0.025 mm/s.

## 3. Results and discussion

### *3.1. Reactive systems as innovative heat source*
After brazing the metallographic samples were cut vertically to the joint, embedded, and finally polished finally with a 1 µm diamond suspension. Then, the joint was examined using means of light microscopy, SEM- and EDS-analyses. In figure 2, characteristic SEM images of the produced joints are shown.

In case of the conventional/conventional combination, almost defect-free joints were observed. Merely a few pores were detected at the interface of the braze metal/diffusion area. As visible in figure 2(a), some manganese sulphides exist next to the brazing area, which may have encouraged the formation of pores. The microstructure of the diffusion area was usually dominated mainly by chromium boride precipitations at the grain boundaries. The hybrid SLM/conventional joints showed a single-sided pore space at the interface of the braze metal/SLM diffusion area (figure 2(b)). Means of

SEM/EDS-analyses verified the presence of pores whereas no influence of oxide inclusions within the SLM microstructure was verifiable since resin (consist of carbon and oxygen) was polished into the pores. Furthermore, the lamellar structure of the SLM part was preserved after brazing and solution annealing, respectively. Thus, the precipitations at the grain boundaries within the diffusion area at the SLM part are located at the grain boundaries and seem to be significantly finer when compared to the conventional parts. Again, the diffusion area within the conventional part was comparable to the completely conventional joint. Consequentially, the samples of the SLM/SLM combination revealed a double-sided pore space and a diffusion area which corresponds with the SLM side of the hybrid joints (figure 2(b)). An irregularly shaped grain geometry within the microstructure is clearly visible in figure 2 (c). The average size of the observed pores and oxides was 10-35 μm as already observed in the hybrid joints. Figure 2(d) illustrates the size ratio between the braze metal (initially 50 μm) and oxide inclusions next to the joint. Thus, a significant influence of the oxides nearby on the formation of single- (SLM/conventional) and double-sided pore spaces must be considered. In contrast to this assumption, the formation of pores could also be a result of greatly enhanced diffusion rates from the braze metal into the SLM base material compared to the diffusion rates into conventional material grades. This formation of voids, the so-called Kirkendall effect, describes higher diffusion rates from the braze metal into the base material than vice versa. As a result, a slowdown of volume is promoted within the braze metal, which results in the characteristic porosity at the interface.

The average thickness of the resulting braze metal was 41.5 μm (s = 2.1) for the conventional/ conventional joints, 42.7 μm (s = 1.3) for the hybrid joints, and 44.1 (s = 1.9) in case of the SLM/SLM joints. Due to the heat treatment, the hardness was reduced to 146.5 $HV_{0.3}$ (s = 4.5) for the conventional material and to 188.2 $HV_{0.3}$ (s = 6.7) for the SLM material compared to the as-delivered/as-build condition. This could be related to the tearing-down of cold work hardenings.

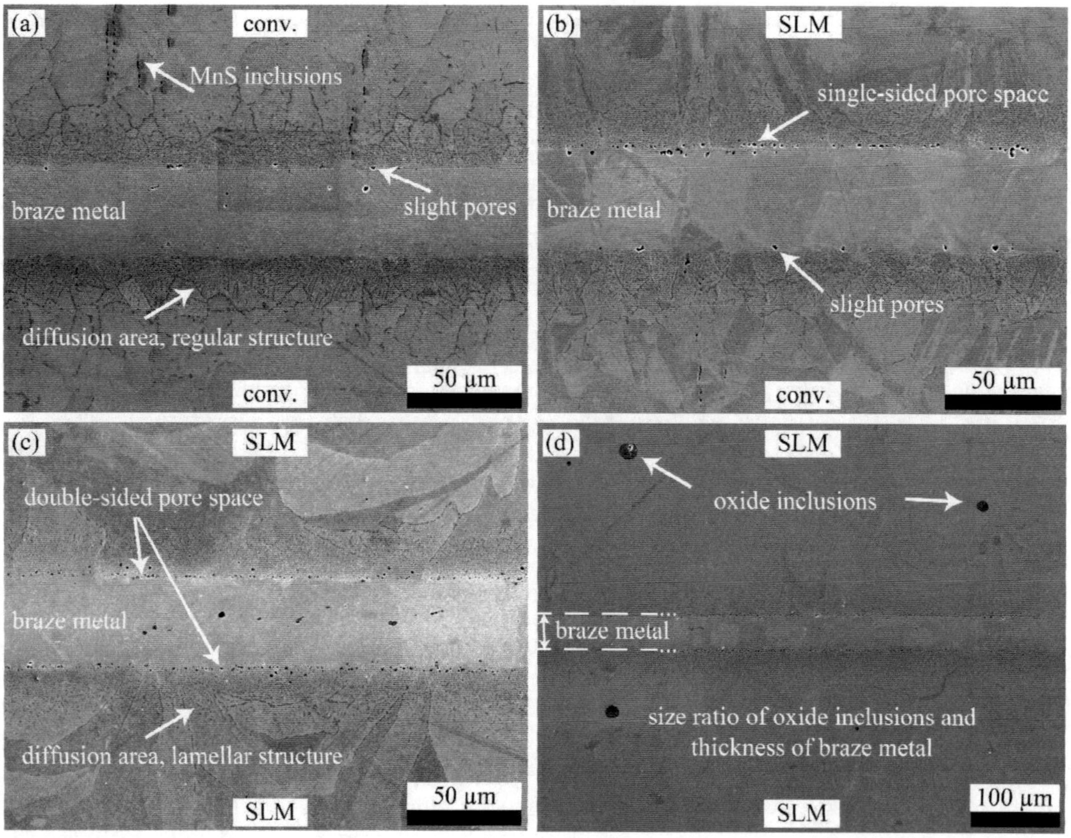


**Figure 2.** Characteristic SEM images of brazed joints: (a) conv./conv., (b) SLM/conv., (c) SLM/SLM, (d) SLM/SLM – size ratio of oxide inclusions and the thickness of braze metal.

EDS-analyses of the specific joints showed no significant difference with regard to the distribution of the main elements Fe, Cr, Ni, and Si along the joint path. Figure 3 illustrates, the EDS-line profiles of the conventional/conventional and SLM/SLM combinations. As a result of the small atomic size and the emerging low quantification quality, boron was excluded from the analyses results.

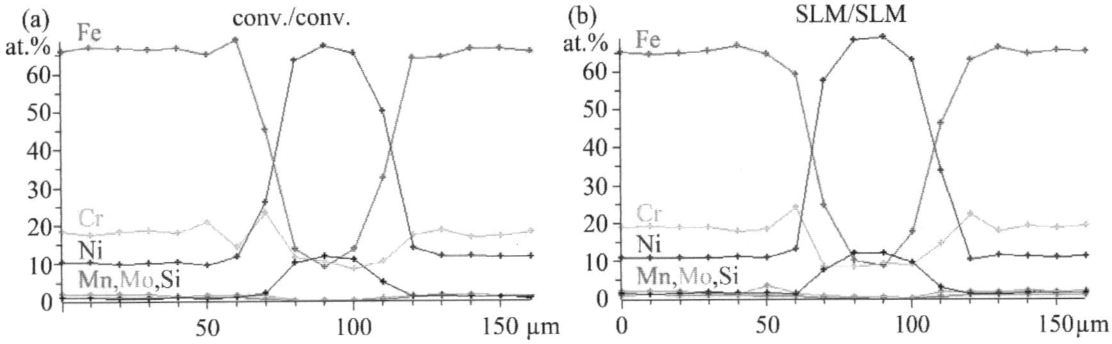


**Figure 3.** EDS-analyses along the joint: (a) conventional/conventional, (b) SLM/SLM.

Nevertheless, the initial proportion of boron in the B-Ni2 brazing foil was 3.2 wt.-%. Therefore, a possibly decisive influence of this element for the formation of the pore spaces at the interface could not be verified with the utilized analysis methods. However, boron has very high diffusion rates, hence, encouraging, the formation of Kirkendall porosity.

*3.2. Analyses of tensile samples*

No visual difference was detectable between the samples after machining. The average tensile strength was 470.8 MPa (s = 27.3) for the conventional/conventional samples and 317.4 MPa (s = 17.8) for the SLM/conventional samples. Thus, the average tensile strength of the SLM/conventional samples was merely 67.4 % compared to the conventional/conventional samples and just 51.1-54.7 % compared to the tensile strength of full SLM material samples in the as-build condition (580-620 MPa) [10]. In every case, the fracture occurred spontaneously without any visible sample deformation. The fractured surfaces were cut in the center and were metallographically prepared. Figure 4 shows a characteristic SEM image of the fracture mirror for the conventional/conventional and the SLM/conventional samples.

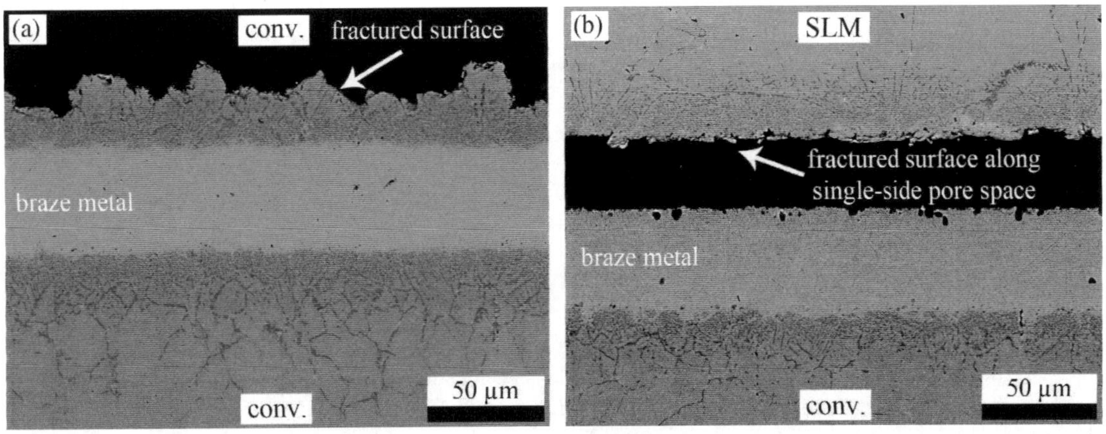


**Figure 4.** Characteristic SEM images of brazed joints, fractured by tensile testing: (a) conv./conv., (b) SLM/conv..

In case of the conventional/conventional samples, the fracture always propagated at the interface diffusion area/base material. In contrast to this, the fracture mirror of the SLM/conventional samples was located at the interface braze metal/diffusion area at the single-sided pore space. Due to the severe formation of pores, the joint strength was significantly reduced.

## 4. Summary and conclusion

Vacuum brazing was successfully performed, using both SLM as well as conventional material grades of AISI 316L. Within the frameworks this work, combinations of conventional/conventional, SLM/conventional, and SLM/SLM samples were brazed. If a faying surface was given by SLM material, a severe pore space was examined at the interface between the braze metal and the diffusion area of the SLM part. Thus, the fracture in SLM/conventional steel brazements was located at the pore space, leading to a joint strength of 317.4 MPa. This corresponds to only 67.4 % of the joint strength obtained with conventional steel, where the fracture propagated through the diffusion area. Therefore, the pores spaces must be considered as the weak point of the SLM/conventional joint. The cause of failure of the pore spaces could not be clearly determined within the present work. In this regard, three different defect mechanisms are conceivable, which may overlap each other. Firstly, there could be an influence of oxide inclusions within the SLM material close to the joint. Secondly, the open porosity of the SLM material at the faying surface can promote gas inclusions between the brazing foil and the steel. In this context, gas inclusions as well as finely dispersed oxides will be solved in the liquid braze metal and finally precipitated at the interface braze metal/diffusion area. Finally, the so-called Kirkendall porosity that results from a high boron diffusion activity into the SLM material, can be expected. A key challenge of future works is to understand pore formation mechanisms and either to develop technologies or to find suitable process parameters to prevent the formation of pores when brazing AISI 316L material grades. Therefore, exceeding research has to be conducted in order to enhance the joint strength of brazed hybrids.

## 5. References

[1] Thompson M K, Moroni G, Vaneker T, Fadel G, Campbell R I, Gibson I, Bernard A, Schulz J, Graf P, Ahuja B and Martina F 2016 Design for additive manufacturing: Trends, opportunities, considerations and constraints *Manufacturing technology* **65** 737-760

[2] Hölker R and 2 A E 2016 Advancements in the manufacturing of dies for hot aluminum extrusion with conformal cooling channels *International journal of advanced manufacturing technology* **83** 1209-1220

[3] Tillmann W, Wojarski L, Lehmert B 2010 The quality of brazed ceramic and cemented carbide joints. A mechanical and metallurgical assessment. *Advances in science and technology* 108 **64** 108-114

[4] Ascari L, Fortunato A, Liverani E, Gamberoni A and Tomesani L 2016 New possibilities in the fabrication of hybrid components with big dimensions by means of selective laser melting (SLM) *Physics procedia* **83** 839-846

[5] Niendorf, T, Brenne F and Schaper M 2014 Lattice structures manufactured by SLM: On the effect of geometrical dimensions on microstructure evolution during processing *Metallurgical and materials transactions B* **45(4)** 1181-1185

[6] Niendorf T, Leuders S, Riemer A, Richard H A, Troester T and Schwarze D 2013 Highly anisotropic steel processed by selective laser melting *Metallurgical and materials transactions B* **44(4)** 794-796

[7] Hinojos A, Mireles J, Reichhardt A, Frigola, P, Hosemann, P, Murr L E and Wicker R B 2016 Joining of Inconel 718 and 316 stainless steel using electron beam melting additive manufacturing technology *Materials and Design* **94** 17-27

[8] Kirchheim A, Zumofen L and Gerig M 2014 Additive Fertigung mit 1.4404 (316L): Prozess- und Charakterisierungsergebnisse. Zentrum für Produkt- und Prozessentwicklung ETH Zürich, 30.09.2014

[9] Cooper D, Thornby J, Blundell N and Gibbon G J 2015 Design and manufacture of high performance hollow engine valves by additive layer manufacturing *Materials and design* **69** 44-55
[10] Niendorf T, Leuders S, Riemer A, Brenne F, Troester T, Richard H A and Schwarze D 2014 Functionally graded alloys obtained by additive manufacturing *Advanced engineering materials* **16** 857-861
[11] Bobzin K, Schlaefer N, Kopp S, Puidokas S, Tillmann W, Wojarski L and Liu C 2010 Systematik zur Auslegung und anwendungsrelevanten Prüfung von Nickel-Basis Lötungen *Proceedings of the 9th International Conference of Brazing, High Temperature Brazing and Diffusion Bonding LÖT 2010*, 15.-17.06.2010, 17-22, Aachen/Germany
[12] Thijs L, Humbeeck J V, Kempen K, Yasa E, Kruth J P and Rombouts M 2012 Investigation on the inclusions in maraging steel produced by selective laser melting. Innovative Developments in Virtual and Physical Prototyping *Proceedings of the 5th international conference on advanced research and rapid prototyping* Leira Portugal, 28.09.-01.10.2011, 297-304
[13] Tillmann W, Yurchenko S, Pfeiffer J, Laemmerhirt I A and Wojarski L 2014 Microstructural investigation of the precipitation hardening alloy Cu12Ni4.6SiB for brazing materials with a high CTE-mismatch. *Conference Proceedings International conference on brazing, soldering and special joining technologies*, Beijing/China 301–305
[14] Schaak C, Tillmann W, Schaper M and Aydinoez M E 2016 Process gas infiltration in Inconel 718 samples during SLM processing *RTeJournal*
[15] Tillmann W, Schaak C, Nellesen J, Schaper M, Aydinoez M E 2015 Functional encapsulation of laser melted Inconel 718 by Arc-PVD and HVOF for post compacting by hot isostatic pressing. *Powder Metallurgy* **58** 259–64
[16] Hartmann T and Nuetzel D 2010 Nickel-chromium-based amorphous brazing foils for continuous furnace brazing of stainless steel *Proceedings of the 9th International Conference of Brazing, High Temperature Brazing and Diffusion Bonding LÖT 2010*, 15.-17.06.2010, 42-47, Aachen/Germany

**Acknowledgments**
The authors acknowledge F. Hengsbach of the department of material science of university Paderborn and Direct Manufacturing Research Center (DMRC) for supporting the SLM material grades.

# Calculation of trajectories and the rate of growth of curvilinear fatigue cracks in isotropic and composite plates

**H Pokhmurska, O Maksymovych, A Dzyubyk and L Dzyubyk**

National University "Lviv Polytechnic", Lviv, 79013, Ukraine
dar.lviv@gmail.com

**Abstract.** The methods of calculating the trajectories and the rate of growth of curvilinear fatigue cracks in isotropic and composite plate structure elements during cyclic loading along straight or curvilinear trajectories are developed. For isotropic and anisotropic materials, the methodes are developed on the basis of the force criterion of destruction with the additional application of the fatigue fracture diagrams. To find the change in the shape of the cracks in the loading process, the step-by-step method was used. At each stage, the direction of the growth of all vertices of cracks and the lengths of their arcs was found on the basis of determining the intensity coefficients of stresses by the method of singular integral equations. The results of calculations of the cracks system growth process are presented.

## 1. Introduction

One of the most common reasons of lamellar elements destruction in cyclic loads is the growth of small initial cracks until they reach critical dimensions. [1,2].

Investigation of the process of growth of fatigue cracks makes it possible to predict the survivability of structures, that is, to evaluate the process from the origin of cracks to complete destruction [1-10]. The criteria for destructive mechanics are used to develop such methods (to find the directions of cracks growth), Fatigue Fracture Diagram (FFD, necessary for finding the speed of peaks) and numerical algorithms for calculating Stress Intensity Factor (SIF) for cracks [11-13]. In calculations in the literature, the most commonly used step-by-step methods, in which the trajectory of cracks growth is described by segments of a straight line (polygonal chain). In this paper, the trajectory of a crack is described by splines of the third degree, which enables simultaneously to control the accuracy of calculations, increase steps and thereby reduce the volume of computations and obtain smooth trajectories of cracks, which are usually observed in practice.

For composite materials, the question of determining the directions of fatigue cracks growth is studied much less than for isotropic materials due to the complication of constructing general kinetic equations of the peaks movement. In the work to find the trajectories of growth of cracks, a force destruction criterion is used, which takes into account the anisotropy of the strength characteristics of the composite material. Algorithms based on the method of singular integral equations are used to calculate stresses at cracks. [11-14].

## 2. Algorithm for calculating the growth process of the system of curvilinear fatigue cracks in plates with a small degree of anisotropy

We consider a plate that at the initial moment of time is weakened by a system of cracks placed along the curves, which is loaded with cyclically variable forces at infinity (for plates of infinite size), concentrated forces, stress, which are applied in accordance with the boundary of the plate and the sides of the cracks.

For isotropic materials, when finding the direction of change in the trajectory of the selected peak of the crack (depending on the conditions) used power, displacement or mixed criteria of destruction. According to them

$$\theta = f(K_I^*, K_{II}^*, C_\theta), \qquad (1)$$

where $\theta$ is the angle measured from the tangent to the crack at its peak, $C_\theta$ is set of additional parameters, which take into account the property of the material and the conditions of operation of the structure, $K_I^*, K_{II}^*$ are characteristic (for example, high or medium) value of SIF in a cycle.

In determining the length of the arc increments were used FFD, which were selected in the form

$$V = F(K_{\max}, K_{\min}, C_V), \qquad (2)$$

where V is the growth rate of the crack, $K_{\max}, K_{\min}$ are respectively equal to the largest and smallest value of K, which characterizes SSS (Stress-Strain State) at the peak of the crack during cyclic loading, $C_V$ are additional parameters that take into account the properties of the material and the conditions of operation of the structure. The relations (1) and (2) are used for isotropic materials. In the same form, relations can also be written for anisotropic materials, while taking into account the strength of anisotropy materials near the peak of cracks.

The calculation algorithm is as follows. Let's consider n – step. We determine the peaks of j cracks by $A_{j,n}$, $B_{j,n}$. In the process of tracking the cracks change their shape. To find SIF using the method of integral equations [12, 13, 14] it is necessary to describe them in an analytical form. For this purpose, on each of the cracks, we will specify the coordinates of the nodal points and, in addition, the angles of the inclination of the tangents at the peaks..

We denote the selected nodes on the crack at the number j at the number of the step n – via $(x_{j,1}^{(n)}, y_{j,1}^{(n)}), (x_{j,2}^{(n)}, y_{j,2}^{(n)}), ..., (x_{j,N_n}^{(n)}, y_{j,N_n}^{(n)})$, where $N_n$ is number of nodal points. Note that the first and last points correspond to the crack tips. We will also give the values of the angles of inclination of tangents at the peaks of the crack $A_{j,n}$, $B_{j,n}$ to the axis Ox, which we will denote by $\varphi_{j,n}^A, \varphi_{j,n}^B$.

On the basis of the given coordinates of the nodal points and the derivatives of the equality of each crack, we describe crack form with the help of a cubic spline, which is continuous along with the first two derivatives and has predetermined inclining tangents at the vertices.Then using known methods of calculation SSS [12, 13, 14] of plates with cracks we find SIF for both vertices of cracks, which we denote by a $K_{j,I}$, $K_{j,II}$.

Using equation (1) we find angles $\theta_{j,n}^A$, $\theta_{j,n}^B$, under which two peaks will grow j crack, which are counted from the corresponding tangent. Let's assume that in the process of destruction, all the peaks of the cracks can move. Denote the coordinates of the peaks of the crack in the next step through $A_{j,n+1}, B_{j,n+1}$ and the distances to which they moved, through $\Delta_{j,n+1}^A$, $\Delta_{j,n+1}^B$, where $\Delta_{j,n+1}^A = |A_{j,n+1} - A_{j,n}|$, $\Delta_{j,n+1}^B = |B_{j,n+1} - B_{j,n}|$.

Based on the ratio (2) we have

$$\Delta_{j,n+1}^A = F_{j,n}(A)\Delta\tau_{n+1}, \quad \Delta_{j,n+1}^B = F_{j,n}(B)\Delta\tau_{n+1}, \qquad (3)$$

where $F_{j,n}(A) = F(K_{\max}(A_{j,n}), K_{\min}(A_{j,n}))$, $F_{j,n}(B) = F(K_{\max}(B_{j,n}), K_{\min}(B_{j,n}))$, $\Delta\tau_{n+1}$ is time of growth.

Here we assumed that the peaks of the crack grew along the straight line.
When calculating, we assume that the distance to which the peak with the highest value of SIF has moved at a certain point is equal to some value, small in size. Then, based on (2) we will get

$$\Delta\tau_{n+1} = \Delta / F_{j,\max} \tag{4}$$

where $F_{j,\max}$ are highest value $F_{j,n}(A),\ F_{j,n}(B),\ j=1,...,J$.

On the basis of the above relations, we find the coordinates of the peaks of the cracks at a new stage $A_{j,n+1}(x^A_{j,n+1}, y^A_{j,n+1}), B_{j,n+1}(x^B_{j,n+1}, y^B_{j,n+1})$, where $x^A_{j,n+1} = x^A_{j,n} + \Delta^A_{j,n+1}\cos\varphi^A_{j,n}$, $y^A_{j,n+1} = y^A_{j,n} + \Delta^A_{j,n+1}\sin\varphi^A_{j,n}$, $x^B_{j,n+1} = x^B_{j,n} + \Delta^B_{j,n+1}\cos\varphi^B_{j,n}$, $y^B_{j,n+1} = y^B_{j,n} + \Delta^B_{j,n+1}\sin\varphi^B_{j,n}$. Next we find $\varphi^A_{j,n+1} = \varphi^A_{j,n} + \theta^A_{j,n}$, $\varphi^B_{j,n+1} = \varphi^B_{j,n} + \theta^B_{j,n}$ – angles of inclination tangent to a crack in its peaks to the axis Ox. The results obtained allow us to proceed to the next stage of tracking the trajectory of the crack. For this purpose, for the description of the crack, we lead nodal points in consideration $(x^{(n+1)}_{j,1}, y^{(n+1)}_{j,1}), (x^{(n+1)}_{j,2}, y^{(n+1)}_{j,2}),...,$ $(x^{(n+1)}_{j,N_{n+1}}, y^{(n+1)}_{j,N_{n+1}})$, where $N_{n+1} = N_n + 2$ ; $x^{(n+1)}_{j,1} = x^A_{j,n+1}$, $y^{(n+1)}_{j,1} = y^A_{j,n+1}$, $x^{(n+1)}_{j,N_{n+1}} = x^B_{j,n+1}$, $y^{(n+1)}_{j,N_{n+1}} = y^B_{j,n+1}$ , $x^{(n+1)}_{j,k} = x^{(n)}_{j,k-1}$, $y^{(n+1)}_{j,k} = y^{(n)}_{j,k-1}$, when $k = 2,...,N_n$.

In this way, we sequentially establish the shape of the crack at the following stages and moments of time, which these stages correspond to relative to the crack.

The time required for crack growth $\tau_n$ (number of cycles), for which peaks of crack number j will reach points $A_{j,n}$, $B_{j,n}$, is determined by the formula $\tau_n = \Delta\tau_1 + \Delta\tau_2 + ... + \Delta\tau_n$.

We divide the entire load process, which we will characterize by the parameter p on the stages. Assume that we have already found the equation of a crack at the end of stage n and denote its equation via $x = \varphi_n(t),\ y = \psi_n(t)$, where t is the parameter that varies within $-1 \le t \le 1$. We will assign the number 0 to the initial crack.

## 3. Determination of cracks growth directions

Consider the general case where both SIF values are different from zero and in the process of loading, the crack grows along the curvilinear trajectories. In this case, it is necessary to choose a criterion for changing the direction of growth of fatigue cracks on steps. In the literature is shown [12], which in many cases can be used for this purpose specified above $\sigma_\theta$ - criterion. Calculations showed, that in the process of tracking the motion of the tops of the crack SIF $K_{II}$ practically is equal to zero. The results obtained in this paper also confirm this conclusion. That is, in this case, the destruction process will pass as the mechanism of normal separation, at which $K_{II} = 0$ . Thus, when applied $\sigma_\theta$ is the criterion for calculating the curvilinear trajectories can FFD be used, which are based on the study of the growth of rectilinear cracks by the mechanism of normal separation.

## 4. Determination of the curvilinear cracks system growth process by force criterion of destruction

In the literature, the development of a single fracture is sufficiently studied. Let's consider a less studied problem about the growth of the system of fatigue cracks.

To find the length of the arc of the crack, it is also necessary to use the equation for the rate of cracks growth. We use the following formula [15]

$$v = v_0 \left( \frac{\Delta K - \Delta K_{th}}{\Delta K_{fc} - \Delta K} \right)^q , \tag{5}$$

where $\Delta K_{th}$, $\Delta K_{fc}$ are respectively, the initial and critical ranges of SIF $\Delta K_I$, $v_0$ and $q$ are constants of the material. Note that when $\Delta K < \Delta K_{th}$ the crack does not grow, but with $\Delta K > \Delta K_{fc}$ there is a fragile (instantaneous) destruction.

Let's rewrite this formula in the form

$$v_0 \Delta N = \left( \frac{1 - k\eta}{k\eta - \varepsilon} \right)^q \Delta l ,$$

where $k = \dfrac{\Delta K}{\Delta K_0}$, $\eta = \dfrac{\Delta K_0}{\Delta K_{fc}}$, $\varepsilon = \dfrac{\Delta K_{th}}{\Delta K_{fc}}$, $\Delta K_0 = p\sqrt{\pi L}$ , $\Delta\ l$ is increase in the length of the trajectory, $\Delta N$ is the number of cycles during which the growth takes place. Note that $\Delta K_0$ is range of SIF for the initial single fracture in the plate, which is under the action of tension perpendicular to it.

**5. Results of calculations**

Calculations are performed at $\varepsilon = 0.12,\ q = 1.26, \eta = \sqrt{\pi} / 3 \approx 0.6$

The case of one-sided stretching of plates with straight and weakly curved cracks is studied in detail. The results of calculations of the crack growing process, located on a circle of radius $R = 0.1$ m at $-p/4 < j < p/4$ for stretching the plates with effort pC, where C – cyclic factor, which varies in each of the load cycles from 0 to 1 (ie, in the cycle, the maximum value of the effort is p, and the minimum value is 0) at an angle $p/6$ to the axis Ox shown in Fig. 1. Calculations were made at $p = 5$ MPa (ie, in the cycle, the maximum value of the effort is p, and the minimum value is 0), $N = 60$, $K = 80$, where $K$ – number of steps. Trajectories of the growth crack are shown on the upper figure; the rate of crack growth (number of cycles), depending on the number of steps, is shown on the middle figure; the SIF value at the top of the crack is on the bottom figure. Here the numbers on charts $0, \mathrm{K}, 4$ correspond to the numbers of the steps: 0 (initial crack), $K/4$, $K/2$, $3K/4$, $K$. On the basis of the calculations performed, it was found that rectilinear and weakly curved cracks in endless plates, which are in the form of one-piece stretch, grow, as a rule, perpendicular to the direction of the efforts.

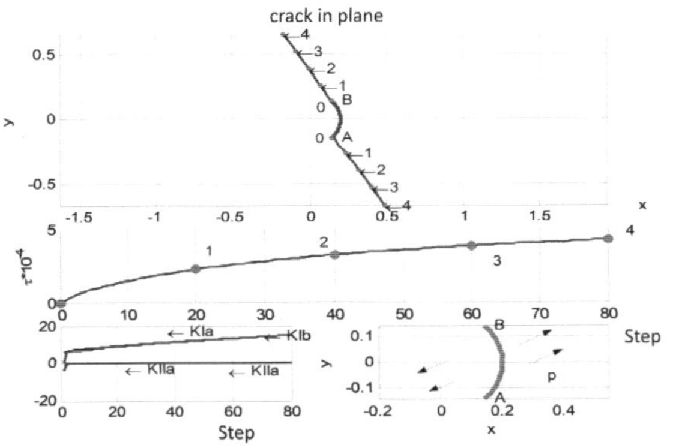


**Figure 1.** Trajectory and speed of crack propagation along the circle arc when the plate is stretched at an angle $p/6$.

Consider a plate with two half-length cracks, parallel to the axis Oy at a distance between them d. Let's assume that the plate is stretched along the axis Ox with efforts pC.

Calculations are performed at 40 steps in the tracking method. On figure 2above shown the calculated trajectories of the two identical cracks movement, the distance between them L, and here the coordinates are attributed to the half-length of the cracks. At the left side below shows the dependence of the value $\tau = v_0 N / L$ from the value of the steps. Note that the characteristic points on the trajectories, near which numbers 1,...4 placed match the step numbers 10, 20, 30, 40. At the right side below is the value $K_m = k\eta K_{fc}$, which characterizes the size of SIF in the process of propagation of cracks, and accepted $K_{fc} = 37 MPa\sqrt{m}$, which takes place for steel 75KhGST( 75ХГСТ).

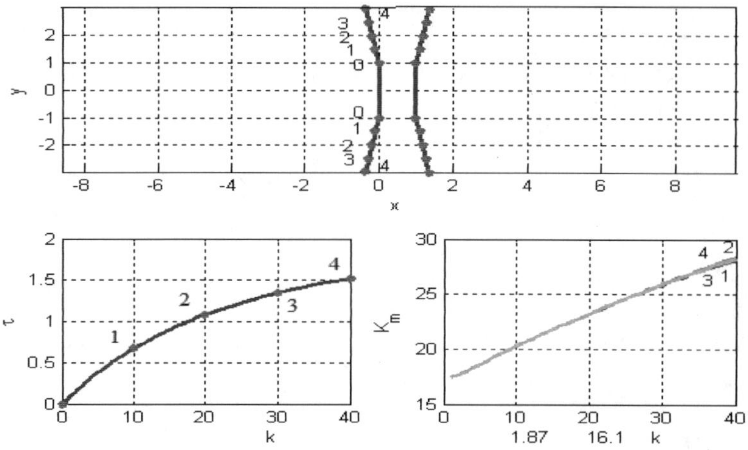

**Figure 2.** A system of two parallel cracks of equal length

To ensure accuracy the number of nodes in the quadrature formula was chosen sufficiently large - up to 30. With less accuracy of calculations, there was a loss of symmetry, which corresponds to the real instability of crack growth in practice. It follows that minor random factors can significantly change the propagation of parallel, closely placed cracks. Note that the calculated trajectories almost coincide with the data of the work [16], in which the problem was solved taking into account the symmetry of the problem, in which instability of calculations does not arise.

Consider the case when the right split has a smaller length - equal $0.9L$. The results of calculations for this case are shown in figure 3 Curves 1, 2 on the lower right figure correspond to SIF in the lower and upper peaks of the left crack, and 3, 4 - for the right cracks.

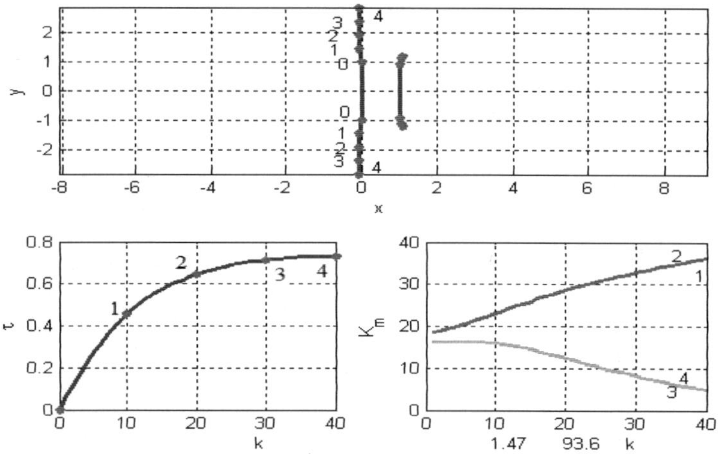

**Figure 3.** A system of two parallel cracks of different lengths

The upper figure shows that a smaller fracture grows at a slower rate than a larger one. From the bottom left figure it follows that a slight change in the length of one of the cracks leads to a significant

reduction in durability (at about two times at $L_2 / L_1 = 0.9$ ). We give relative values of SIF $F_{I,II} = K_{I,II} /(p\sqrt{\pi L})$ for initial cracks: 0.8309, 0.0970 (left crack) and 0.7554, - 0.1331 (right crack).
Now assume that in the above system the center of the right-handed crack is shifted - it is at the point $(L, 0.1L)$. The results of calculations for this case are shown in figure 4.

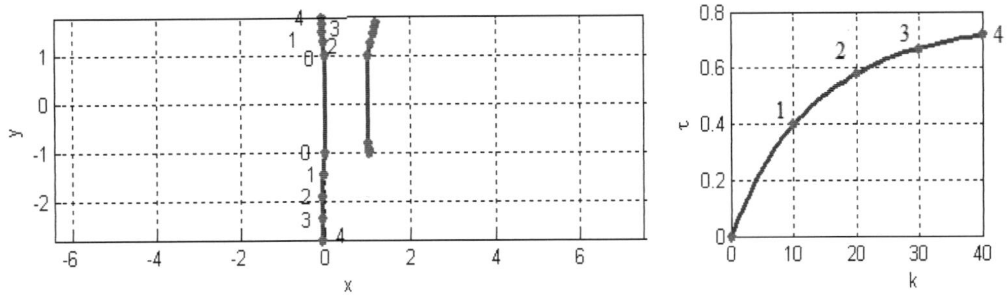


**Figure 4.** System of two parallel shifted cracks of different lengths

From the upper figure it can be seen that the upper peaks of the cracks grow at almost the same speed (This is confirmed by the fact that curves 2 and 4, which describe the SIF in these peaks are close), with the trajectories of their movement somewhat removed from each other. The maximum speed has a lower peak of a longer crack, and closer to it peak of a smaller crack is practically immobile. The durability of the crack system, considered in the last two cases, is practically the same.

*5.1. System of four parallel cracks*
We will initially calculate the SIF for the system parallel to the axis Oy our equidistant cracks half length of *L*, stretching by the forces of *pC* in the direction of the axis *Ox*. The calculation results of the relative values of the SIF $F_{I,II} = K_{I,II} / p\sqrt{\pi L_j}$ for the average (j=1) and for the right (j=2) top peaks of the cracks are shown in Table 1.

**Table 1.** Relative SIF for a system of four cracks

| | central crack | | side crack | |
|---|---|---|---|---|
| d/L | $F_I$ | $F_{II}$ | $F_I$ | $F_{II}$ |
| 0.25 | 0.441 | -0.002 | 0.675 | 0.147 |
| 0.50 | 0.532 | 0.000 | 0.724 | 0.113 |
| 0.75 | 0.606 | 0.001 | 0.766 | 0.085 |
| 1.00 | 0.672 | 0.002 | 0.805 | 0.062 |
| 1.50 | 0.778 | 0.002 | 0.867 | 0.033 |
| 2.00 | 0.848 | 0.001 | 0.909 | 0.018 |
| 2.50 | 0.892 | 0.001 | 0.935 | 0.011 |
| 3.00 | 0.920 | 0.001 | 0.952 | 0.007 |
| 4.00 | 0.952 | 0.000 | 0.971 | 0.003 |
| 5.00 | 0.968 | 0.000 | 0.981 | 0.002 |

The table shows that the SIF values for the side cracks are much larger than for the central. In addition, for central cracks the value $K_{II}$ practically equal to zero. From here it is clear that side cracks will grow faster, and they will grow in the direction opposite to the central cracks.
Tables 2 and 3 show similar results for a case where the half-length of the central cracks is equal $L_1 = 1.1L$ and $L_1 = 1.2L$, and side cracks - $L_2 = L$.

**Table 2.** Relative SIF for a system of four cracks (the half-length of the central cracks is equal $L_1 = 1.1L$ and side cracks is $L_2 = L$.)

| d/L | central crack $F_I$ | central crack $F_{II}$ | side crack $F_I$ | side crack $F_{II}$ |
|---|---|---|---|---|
| .25 | 0.546 | 0.035 | 0.574 | 0.167 |
| 0.50 | 0.578 | 0.018 | 0.671 | 0.126 |
| 0.75 | 0.628 | 0.013 | 0.729 | 0.095 |
| 1.00 | 0.681 | 0.011 | 0.775 | 0.070 |
| 1.25 | 0.731 | 0.009 | 0.814 | 0.051 |
| 1.50 | 0.775 | 0.007 | 0.846 | 0.037 |
| 2.00 | 0.842 | 0.004 | 0.893 | 0.021 |
| 2.50 | 0.886 | 0.003 | 0.924 | 0.012 |

**Table 3.** Relative SIF for a system of four cracks (the half-length of the central cracks is equal $L_1 = 1.2L$ and side cracks is $L_2 = L$.)

| d/L | central crack $F_I$ | central crack $F_{II}$ | side crack $F_I$ | side crack $F_{II}$ |
|---|---|---|---|---|
| 0.25 | 0.619 | 0.074 | 0.462 | 0.173 |
| 0.50 | 0.617 | 0.038 | 0.613 | 0.136 |
| 0.75 | 0.649 | 0.026 | 0.688 | 0.104 |
| 1.00 | 0.690 | 0.020 | 0.744 | 0.077 |
| 1.25 | 0.733 | 0.016 | 0.788 | 0.057 |
| 1.50 | 0.772 | 0.013 | 0.825 | 0.042 |
| 2.00 | 0.836 | 0.008 | 0.877 | 0.024 |
| 2.50 | 0.879 | 0.005 | 0.912 | 0.014 |

From Table 2 it can be seen that the character of the SIF distribution is the same as for the cracks of the same length. At $L_1 = 1.2L$ SIF values for central crack dominate the SIF for the side crack. It follows that in the latter case faster growth of central crack should be expected.

Let's now calculate the process of fatigue destruction of the plate. figure 5 shows the results of calculations for four cracks of equal length L, the distance between which is equal to L.

Similar results of calculations for cases where side cracks have a half length of *L*, and average *0.9L* and *0.8L* shown in figure 6, 7.

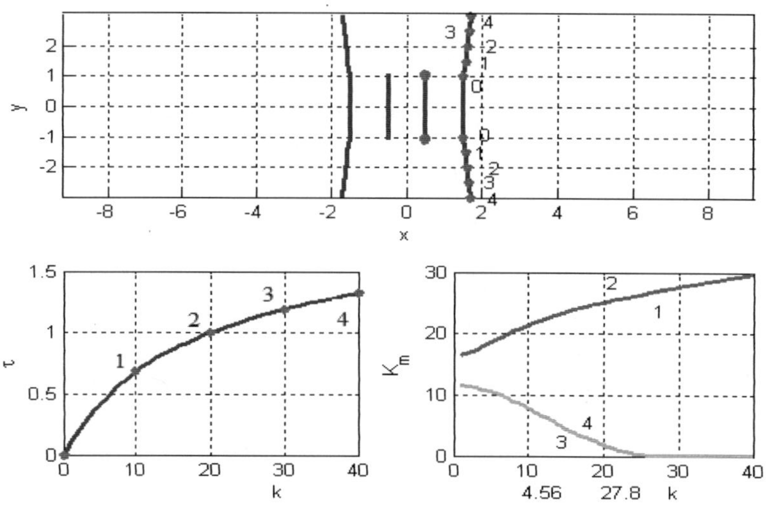


**Figure 5.** The system of four parallel cracks of the same length

IOP Conf. Series: Materials Science and Engineering **373** (2018) 012024 doi:10.1088/1757-899X/373/1/012024

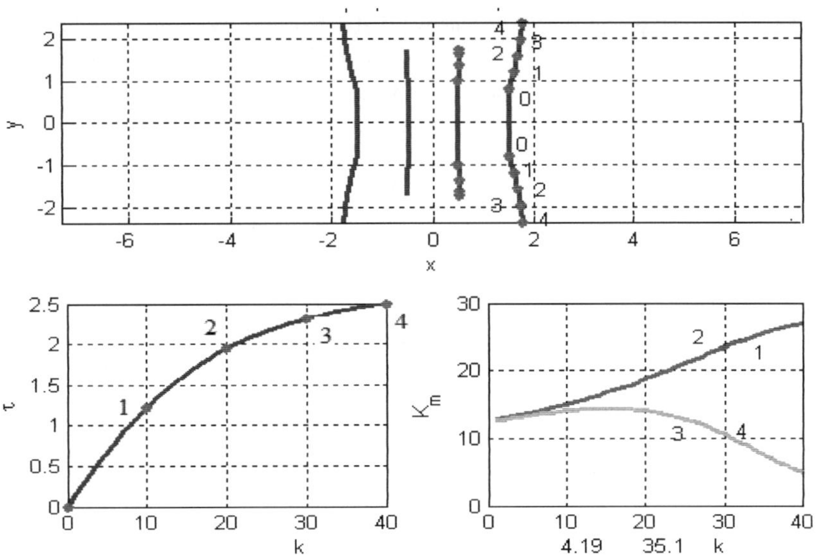


**Figure 6.** System of four parallel cracks at $L_1 = 0.9L$

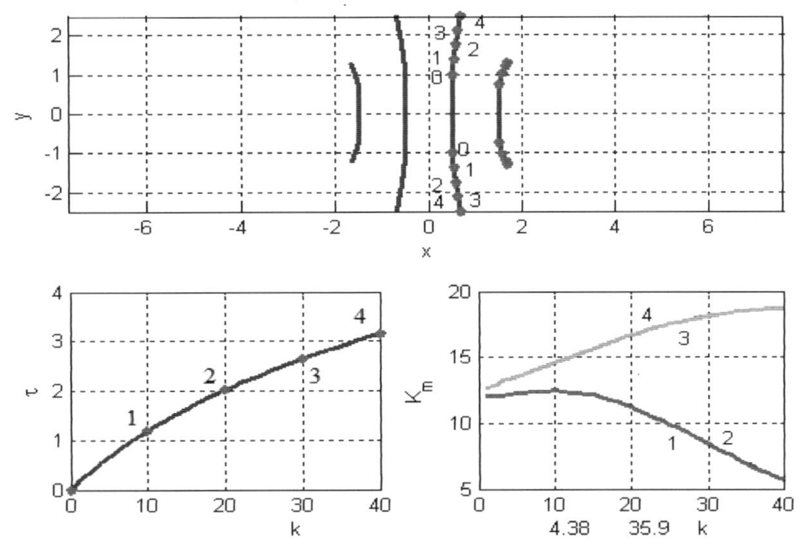


**Figure 7.** System of four parallel cracks at $L_1 = 0.8L$

*5.2. Consider a composite orthotropic plate with a crack*

The calculation of the fatigue cracks growth process is complicated, since the direction of growth is determined not only on the basis of SIF, but also taking into account the placement of a crack relative to the orthotropic axes. Building a FFD for composite materials is also a difficult task. Therefore, we will restrict ourselves to the tasks in which the crack growth trajectory can be found without using FFD.

Consider the crack, the tangent to which at its peak $A$ inclined at an angle $\varphi$ to the axis Ox. Here and then the Ox axis is chosen so that along with it the module of elasticity of the material was maximal. We will use the polar coordinate system $(r,\theta)$ with center at the top of the crack so that the angular coordinate $\theta=0$ coincided with the tangent to the crack at its peak (figure 8).

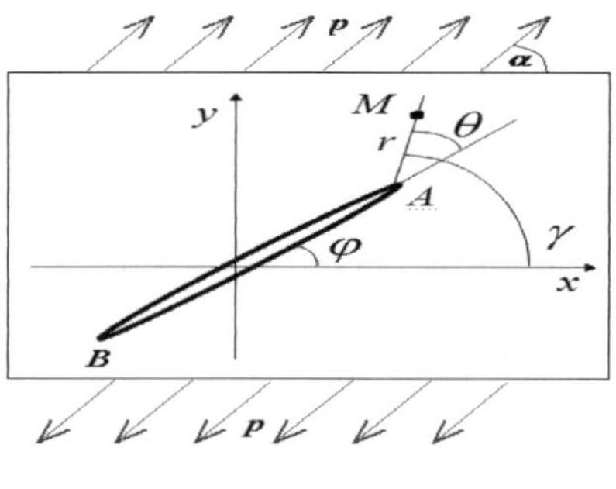


**Figure 8.** Load scheme of the plate with a crack

We will consider the known SIF for this peak, which we denote by $K_I, K_{II}$.
For stresses in the neighbourhood of the peak of the crack, we have asymptotic relations [12,16]

$$\sigma_\theta(r,\theta) \approx \frac{K_{I\theta}(\theta,\varphi)}{\sqrt{2\pi r}},\ \tau_{r\theta}(r,\theta) \approx \frac{K_{II\theta}(\theta,\varphi)}{\sqrt{2\pi r}}, \tag{6}$$

where

$$K_{I\theta}(\theta,\varphi) = K_I f_{11}(\theta+\varphi,\varphi) + K_{II} f_{12}(\theta+\varphi,\varphi),$$
$$K_{II\theta}(\theta,\varphi) = K_I f_{21}(\theta+\varphi,\varphi) + K_{II} f_{22}(\theta+\varphi,\varphi), \tag{7}$$
$$f_{11}(\gamma,\varphi) = \mathrm{Re}\left[\left(g_1\delta_2(\gamma)\sqrt{\delta_2(\gamma)/d_2} - g_2\delta_1(\gamma)\sqrt{\delta_1(\gamma)/d_1}\right)/(s_2-s_1)\right],$$
$$f_{12}(\gamma,\varphi) = -\mathrm{Re}\left[\left(d_1\delta_2(\gamma)\sqrt{\delta_2(\gamma)/d_2} - d_2\delta_1(\gamma)\sqrt{\delta_1(\gamma)/d_1}\right)/(s_2-s_1)\right]$$
$$f_{21}(\gamma,\varphi) = \mathrm{Re}\left[\left(g_1\gamma_2(\gamma)\sqrt{\delta_2(\gamma)/d_2} - g_2\gamma_1(\gamma)\sqrt{\delta_1(\gamma)/d_1}\right)/(s_2-s_1)\right],$$
$$f_{22}(\gamma,\varphi) = -\mathrm{Re}\left[\left(d_1\gamma_2(\gamma)\sqrt{\delta_2(\gamma)/d_2} - d_2\gamma_1(\gamma)\sqrt{\delta_1(\gamma)/d_1}\right)/(s_2-s_1)\right],$$

$\delta_j = \delta_j(\gamma) = \cos\gamma + s_j\sin\gamma$, $\gamma_j = \gamma_j(\gamma) = \sin\gamma - s_j\cos\gamma$, $g_j = \gamma_j(\varphi)$, $d_j = \delta_j(\varphi)$, $\gamma = \varphi + \theta$, $s_{1,2}$ are the roots of the characteristic Lehnitsky equation [11].

Assume that the fracture toughness is known $K_C(\varphi)$, which is determined on the basis of the stretching of the plate in a direction perpendicular to the crack. Consider the fiberglass ЕФ 32-301, which is made of fiberglass АСТТ. For functions $K_C(\gamma)$ use experimental data [17] (table 4).

**Table 4.** Values $K_{Ic}$ (MPa · m$^{1/2}$) depending on the angle $\gamma$ (angle between the crack and the direction with maximum stiffness of the material)/

| $\gamma$, radian | 0 | $\pi/6$ | $\pi/4$ | $5\pi/16$ | $\pi/2$ |
|---|---|---|---|---|---|
| ЕФ 32-301 | 13,0 | 16,79 | 18,66 | 22,74 | 25,55 |

Function $K_C(\gamma)$ (MPa m$^{1/2}$) based on the data in the table for $-\pi \le \gamma \le \pi$ we will describe in the form

$$K_C(\gamma) = a_0 + a_1 \sin^2 \gamma + a_2 \sin^4 \gamma + a_3 \sin^6 \gamma .$$

The coefficients in this formula, found by the method of least squares, will be: $a_0$ = 25.6074, $a_1$ = -11.2031, $a_2$ = -6.0858, $a_3$ = 4.7243.

The crack will grow at an angle $\theta$, at which value $\sigma_\theta(r,\theta)/K_C(\gamma)$ is the maximum. Based on (7) then we get that angle $\theta = \gamma - \varphi$ is such that the value is the maximum.

$$I(\gamma) = \frac{K_I f_{11}(\gamma,\varphi) + K_{II} f_{12}(\gamma,\varphi)}{K_C(\gamma)},$$

Calculations of the trajectory growth of the initially rectilinear a half-longitudinal fracture a, inclined at an angle $\varphi = 0^0$ when stretching the plate with the forces acting at angles $\omega = 18^o j$, $j = 1,...,9$ to the axis Ox. Calculated trajectories are shown in figure 9 at which angle $\omega$, is indicated at the top and here the coordinates are assigned to the half-length of the crack a.

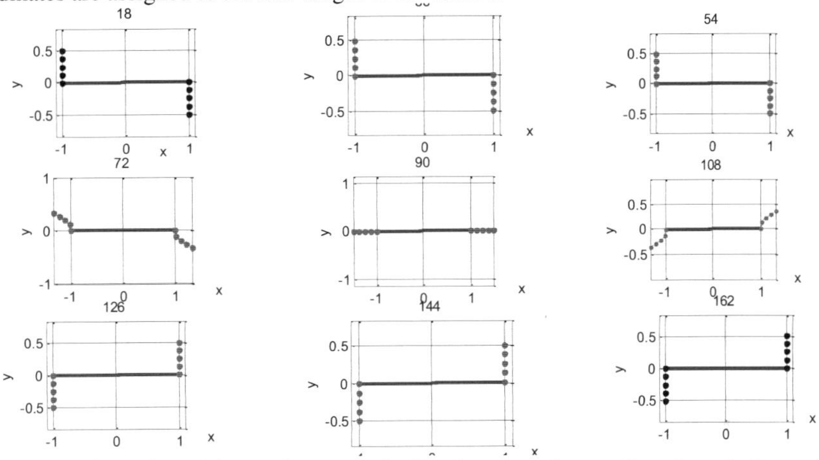


**Figure 9.** Trajectories of the crack growth, inclined at an angle $\varphi = 0$ to the axis Ox

Similar data at the angle of crack inclination $\varphi = 30^o$ shown in figure 10.

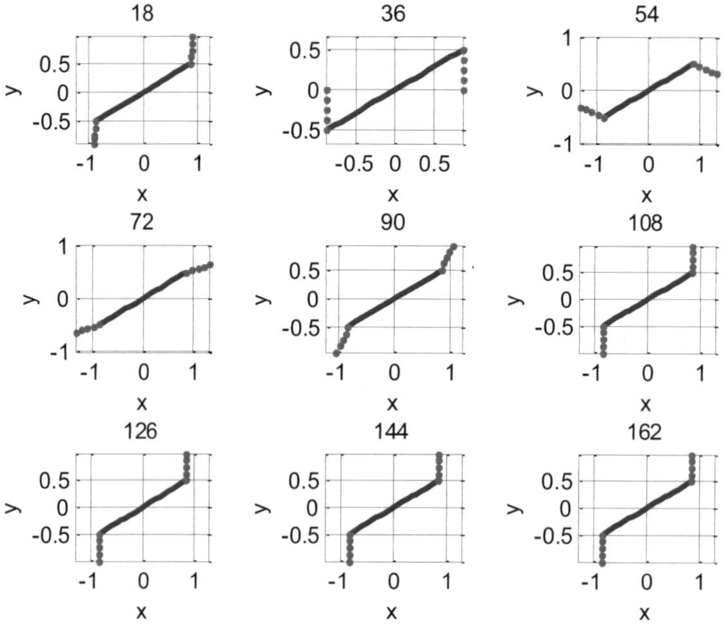


**Figure 10.** Trajectories of the crack growth, inclined at an angle $\varphi = 30^o$ to the axis Ox

Figure 9-10 shows that the cracks mainly grow along the straight lines.

## 6. Conclusions

The method of calculating the durability of isotropic and composite plate elements of constructions under cyclic loading is developed on the basis of the study of the growth process of fatigue cracks along the curvilinear trajectories. Calculations are made on the basis of singular integral equations, force criterion of destruction, diagrams of fatigue failure. The characteristic features of fatigue cracks system growth in isotropic materials are defined. In particular, in the system of four identical cracks, central cracks do not propagate. For longer *(1,1L)* central cracks comparatively to side cracks at the initial stage all cracks grow at almost the same speed, however, further central cracks stop their growth. Only at significantly shorter lengths of the side crack, the growth rate in the central cracks is greater than in the side one, and in the process of propagation the side cracks almost stop their growth.
The curvilinear trajectories of the growth of the initially rectilinear crack in the composite plate are investigated depending on the direction of the applied load.

## 7. References

[1]. Maligno A. R., Soutis C., Silberschmidt V. V. 2013 An advanced numerical tool to study fatigue crack propagation in aluminium plates repaired with a composite patch *Engineering Fracture Mechanics.* vol. 99. – pp. 62-78.

[2]. Richard H. A. et al. Fatigue crack growth in real structures *International Journal of Fatigue.* – 2013. – vol. 50. – pp. 83-88.

[3]. Ooi E. T. et al. 2015 Adaptation of quadtree meshes in the scaled boundary finite element method for crack propagation modelling *Engineering Fracture Mechanics.* – vol. 144. – pp. 101-117.

[4]. Khoei A. R., Yasbolaghi R., Biabanaki S. O. R. 2015 A polygonal finite element method for modeling crack propagation with minimum remeshing *International Journal of Fracture.* – vol. 194. – №. 2. – pp. 123-148.

[5]. Hedayati E., Vahedi M. 2014 Using Extended Finite Element Method for Computation of the Stress Intensity Factor, Crack Growth Simulation and Predicting Fatigue Crack Growth in a Slant-Cracked Plate of 6061-T651 Aluminum *World Journal of Mechanics.* vol. 4. – №. 01. – pp. 24.

[6]. Yu Q. Q. et al. 2014 Crack propagation prediction of CFRP retrofitted steel plates with different degrees of damage using BEM *Thin-Walled Structures.* –. vol. 82. – pp. 145-158.

[7]. Ooi E. T., Yang Z. J. 2011 Modelling dynamic crack propagation using the scaled boundary finite element method *International journal for numerical methods in engineering.* – vol. 88. – №. 4. – pp. 329-349.

[8]. Yan X. A 2006 Boundary element modeling of fatigue crack growth in a plane elastic plate Mechanics Research Communications. – vol. 33. – №. 4. – pp. 470-481.

[9]. Singh I. V. et al. 2012 The numerical simulation of fatigue crack growth using extended finite element method *International Journal of Fatigue.* – vol. 36. – №. 1. – pp. 109-119.

[10]. Alshoaibi A. M. 2013 Adaptive finite element modeling of fatigue crack propagation *Int. J. Mater. Sci. andApplications.* – vol. 2. – №. 3. – pp. 104-108.

[11]. Maksymovych, O.,Illiushyn O. 2017 Stress calculation and optimization in composite plates with holes based on the modified integral equation method *Engineering Analysis with Boundary Elements*, vol. 83, pp. 180-187

[12]. Bozhydarnik, V.V., Maksymovych, O.V. 2010 Determination of the stressed state near edge cracks in a plate containing a hole of complex shape *Materials Science*, vol. 46 (1), pp. 16-26.

[13]. Maksymovych O., Pasternak Ia., Sulym H., Kutsyk S. 2014 Doubly periodic cracks in the anisotropic medium with the account of contact of their faces *Acta mechanica et automatica.* – Vol. 8, No. 3. – pp. 160–164.

[14]. V. Bozhydarnyk · O. Maksymovych 2001 Elastic Equilibrium of an Anisotropic Half Plane with Periodic System of Holes and Cracks *Materials Science*, 37 (6), pp. 857-865.

[15]. Yarema S.Ya. 1987 On the foundations and some problems of the mechanics of fatigue failure / *Physico-chemical mechanics of materials.* - № 5. - pp. 17-29.

[16]. Savruk M.P. 1981 Two-dimensional elasticity problems for bodies with cracks / K.: *Scientific. thought*- p. 324.

[17]. VV Bozhydarnik, O.E. Andreyuk, G.T. Sulim 2007 Mechanics of fracture, durability and durability of unevenly reinforced composites Lutsk: "Supermarket", vol. 1. - P. 424, vol. 2. - P. 410.

# Dynamic behavior of geometrically complex hybrid composite samples in a Split-Hopkinson Pressure Bar system

**M Pouya[1], S Balasubramaniam[1, 2], S Sharafiev[1] and M F-X Wagner[1]**

[1]Technische Universität Chemnitz, Institute of Materials Science and Engineering, 09125 Chemnitz, Germany
[2]now at Technische Universität Chemnitz, Department of Electrical Engineering and Information Technology, 09126 Chemnitz, Germany

**Keywords:** high strain rate deformation, Split-Hopkinson Pressure Bar (SHPB), Finite Element Method (FEM), geometry effects, composite materials

**Abstract.** The interfaces between layered materials play an important role for the overall mechanical behavior of hybrid composites, particularly during dynamic loading. Moreover, in complex-shaped composites, interfacial failure is strongly affected by the geometry and size of these contact interfaces. As preliminary work for the design of a novel sample geometry that allows to analyze wave reflection phenomena at the interfaces of such materials, a series of experiments using a Split-Hopkinson Pressure Bar technique was performed on five different sample geometries made of a monomaterial steel. A complementary explicit finite element model of the Split-Hopkinson Pressure Bar system was developed and the same sample geometries were studied numerically. The simulated input, reflected and transmitted elastic wave pulses were analyzed for the different sample geometries and were found to agree well with the experimental results. Additional simulations using different composite layers of steel and aluminum (with the same sample geometries) were performed to investigate the effect of material variation on the propagated wave pulses. The numerical results show that the reflected and transmitted wave pulses systematically depend on the sample geometry, and that elastic wave pulse propagation is affected by the properties of individual material layers.

## 1. Introduction

Hybrid composite structures can be built of different macro-scale layers to combine the beneficial properties of single components, e.g., strength, stiffness, fatigue or corrosion resistance [1, 2]. The contact interfaces of the joined materials play a key role in determining the mechanical behavior of such composites, especially at high strain rates [3, 4]. The interfaces are also important in composite structures with complex shapes, and a detailed experimental characterization is often needed to fully understand their mechanical behavior [5]. The present study is motivated by a need for novel, modified sample geometries that are optimized with respect to homogeneous stress distributions at hybrid interfaces, where wave pulse propagation and reflection typically produce complex stress gradients in conventional samples. Sample preparation of such composite materials, particularly with a reproducible orientation of the interfaces, and the corresponding experimental work, can be quite difficult and costly. Numerical simulations, in contrast, provide a novel tool at reduced costs. They moreover reduce the complexity of mechanical testing and they allow analyzing more details (like local stress and strain distributions) during dynamic testing.

In this paper, the Split-Hopkinson Pressure Bar (SHPB) technique is used to experimentally determine the dynamic behavior of several complex sample geometries under dynamic compressive loading (which produces a shear-dominated deformation of the interfaces that are oriented parallel to the macroscopic loading direction). The SHPB system allows determining the deformation behavior of materials at strain rates from $10^2$ to $10^4$ $s^{-1}$. A careful design of this system and high frequency signal acquisition are required to evaluate and analyze the resulting material behavior at these high strain rates [6]. First, a series of experiments is performed with different sample geometries made of monomaterial steel. To complement the experimental studies, the Finite Element Method (FEM) is employed to explicitly simulate the SHPB experiment. The different sample geometries are simulated and the resulting SHPB signals are directly compared to those from the experiments. Finally, the same sample geometries are simulated considering different composite layers of steel and aluminum to determine the effect of both the geometry and of material variations on elastic wave pulse propagation.

## 2. Experimental and numerical methods

The SHPB system was used to experimentally analyze geometry effects on the dynamic behavior of samples with different, complex shapes. To perform systematic mechanical tests, five different sample geometries made of uniform monomaterial steel were prepared. Figure 1 presents the experimental set-up of the SHPB system (I), and the different sample geometries placed and tested in this set-up (II). The physical mechanism used in SHPB testing is the propagation of a stress wave pulse along a series of long, thin bars. The setup consists of a gas launcher and four bars including the striker, incident and transmitter bars beside a shock absorber bar at the end. The striker bar is accelerated to a pre-defined velocity with the help of the gas launcher (which is charged by a pressurized gas) to hit the incident bar and thus to create an elastic stress pulse. The stress pulse then propagates at the material's speed of sound through the incident bar. When it meets the sample (placed between the incident and transmitter bars; highlighted by a black circle in figure 1-I), part of the pulse is reflected back to the incident bar, due to a change in cross section and material. In fact, whenever an elastic wave encounters interfaces or changes of the geometrical boundary conditions, partial wave reflection can occur – this also affects the stress state at internal material interfaces [7]. The other part of the signal is passed on through the sample and to the transmitter bar (there is again some reflection at the interface between sample and transmitter bar) and is finally stopped at the shock absorber. In the experimental setup used here, two additional protectors (i.e., thin plates of the same material) are placed at both sides of the sample to prevent damage at the end and front faces of the incident and at transmitter bars. A measurement of the propagated wave pulses, and therefore an analysis of the mechanical response (in terms of force and uniaxial displacement as a function of time) of the material sample, is possible by means of the carefully designed strain gauges, marked using four red arrows in figure 1-I. Not shown is a pulse shaper that is fixed between the striker and incident bars in the experimental SHPB setup, with the aim to reduce superimposed wave pulse oscillations during dynamic testing. Further details on the experimental setup and on the effect of the pulse shaper are, for instance, given in reference [8].

The sample geometries studied here were designed to ultimately serve for mechanical testing of hybrid composite samples consisting of different layers of materials. To study the effect of the unconventional geometries on wave propagation, we first experimentally considered samples produced uniformly from C45 steel (figure 1-II). The geometries start from a simple cube shape with an edge length of 9 mm (geometry "a"), and are then evolved systematically by cutting the front and end faces of the samples in different ways (geometries "b" to "e"). The rationale behind these geometrical variations is that ultimately, interfaces between different material layers are supposed to be subjected to a predominant shear loading. Shear testing of materials is challenging in general [9], and to achieve this in a composite sample, the interfaces can be oriented parallel to the (macroscopically compressive) loading direction as indicated schematically in figure 1-IIb.

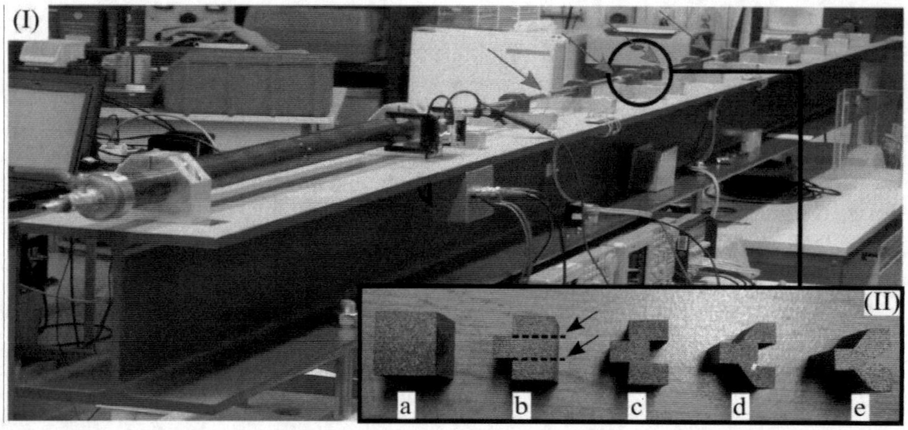


**Figure 1.** (I) Experimental setup of the SHPB system for dynamic testing at high strain rates up to $10^4$ $s^{-1}$. The sample is located between the incident and transmitter bars (black circle); red arrows mark positions of strain gauges. (II) Five different sample geometries ("a" - "e") made of steel were placed in the SHPB setup in the present study to investigate geometry effects on signal propagation during high strain rate testing.

A complementary, explicit FE model of the SHPB system was developed for careful numerical analysis of wave propagation during SHPB testing of the different sample geometries. The fundamental structure of the model is shown schematically in figure 2. The SHPB model consists of the four bars, which are assigned the same material properties (corresponding to steel: E = 210 GPa; ν = 0.28) and which only deform elastically throughout the whole experiment. A constant diameter of D = 16 mm (similar to the actual experimental setup) was set for all the four bars. The five different sample geometries were positioned between the incident and transmitter bars. The simple elastic-plastic mechanical behavior of the material samples was defined by a piecewise fit of the experimentally obtained hardening curve. All sample geometries were created using linear brick elements (C3D8 type) resulting in a relatively large total number of elements (because of the sharp radii present in some samples) of 35,000 – 46,000.

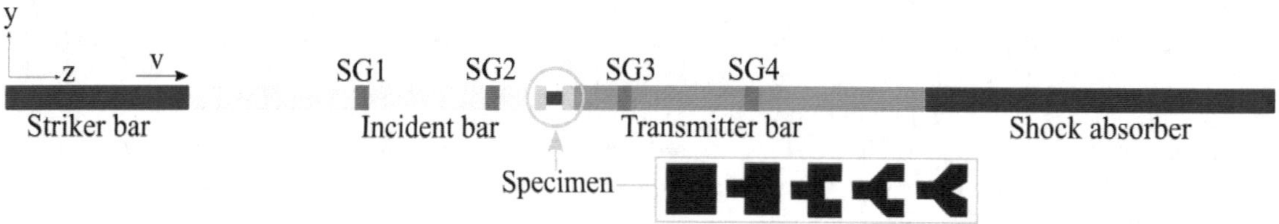


**Figure 2.** Schematic representation of the simulated SHPB setup. The four bars are highlighted using different colors; the velocity applied to the striker bar is denoted as v. The strain gauges (SGs) are used to measure the propagating elastic wave pulses during dynamic loading. Different specimen geometries are positioned in the SHPB system to determine geometry effects on the contact interfaces of mono- and composite materials with complex shapes.

In all simulations, the striker bar is accelerated to a predefined velocity (v = 8 $ms^{-1}$) to hit the incident bar and thus to create an elastic stress wave pulse. The pulse propagates through the whole length of the SHPB system (in z direction as marked in figure 2) and the corresponding elastic deformation is measured (both in the actual experiments and in our simulations) in four different locations by strain gauges (SGs). The SGs 1 and 4 are mounted in the middle of the incident and transmitter bars, respectively, and SGs 2 and 3 are positioned near the front and end faces of the material sample. Two protectors were also placed at both sides of the sample, similar to the experimental set up (green in figure 2). The effect of a pulse shaper was neglected during the FE simulations. More details on the simulated SHPB setup, specifically on material parameters and boundary conditions, can be found in reference [10].

Finally, an additional series of simulations was performed to numerically assess the effect of material variations and combinations at both sides of the interfaces (as shown schematically in figure 1-IIb) in hybrid composite samples on wave reflection. The same sample geometries were studied, now consisting of a central layer of the C45 steel and two outer layers of technically pure aluminum (E = 70 GPa, ν = 0.34). Similar to the initial simulations, the mechanical response was described by simple fits of the corresponding hardening behavior. We note that the focus of these simulations was on wave propagation as opposed to interface damage mechanisms and, therefore, we did not study interface degradation (which would e.g. require a cohesive zone modeling approach).

## 3. Results and discussion

The numerical results shown in figure 3 illustrate the initiation of the elastic wave pulse and its propagation through the total length of the SHPB system. The individual bars are separated using black dashed lines. The elastic signal is represented by means of the von Mises stress ($\sigma_{vm}$) distribution at different time steps (given in milliseconds, ms) of the FE simulations. At the beginning of the simulation, the striker bar hits the incident bar at a known velocity and thus creates an elastic stress pulse between the striker and input bars. The initiated pulse is well presented in figure 3, after 0.27 ms. As expected from the fundamental physics of SHPB testing [8], this wave pulse grows until it has reached a total length equal to double the length of the striker bar, as illustrated after 0.31, and 0.36 ms. The signal is then completely transferred to the incident bar (as presented after 0.46 ms). Once the wave pulse reaches the sample, it encounters a change of boundary conditions (both in terms of sample dimensions and material properties), and therefore a partial reflection of the wave occurs at the specimen's front face. The other part of the wave pulse runs through the material sample towards the transmitter bar. Additional wave reflection occurs between the end and front faces of the material sample and the transmitter bar, respectively. The presence and superposition of the reflected and transmitted elastic waves are evident in figure 3 after 0.60 ms, where von Mises stress distribution clearly exhibits fluctuations along the different bars of the SHPB system.

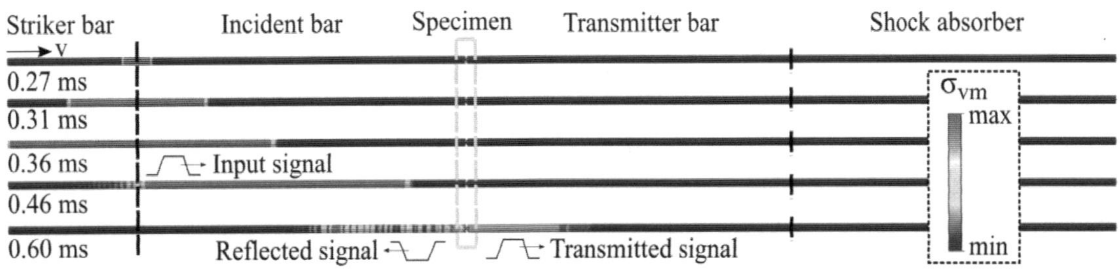


**Figure 3.** The von Mises stress distributions at different time steps (time given in ms on the left) of the FE simulations represent the initiation and propagation of an elastic wave pulse in the SHPB system. The input wave pulse moves through the length of the incident bar, parts of this pulse are reflected at the specimen front and end faces while the other part moves through the transmitter bar.

Figure 4 presents the experimental (left) and simulated (right) elastic wave pulse signals (force as a function of time) for testing of the steel samples. The input and reflected signals were measured using SG1 and the transmitted signal by SG3. A shift in the transmitted signals with time compared to the input signals is directly related to the location of SG3 in the SHPB setup. Although the numerical signals exhibit more pronounced oscillations due to the absence of a pulse shaper, they agree well with experiment. Note that in contrast to the idealized simulations, the strain rate (or the applied velocity) in the series of experiments is subjected to some variations. Therefore the experimental input signals (i.e., the wave pulses observed during the first 5 to 6 ms) of the five separate tests differ somewhat in terms of the maximum force. The samples were deformed plastically during SHPB experiments and no fracture was observed. Obviously, this is in line with the simple simulation approach used in the present study.

According to figure 4, both the experimentally and numerically obtained reflected and transmitted signals exhibit a systematic change with respect to the sample geometry. Geometry "a" transmits almost the entire input signal. A minor reflection dip occurs at the sample front face because of the local change of boundary conditions (material and dimensions). Compared to geometry "a", a larger part of the input signals is reflected when testing the other geometries ("b", "c", "d" and "e"). This indicates more pronounced wave reflections (and a more pronounced decay of the transmitted signals) at the faces of the samples, the more front and end faces are changed from the initial cube geometry. Additionally, an exact analysis of the resulting reflected and transmitted signals indicates a change in the shape of the reflected signals. It could be seen from figure 4 that, except for the geometry "a" where the tangential angle or slopes of the reflected/transmitted signals are almost the same as that of the input signal, the reflected/transmitted signals of the other geometries exhibit a finite slope with time. This slope is likely related to the increase of cross-sectional area as the elastic wave pulse travels through the length of the material samples "b" to "e".

In summary, these FE results demonstrate that the different specimen geometries studied here have a distinct effect both on the transmitted and reflected signals. The numerical simulations capture various subtle details of the SHPB experiment (for instance, the overshoot dips and peaks of the reflected signals shortly after 6 and 8 ms, respectively).

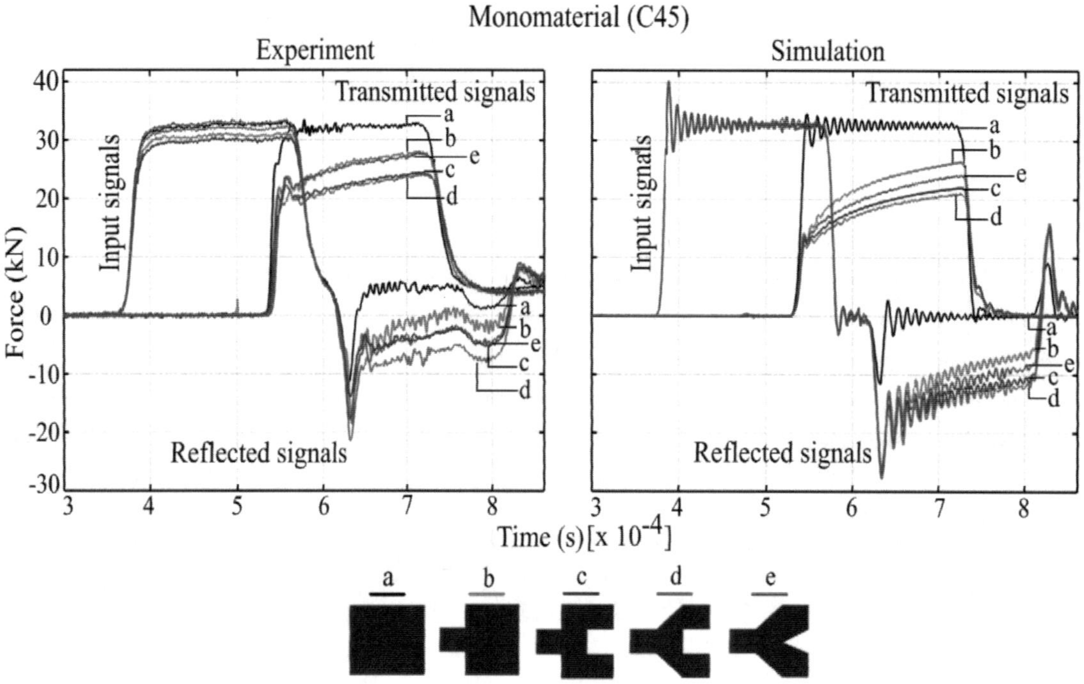


**Figure 4.** Experimental (left) and simulated (right) elastic pulse signals for different sample geometries made of steel (C45). The numerical input, reflected and transmitted signals measured in SGs 1 and 3 are in a good agreement with experimental results. Cross-sectional changes clearly affect wave propagation and reflection: samples "b" to "e" exhibit finite slopes of the reflected and transmitted signals.

The results from the monomaterial simulations, particularly the good agreement with the experiments, encouraged the next series of the FE simulations using different composite layers of C45 steel and Aluminum (Al), figure 5. The same sample geometries “a” to “e” were used for a direct comparison between the signals of monomaterial and hybrid samples. Figure 5 presents the resulting input, reflected and transmitted signals of the five composite samples under dynamic loading. The input signals presented in this figure are exactly the same as those of monomaterial simulations presented in figure 4. The overall profiles of the reflected and transmitted signals exhibit a quite similar shape as those of the monomaterial. In contrast to the monomaterial results, a large part of the input signals is reflected because the material variation introduces new contact interfaces even in the cube-shaped sample (geometry “a”). This also results in a considerable wave reflection and a significant slope of the transmitted and reflected signals for sample geometry “a”.

The effect of the different sample geometries can be rationalized by considering the impact of sudden changes in cross-sectional area (highlighted by dashed lines in figure 5) as the wave pulse propagates through each sample. In sample “b”, two distinct jumps occur – one when the cross-section is suddenly increased by adding the outer Al layers, and the other at the free surface (end face of the sample). In sample geometries “c” and “d”, three sudden changes occur; the additional change is related to the free surface at the end of the central steel layer. These samples clearly exhibit the most pronounced amount of signal reflection. Finally, sample “e” is characterized by a soft transition of cross-sectional area (between the dashed lines) and this explains why a larger part of the signal is transmitted. While, from a fracture mechanics point of view, one would expect the geometry of the front faces of the different samples to affect crack initiation, our numerical results demonstrate that the geometrical design of the end faces of such complex hybrid samples also play a key role in determining wave pulse reflection.

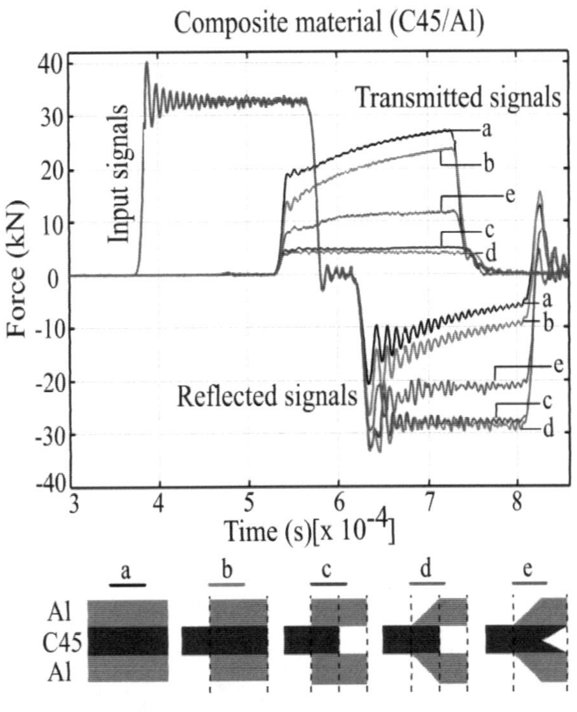


**Figure 5.** Numerical SHPB signals for the five specimen geometries (“a” – “e”) defined using different layers of steel and aluminum. There clearly is an effect of material variation on the propagated signals because hybrid samples contain additional internal interfaces. The reflected and transmitted signals also indicate a more pronounced, systematic dependence on the sample geometry.

IOP Conf. Series: Materials Science and Engineering **373** (2018) 012025 doi:10.1088/1757-899X/373/1/012025

We note in closing that, in geometries “a” and “b” with a constant cross-sectional area at the end faces, the mechanical behavior of steel with a higher yield stress strongly affects the SHPB signal. In geometries “c” and “d”, which consist only of Al at their end faces, the SHPB signal is instead dominated by the mechanical response of the much softer Al. It is likely that a modified stacking order of the different materials would further affect to dynamic signals. Obviously, the response of hybrid samples to dynamic testing also depends on the mechanical behavior of the different materials; the complex interplay between different material combinations and geometries warrants further experimental and numerical studies.

## 4. Summary

The Split-Hopkinson Pressure Bar system was used to study geometry and material effects on hybrid composite materials under dynamic loading. Experiments were performed on monomaterial C45 steel samples with different complex shapes. Complementary Finite Element simulations were performed both on monomaterial samples and on virtual samples consisting different layers of C45 steel and pure aluminum. The experimental and numerical measurements in the Split-Hopkinson Pressure Bar setup for both monomaterial and composite samples with complex geometries indicate a systematic influence of sample geometry and material combination on the reflected and transmitted signals. Free surfaces, changes of cross-sectional area, and internal interfaces between different material layers contribute to varying degrees to partial elastic wave pulse reflection. Particularly in hybrid composites made of different material layers, a substantial part of the input signal is reflected at the interfaces. The overall shape of the reflected signals, and likely internal stress and strain states in the different samples, thus depend on the material combination and on individual specimen geometry.

## 5. References

[1] Ashby M F, Bréchet Y J M 2003 *Acta Mater.* **51** 5801–5821
[2] Henning F, Moeller E 2011 *Handbuch Leichtbau* (Carl Hanser Verlag, München)
[3] Kießling R, Ihlemann J, Riemer M, Drossel W G, Scharf I, Lampke T, Sharafiev S, Pouya M, Wagner M F-X 2016 *IOP Conf Ser.: Mat. Sci. Eng.* **118** 012017
[4] Kießling R, Ihlemann J, Riemer M, Drossel W G, Scharf I, Lampke T, Sharafiev S, Pouya M, Wagner M F-X 2017 *Proc. CRIP* **66** 289–293
[5] Carrado A, Faerber J, Niemeyer S, Ziegmann G, Palkowski H 2011 *Compos. Struct.* **93** pp 715–721
[6] Elibol C, Wagner M F-X 2015 *Mat. Sci. Eng.* A **643** 194-202
[7] Meyers M A 1994 *Dynamic Behavior of Materials* (John Wiley & Sons, New York)
[8] Halle T 2009 *Habilitation Thesis* TU Chemnitz
[9] Yin Q, Zillmann B, Suttner S, Gerstein G, Biasutti M, Tekkaya A E, Wagner M F-X, Merklein M, Schaper M, Halle T, Brosius A 2014 *Int. J. Sol. and Struc.* **51** 1066–1074
[10] Pouya M, Winter S, Fritsch S, Wagner M F-X 2017 *IOP Conf. Ser.: Mat. Sci. Eng.* **181** 012022

IOP Conf. Series: Materials Science and Engineering **373** (2018) 012026 doi:10.1088/1757-899X/373/1/012026

# Characteristics of joining and hybrid composite forging of aluminum solid parts and galvanized steel sheets

**V Wesling[1], K Treutler[1], T Bick[1], M. Stonis[2], J Langner[2] and M Kriwall[2]**

[1] Institute of Welding and Machining (ISAF), Clausthal University of Technology, Germany
[2] IPH – Institut für integrierte Produktion Hannover, Germany

e-mail: office@isaf.tu-clausthal.de

**Abstract.** In lightweight construction, light metals like aluminum are used in addition to high-strength steels. However, a welded joint of aluminum and steel leads to the precipitation of brittle, intermetallic phases and contact corrosion. Nevertheless, to use the advantages of this combination in terms of weight saving composite hybrid forging has been developed. In this process, an aluminum solid part and a steel sheet were formed in a single step and joined at the same time with zinc as brazing material. For this purpose, the zinc was applied by hot dipping on the aluminum in order to produce a connection via this layer in a forming process, under pressure and heat. Due to the formed intermediate layer of zinc, the formation of the Fe-Al intermetallic phases and the contact corrosion are excluded. By determining the mathematical relationships between joining parameters and the connection properties the strength of a specific joint geometry could be adjusted to reach the level of conventional joining techniques. In addition to the presentation of the joint properties, the influence of the joining process on the structure of the involved materials is also shown. Furthermore, the failure behavior under static tensile and shear stress will be shown.

## 1. Introduction

To save weight in constructions and to reduce the consumption of resources during production process and operation multi-material designs are being used more and more frequently. The components are adapted to the loads, so the best material is used at the point of loading. As aluminum has a low density and steel a high degree of strength these two materials are particularly suitable for a multi-material design. However, this poses some challenges for the joining technique because the used materials have different mechanical and metallurgical properties. As soon as aluminum and iron come into contact in the molten state they form brittle intermetallic phases. Furthermore, the difference in electronegativity of the two materials leads to contact corrosion. To join these two materials despite their properties mechanical joining processes like riveting, clinching, etc. or gluing are frequently used. But also connections with metallic continuity are possible as several studies have shown. Meco et al. investigated the influence of the intermetallic phases and the width of the heat affected zone as a function of the temperature during the laser beam welding process of aluminum and steel. They showed that a narrow heat-affected zone with a low proportion of intermetallic phases is as damaging to the strength as large pronounced phases themselves, [1]. Kashani et al. showed that the strength of the compound increases when the size of the intermetallic phases is less than 10 microns, [2]. Zhang et al. investigated the influence of zinc in the grinding of aluminum and steel. They found that the zinc from the coating of the steel sheet has a positive influence on the formation of the intermetallic phases and thus on the strength, [3].

IOP Conf. Series: Materials Science and Engineering **373** (2018) 012026 doi:10.1088/1757-899X/373/1/012026

Engelbrecht et al. developed parameters for a laser soldering process to avoid a melting of both materials. In this case, aluminum and steel should also be isolated from each other as it can also lead to the formation of intermetallic phases in the solid state, [4]. Reisgen et al. also examined the formability of an arc-blended mixed aluminum / steel compound, [5]. In addition, hybrid composite forging has been developed, [6–8]. In this process, the two materials are joined to each other without transition to the molten state in a material-tight manner. The peculiarity of the method is that only one process step for forming and producing the connection is necessary. So far, this was done separately. Such a connection shows figure 1. Figure 1b) clearly shows the mechanical clamping of the created joint.

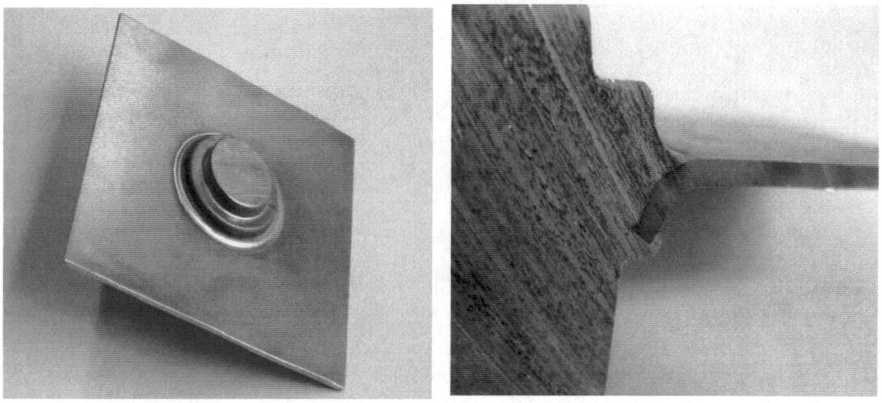

a) b)

**Figure 1**: a) hybrid forging joint; b) micrograph of the joint

This process is now transferred to the joining of aluminum solid parts with galvanized steel sheets. Cause zinc has a positive influence on the formation of the phases, an intermediate layer was considered to realize the compound. In the first step, the aluminum solid parts were coated with zinc. In the second step, the galvanized aluminum was bonded to the galvanized sheet steel under pressure and temperature.

**Table 1**: Material properties of AlMg4,5Mn and Dx54+Z

For the experiments the aluminum alloy AlMg4,5Mn and steel grade DX54 + Z (thickness 1 mm) is used. The material characteristics are listed in table 1. For the joining process, the yield strength and compressive strength of the AlMg4.5Mn were additionally determined.

| | tensile strength | Yield strength | Young's Modulus | Breaking elongation | compressive yield strength |
|---|---|---|---|---|---|
| AlMg4,5Mn | 329 MPa | 223 MPa | 92.400 MPa | 14,36 % | 328 MPa |
| DX 54+Z | 260 MPa | 120 MPa | 210.000 MPa | 36 MPa | - |

**Table 2**: chemical composition of AlMg4,5Mn

| Al [%] | Mg [%] | Mn [%] | Fe [%] | Si [%] | Cr [%] | Zn [%] | Cu [%] | Ni [%] |
|---|---|---|---|---|---|---|---|---|
| 94,3 | 4,489 | 0,515 | 0,3042 | 0,1555 | 0,0796 | 0,0470 | 0,0403 | 0,0020 |

The chemical composition of the aluminum material is listed in table 2. This material has a good weldability and corrosion resistance. Therefore, it is often used in apparatus, container and vehicle. Due to the good low temperature properties and seawater resistance this alloy is also used in shipbuilding.
The DX 54 is a hot-dip galvanized deep-drawing steel that is mainly used in vehicle construction.

## 2. Hot dipping of Aluminum

To apply the zinc layer on the aluminum stud the process of hot dipping steel was transferred. The bolts were left in the furnace for 20 minutes at 420 °C in the zinc melt. The molten zinc used corresponds to a near-eutectic zinc-aluminum alloy to avoid melting of the bolt edges due to the low-melting eutectic. The classification of the melt in the phase diagram is shown in figure 2. It is composed of 17% aluminum and 83% zinc and is therefore a hypoeutectic alloy. The melting temperature of this composition was found to be 378 °C. Figure 2 also shows micrographs of the hypereutectic and eutectic composition in the cut. The layer thickness of the zinc coating on the aluminum is between 50 and 150 microns. Below this layer, a diffusion layer is formed during coating in which the zinc content decreases with increasing distance to the surface. **Figure 3** shows the aluminum bolt with the zinc layer on the left. The cross section of this layer can be seen in the right part of the figure.

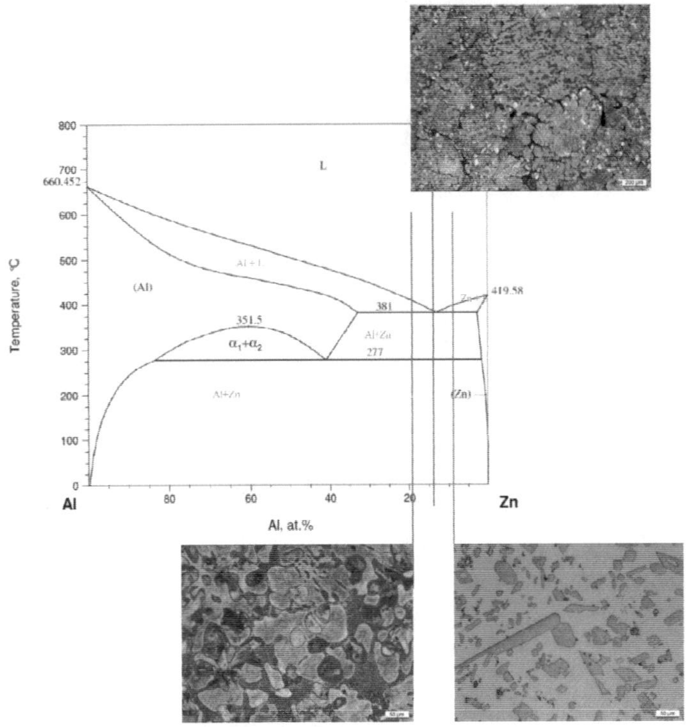


**Figure 2**: Phasediagram of Al-Zn; Source: Materials Science International Services GmbH

The layer thickness can be adjusted by stripping the liquid melt after removal from the zinc bath. The investigation of the coating in the scanning electron microscope with the energy-dispersive X-ray spectrometry (EDX) clearly showed that a nearly pure zinc layer was formed, which increasingly grows in the course of the bolt with aluminum.

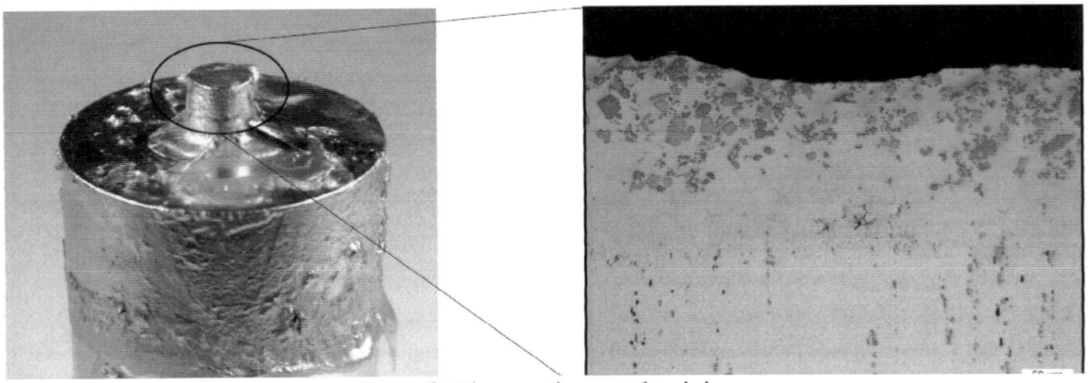

**Figure 3**: Zinc coating on aluminium

## 3. Joining of the aluminum and the galvanized steel sheets

The joining tests were carried out with a tensile compression test bench in an oven. The experimental setup is shown in figure 4. The aluminum studs were hung from the upper punch in a die and heated by natural convection for 20 minutes before the joining process was performed.

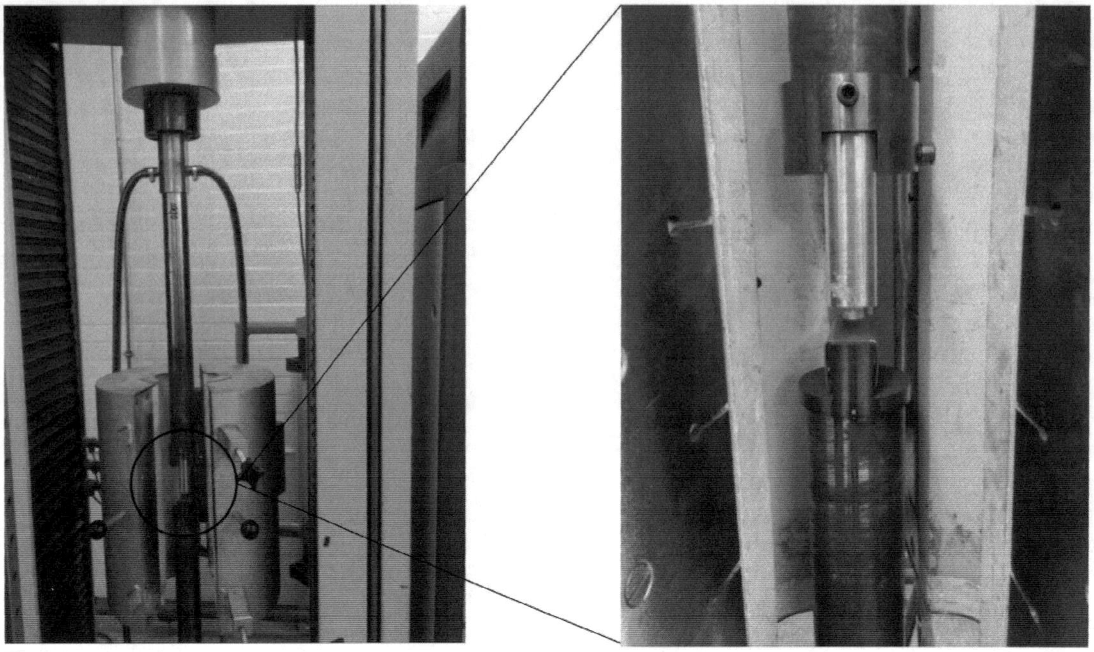

**Figure 4**: Testing setup

On the lower punch, the produced steel sheet samples were placed on a die. The sheets were also preheated. After the preheating time, the punches were used to approach the predetermined force at the specified speed. The parameters were varied with the aid of a D-optimal test plan to make an adequate statement about the process window with the lowest possible number of tests. The parameter field is shown in table 3.

**Table 3**: Joining parameters

| Oven temperature in °C | Specimen temperature in °C | Punching speed in mm/min | Punching force in kN |
|---|---|---|---|
| 375 - 500 | 220-375 | 5-18 | 15-40 |

The head tensile strengths of the individual samples were subsequently determined by tensile testing.

## 4. Results

The applied force and the introduced heat lead to a metallic continuity of the zinc layers of both specimen halves. Figure 5 shows a cross section of the joining area. The joint of the two materials over the zinc layer is clearly recognizable here. The diffusion layer is also visible, which is located in transition of the zinc layer to the aluminum base material.
The base material of the DX54 + Z has not got influence from the zinc during the joining process. The structure has been completely preserved compared to original state.
The chemical composition of the joining area is shown qualitatively in the SEM image in figure 6, clearly to see is the abrupt decrease in iron content and the immediate increase in zinc and aluminum content. The proportion of aluminum in the connection zone is only slightly lower than the zinc content, but much lower than in the case of the parent material. The amount of iron in this area is close to zero. This shows that no diffusion of the iron has taken place and the associated formation of the brittle intermetallic phases has been prevented. The composition of the chemical elements in the joining zone suggests that a cohesive zinc-zinc compound was formed which was formed from the two zinc layers on the base materials. The lower melting point of the zinc also prevents the formation of intermetallic phases as the base materials do not have to be melted.
Furthermore, the resulting intermediate layer leads to the shielding of the materials, whereby the contact corrosion is also prevented.
The geometry of the tensile specimens produced in the joining process is shown in figure 7.
The cross sections in figure 7 show the used materials before the joining process. The two zinc layers are clearly visible. While the zinc layer on the steel sheets is a few microns thick the layer thickness on the aluminum is between 50 and 150 microns.

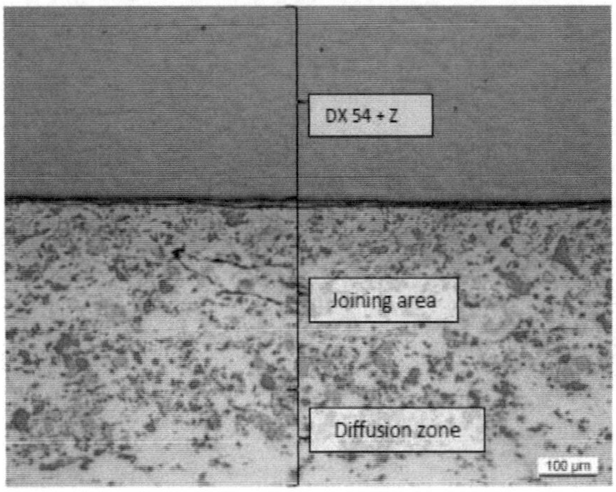


**Figure 5**: Joining area

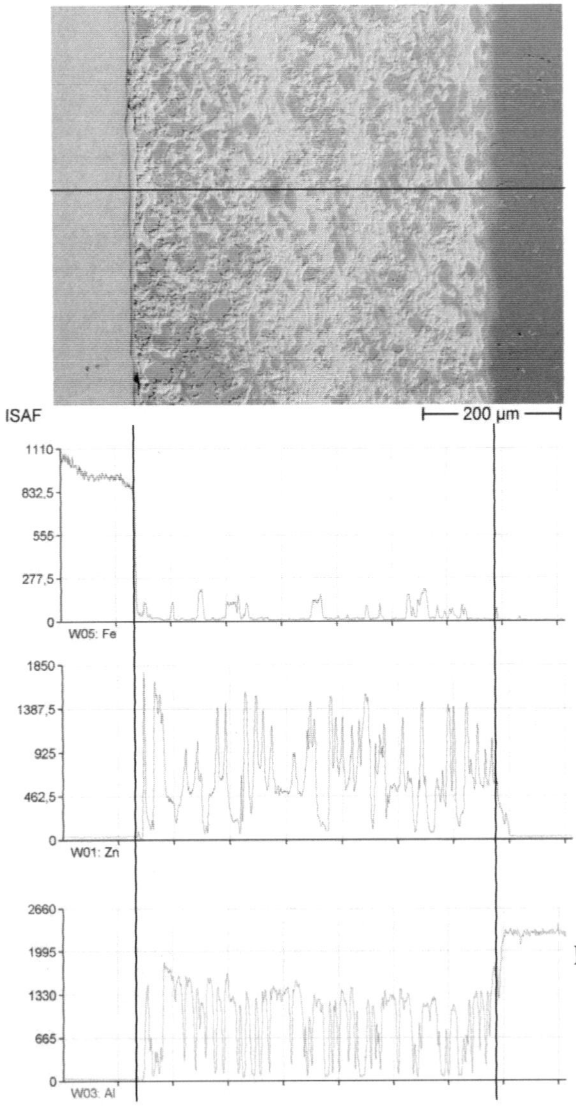


**Figure 6**: SEM-EDX analysis of the join-

IOP Conf. Series: Materials Science and Engineering **373** (2018) 012026 doi:10.1088/1757-899X/373/1/012026

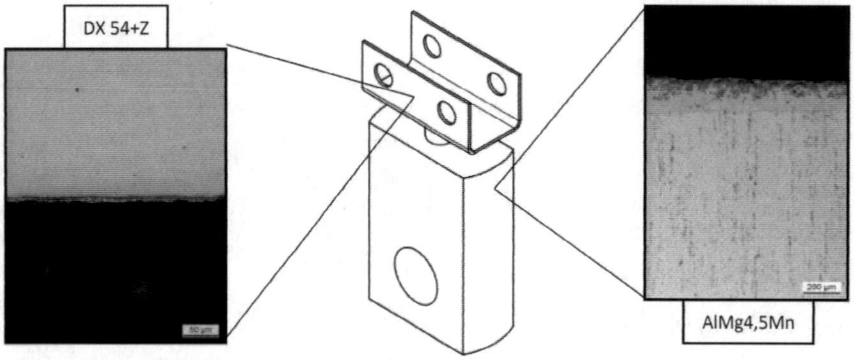


**Figure 7**: Specimen geometry and micrographs of the used materials

The force-displacement curves determined in the tensile test for the samples, which were joined at oven temperatures between 410 and 500 ° C, are shown in figure 8. The measured surface temperatures of the specimens before the joining process were between 220 and 375 °C. This corresponds approximately to the melting temperature of the near-eutectic molten zinc used for coating. As expected, the zinc layer was returned to the molten state at this temperature. This means that under the given process conditions no metallic continuity could be generated as no liability or solidification had occurred at removal from the oven. Cooling in the oven would not be expedient because the determined process parameters and relationships should be transferred to a forging process. Despite of this fact, the temperature-controlled cooling will be part of subsequent investigations.

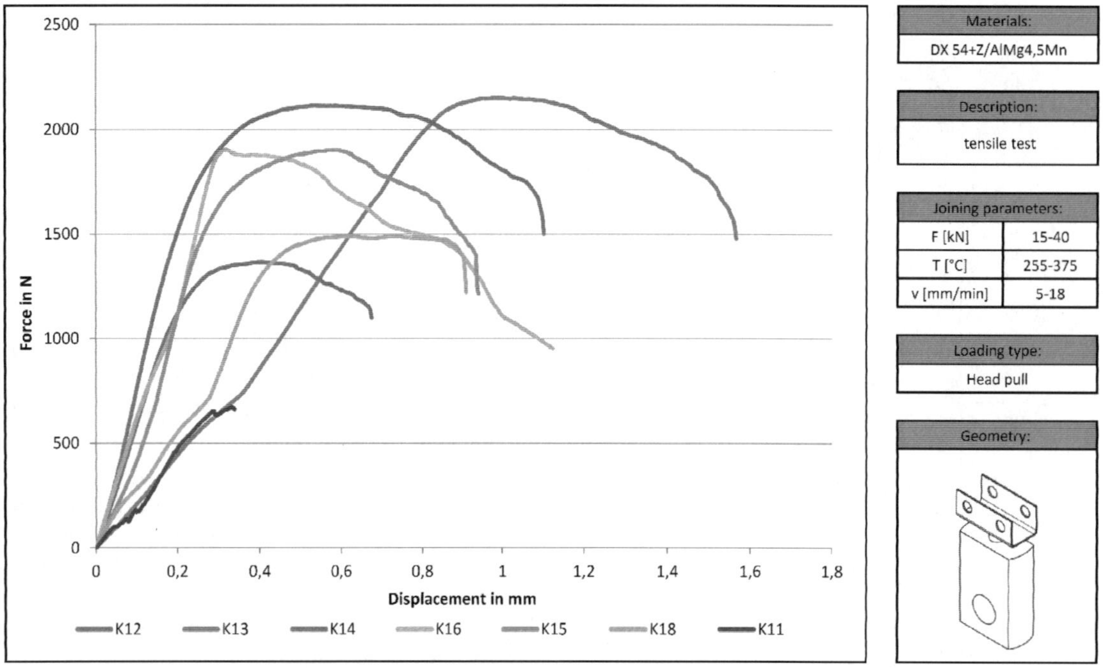


**Figure 8**: Force-displacement curves of specimens

The highest force is borne by the specimens joined at a temperature of 330 °C, a force of 27.5 kN and a punch speed of 11.5 mm/min. The measured specimen temperature on surface of the zinc was about 330 °C.

IOP Conf. Series: Materials Science and Engineering **373** (2018) 012026 doi:10.1088/1757-899X/373/1/012026

The tensile tests were carried out after complete cooling of the samples. The fracture surfaces were also subjected to an EDX analysis after the experiments to obtain the distribution of the elements over the attachment surface and thus to be able to characterize the metallic continuity. Figure 9 shows the analysis

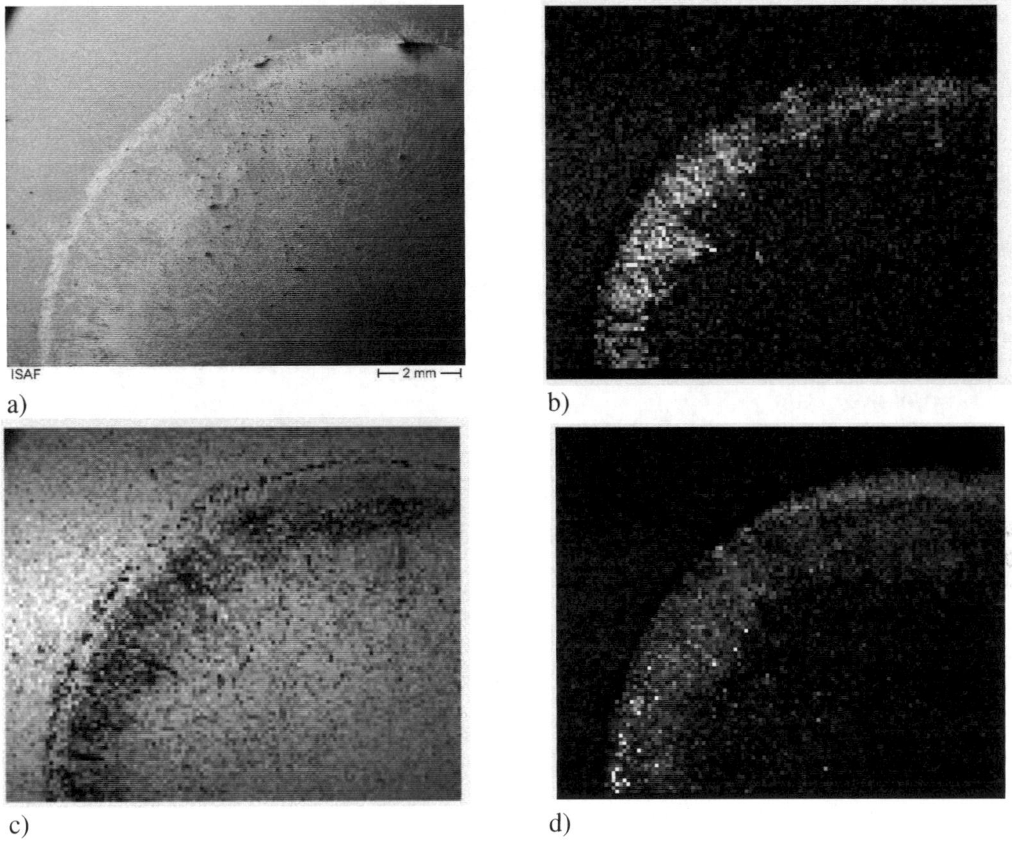


**Figure 9**: a) REM picture of the fracture area; b, c, d) Elementanalysis of Fe, Zn, Al

of the fracture surface on the steel side with the SEM. The distribution of the elements iron, zinc and aluminum (figure 9 b-d) clearly shows the transition of the materials in the attachment zone. Thus, a cohesive connection has been created which withstands forces of up to 2.1 kN.
The evaluation of the degree of deformation as a function of the joining parameters resulted, as expected, in a proportional relationship. The tensile strength, however, shows no correlation as a function of the joining parameters. However, it was assumed that, as with all thermal joining processes, a minimum energy must be introduced to produce a metallic continuity. This energy is composed on the one hand of the introduced heat and on the other hand of the forming energy. The heat energy was calculated by the mass of the bolts, the heat capacity of the aluminum and the temperature difference. The contribution of the deforming energy goes back to the generated plastic deformation. The transformation energy or deformation work is converted up to 95% into heat, which implies a connection between temperature change at the joint and compression, [9]. Therefore, the deformation energy was derived from the very good correlation of the compression ratio and the joining parameters and calculated using the integral of the force-displacement data from the joining process.

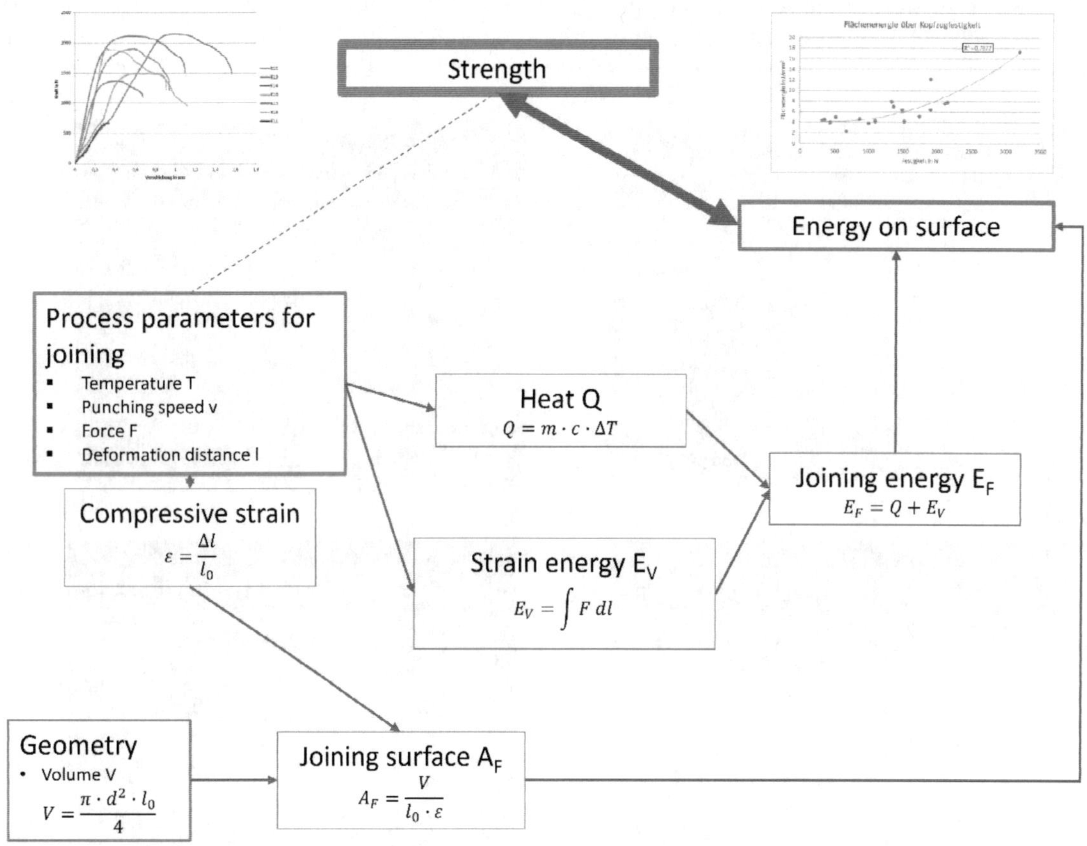


**Figure 10**: Analysis of correlation

The actual joining surface was calculated by means of the compression ratio from the volume of the aluminum bolt. The actual joining surface is used to normalize the introduced total joint energy, in the sum of deformation work and heat energy. The normalized energy values were plotted against the

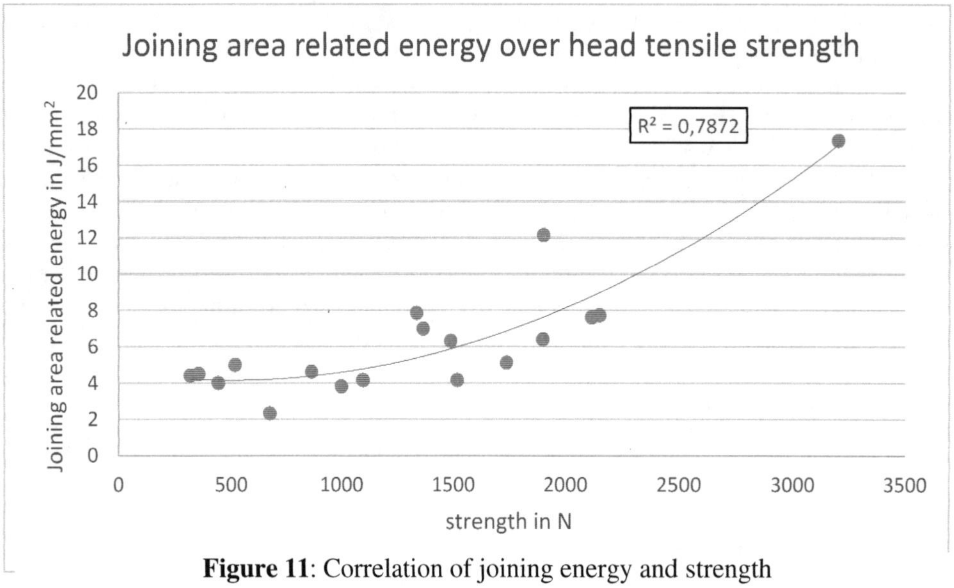


**Figure 11**: Correlation of joining energy and strength

determined head tensile strength values and are shown in figure 10. Furthermore, the degree of normality $R^2 = 0.7872$ was calculated via the trend course, in the form of a quadratic relationship. The established and described relationships are shown in figure 11.

## 5. Conclusions of joining tries

Within the scope of the presented work two significant results have been developed so far.
First, aluminum was galvanized using the hot dip process and the parameters for this process were determined. The influence of the composition of the molten bath was checked and improved by an iterative process. The adhesion of the layer is dependent on the immersion time and the temperature. The layer thickness can be adjusted after removal by stripping. Analysis by SEM and micrographs revealed that during galvanizing, a diffusion layer in the aluminum sets, thereby eliminating the oxide layer.
As a second result, parameters were developed for producing a material connection between the zinc-plated aluminum bolt and the galvanized steel sheet. This could also be detected by SEM and micrographs. This analysis also showed that the two base materials were hardly affected.
The determined strength values are comparable with the values of punctiform geometries produced with conventional joining methods. The comparison of the values from the literature with the data obtained in the experiment (figure 12) shows that the head tensile strength of the compound corresponds to the thermal joining method, [10–12]. Furthermore, it is above the level achievable with mechanical joining methods.

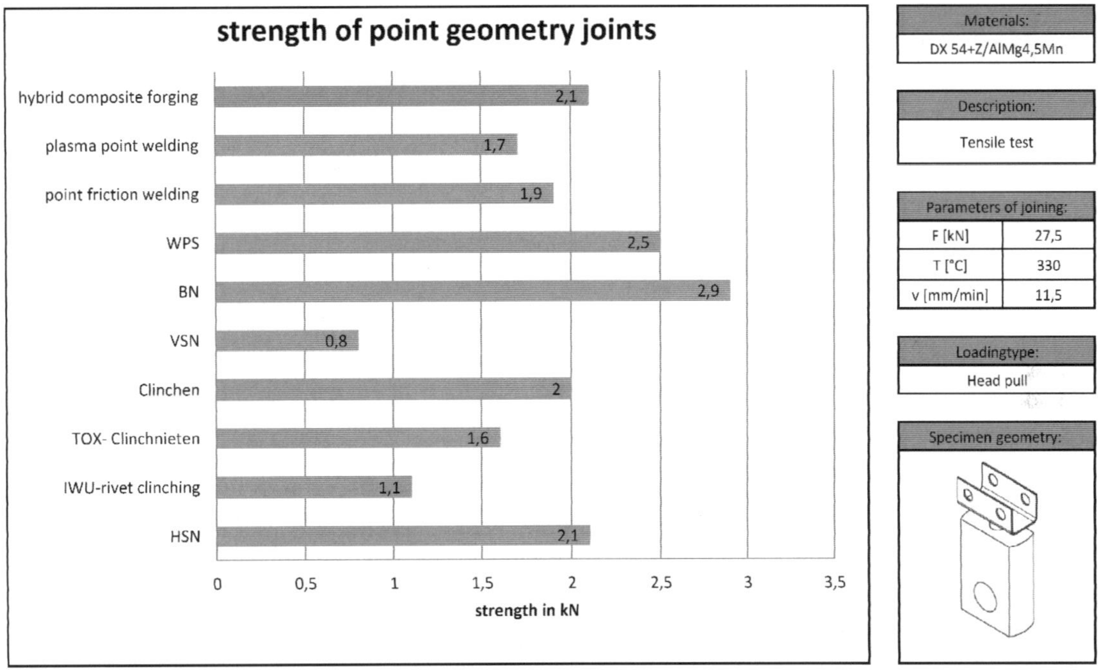


**Figure 12**: Comparison of ultimate strength of different joining techniques

The determined relationships clearly show that a minimum energy must be applied to create a metallic continuity. With increasing joining energy and a concomitant increase in the compression ratio an increase in strength can be achieved to a certain extent. An in-depth evaluation and validation of these relationships should enable the economical production of the joint in industrial production in the medium term.

## 6. Investigation of parameters for hybrid composite forging by FE-simulations

The research on the forging operation of hybrid composite forging is made subsequently to the galvanization of the aluminum part. To identify suitable parameters for hybrid composite forging finite-element simulations were conducted with FORGE NxT 2.1. A cylindrical aluminum bulk part, material AlMg4,4Mn0,7Cr, with a diameter of 10 mm and a hight of 20 mm and a steel sheet, material DD11, with a diameter of 30 mm and a height of 1 mm are the connecting partners of the forging process. The temperature of the forging tools was set to 250 °C. As friction conditions the "Coulomb limited Tresca model" was chosen with high friction conditions ($\mu = 0{,}3$).

On the bulk aluminum part the top layer is defined as a different type of aluminum to simulate the effect of the zinc layer as a first appraoch, as material data for Zinc was currently not available in sufficient quality. Furthermore, simulations above the melting point of zinc lead to simulation errors as it gets liquid. The material description for the hot forming operation is based on the flow curve equation by HENSEL and SPITTEL.

To identify suitable parameters for the forging process different temperatures of aluminum and steel are investigated. In addition, the velocity of the presses is a parameter that was varied between 3 mm/s and 300 mm/s, which represent a spectrum of different industrial presses. Table 4 shows the different investigated parameters.

**Table 4**: Used simulation parameters for investigation of the joining zone

| Temperature aluminum (°C) | Temperature steel (°C) | Velocity (mm/s) |
|---|---|---|
| 450 | 20 | 3 |
| 500 | 200 | 300 |
| 550 | 400 | |

The temperature distribution within the aluminum in the joining zone after forming was investigated. At the forging process with v = 300 mm/s the highest temperature is above 550 °C, for v = 3 mm/s the highest temperature is below 312 °C (see figure 13 and figure 14) Thus, the higher press velocity leads to a more irregular temperature distribution and to significant higher temperature in the joining zone (figure 13) than the low press velocity.

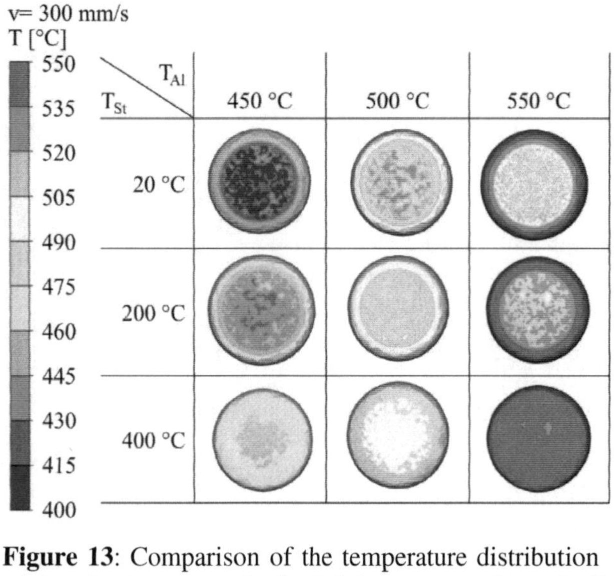


**Figure 13**: Comparison of the temperature distribution of the aluminum part in the joining zone at the end of forming with 300 mm/s at different temperatures of aluminum and steel at the start of the forming process

Figure 13 shows that especially with a bigger difference between the starting temperatures of aluminum and steel the temperature distribution is increasingly irregular. This effect can be observed when the results of the simulations with $T_{St} = 20$ °C are compared to the results of the simulations with $T_{St} = 400$ °C. The simulations with $T_{St} = 400$ °C show a more homogenous temperature distribution.

As seen in figure 14 a lower press velocity leads to a more homogenous temperature distribution in the joining zone. This can be seen when the results in figure 13 are compared with the results in figure 14.

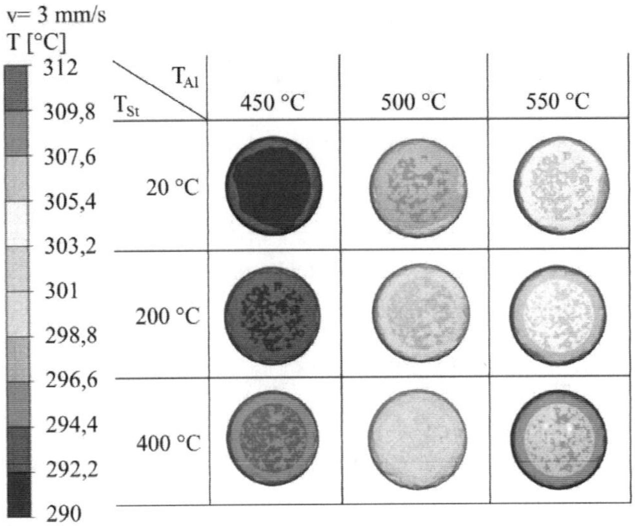


**Figure 14**: Comparison of the temperature distribution of the aluminum part in the joining zone at the end of forming with 3 mm/s at different temperatures of aluminum and steel at the start of the forming process

In simulations in which the low press velocity was used, the temperature of the tools has significant influence on the temperature of the connecting partners. In figure 14 the figure of temperature distribution of the simulation $T_{Al}$ = 450 °C and $T_{St}$ = 400 °C shows that the temperature of aluminum at the end of forming is below 296.6 °C. This decrease in temperature occurs due to heat transfer to the dies and to the ambient medium and is more present when a low press velocity is used.

In general the simulations showed, that the temperature decreases from the outer areas to the middle of the joining zone. The irregular temperature distribution in the joining zone is a difference between simulations with a defined layer of different material on the bulk aluminum part and simulations with only bulk aluminum (figure 15).

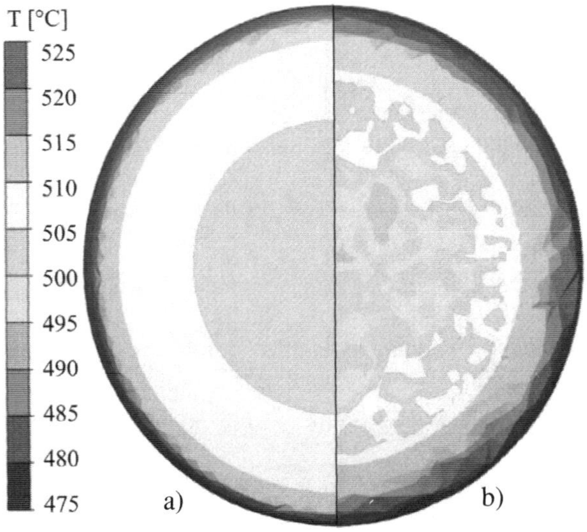


**Figure 15**: Temperature distribution in the joining zone of the aluminum part a) without defined layer of different material and b) with a layer of different material in the joining zone with T-Al: 500 °C, T-St 400 °C and press velocity of v = 300 mm/s

In simulations without a defined layer of different material the temperature decreases form the outside of the joining zone to the middle of the joining zone. Figure 15 shows the comparison of the temperature distribution of a simulation without (figure 15 a) and with a defined layer of different material in the joining zone (figure 15 b) for a forming simulation with the parameters $T_{Al}$ = 500 °C, $T_{St}$ = 400 °C and a press velocity of 300 mm/s. The irregularities in the temperature distribution in the simulations with a defined layer of different material might occur due to problems with remeshing of the top layer. The real temperature distribution can be investigated with experimental tests. An irregular temperature distribution may lead to an inhomogeneous joint.

In the joining zone a homogenous temperature distribution shall be attained, so the joint is homogenous as well. The temperature distribution can have an influence on the distribution of elements in the joining zone as shown in figure 9. To achieve a homogenous temperature distribution lower temperature difference between aluminum part and steel sheet will be investigated in further simulations. For example simulations where the temperature of aluminum part and steel sheet are below the melting point of zinc will be made. Especially the temperatures shown in table 3 will be examined. Furthermore, a variation of press velocity between 3 mm/s and 300 mm/s will be investigated, e.g. 50 mm/s, 100 mm/s and 150 mm/s. Results of simulations with these parameters are expected to lead to homogenous temperature distribution in the joining zone that will investigated in experimental trials and tested further.

## 7. References

[1] S. Meco, L. Cozzolino, S. Ganguly et al., 2017 "Laser welding of steel to aluminium: Thermal modelling and joint strength analysis," *Journal of Materials Processing Technology*, vol. 247, pp. 121–133
[2] H. T. Kashani, P. Kah, and J. Martikainen 2015 "Laser Overlap Welding of Zinc-coated Steel on Aluminum Alloy," *Physics Procedia*, vol. 78, pp. 265–271,.
[3] G. Zhang, W. Su, J. Zhang et al., 2011 "Friction Stir Brazing: A Novel Process for Fabricating Al/Steel Layered Composite and for Dissimilar Joining of Al to Steel," *Metallurgical and Materials Transactions A*, vol. 42, no. 9, pp. 2850–2861
[4] L. Engelbrecht, O. Meier, A. Ostendorf et al., 2006 "Einflüsse auf die mechanischen Eigenschaften lasergelöteter Mischverbindungen aus Stahl und Aluminium," *Materialwissenschaft und Werkstofftechnik*, vol. 37, no. 3, pp. 272–278
[5] U. Reisgen, L. Stein, M. Steiners et al., 2010 "Lichtbogenfügen umformbarer Stahl-Aluminium-Mischverbindungen. Arc joining of ductile steel-aluminium-dissimilar-metal-welds," *Materialwissenschaft und Werkstofftechnik*, vol. 41, no. 11, pp. 951–960
[6] K.-G. Kosch, C. Frischkorn, A. Huskic et al., 2012 "Effizienter Leichtbau durch belastungsangepasste und anwendungsoptimierte Multimaterial-Schmiedebauteile," *UTF Science*, pp. 1–17
[7] B.-A. Behrens and K.-G. Kosch, 2011 "Development of the heating and forming strategy in compound forging of hybrid steel-aluminum parts," *Materialwissenschaft und Werkstofftechnik*, vol. 42, no. 11, pp. 973–978
[8] B.-A. Behrens and F. Holz, 2008 "Verbundschmieden hybrider Stahl-Aluminium Bauteile," *Materialwissenschaft und Werkstofftechnik*, vol. 39, no. 9, pp. 599–603
[9] F. Walther and D. Eifler, 2008 "PHYBAL — Kurzzeitverfahren zur Berechnung der Lebensdauer metallischer Werkstoffe auf der Basis physikalischer Messgrößen," *Materials Testing*, vol. 50, no. 3, pp. 142–149
[10] M. Reinstettel, 2008 *Laboruntersuchung zur Prozessstabilität beim Niet-Clinchen*, Verl. Wiss. Scripten, Zwickau
[11] L. Budde, Hahn, and O., 1991 *Grundsatzuntersuchungen zum Festigkeitsverhalten von Durchsetzfügeverbindungen aus Stahl*, Franfurt am Main
[12] O. Hahn and D. Gieske, 1995 *Ermittlung fertigungstechnischer und konstruktiver Einflüsse auf die ertragbaren Schnittkräfte an Durchsetzfügeelementen*, Franfurt am Main

**Acknowledgements**
Special thanks goes to the German Research Foundation (DFG), which has made this work possible through financial support.